W9-BAJ-374

Planning and Ecology

Planning and Ecology

EDITED BY

R.D. Roberts

Lecturer in Biology
and Director, Industrial and Applied Biology Group
University of Essex

and

T. M. Roberts

Research Officer, Biology Section
Central Electricity Research Laboratories
CEGB
Leatherhead

LONDON NEW YORK
CHAPMAN AND HALL

First published in 1984 by
Chapman and Hall Ltd
11 New Fetter Lane, London EC4P 4EE
Published in the USA by
Chapman and Hall
29 West 35th Street, New York NY 10001
First paperback edition 1986

Printed in Great Britain by the
University Press, Cambridge

ISBN 0 412 28470 7

British Library Cataloguing in Publication Data
Roberts, R. D.
Planning and ecology.
1. City planning 2. Regional planning
3. Ecology
I. Title II. Roberts, T. M.
711 HT 166
ISBN 0 412 28470 7

Library of Congress Cataloging in Publication Data
Planning and ecology.
Bibliography: p.
Includes index.
1. Environmental impact analysis. 2. Land-use–Planning. 3. Ecology. 4. Environmental policy.
I. Roberts, R. D. II. Roberts, T. M.
TD194.6.P63 1984 333.7 83-23170
ISBN 0 412 28470 7

Contents

List of Contributors

J. M. Baker, BSc, PhD, MIBiol, FInstPet, Research Director, Field Studies Council

R. M. Bell, BSc, PhD, Senior Advisory Officer, Environmental Advisory Unit, University of Liverpool

T. J. Bines, BSc, PhD, England Headquarters, Nature Conservancy Council

A. G. Booth, DipEng, DipTP, FRTPI, FFB, FRSA, Town Planning Consultant, Formerly County Planner, Essex County Council and President of the Royal Town Planning Institute

A. D. Bradshaw, MA, PhD, FRS, Holbrook Gaskell Professor of Botany, University of Liverpool. President of the British Ecological Society

R. G. H. Bunce, BSc, PhD, Principal Scientific Officer, Institute of Terrestrial Ecology, Merlewood Research Station

B. D. Clark, BA, MA, Senior Lecturer in Geography, University of Aberdeen

J. Corlett, MSc, FIBiol, Associate Consultant, Natural Environment Research Council

D. M. Colwill, BSc, Principal Scientific Officer, Transport and Road Research Laboratories, Department of the Environment

A. Currie, MIBiol, Assistant Regional Officer, Nature Conservancy Council

W. G. Dallas, BA, Agr(Forest)B, MF, Conservation Manager, Tara Mines Ltd, Ireland

A. W. Davison, BSc, PhD, Lecturer, Plant Biology Dept, University of Newcastle upon Tyne

G. Diprose, BSc, PhD, Environmental Section, Research and Development Dept, ICI, Mond Division, Northwich, Cheshire

J. P. Doody, BSc, PhD, Chief Scientists' Team, Nature Conservancy Council

R. W. Edwards, DSc FIBiol, FIWPC, Professor of Applied Ecology and Head of Department of Applied Biology, The University of Wales, Institute of Science and Technology

W. R. Effer, BSc, PhD, Manager of Environmental Studies and Assessments, Ontario Hydro, Canada

I. H. Findlay, Warden, Nature Conservancy Council

K. M. Gammon, BFC, CEng, FIC, FBIM, Generator Development Engineer, Corporate Strategy Department, CEGB

D. A. Goode, BSc., PhD., Assistant Chief Scientist, Nature Conservancy Council

B. H. Green, BSc, PhD, Senior Lecturer in Ecology and Conservation, Wye College

R. Grove-White, BA, Director, Council for the Protection of Rural England

D. Hammerton, BSc, FIBiol, FIWES, AMBIM, Director, Clyde River Purification Board

J. F. Handley, BSc, PhD, Resource Officer, Merseyside County Council

O. W. Heal, BSc, PhD, Senior Officer, Institute of Terrestrial Ecology, Merlewood Research Station

K. Hiscock, BSc, PhD, Deputy Director, Field Studies Council

M. W. Holdgate, BSc, PhD, CB, Director General of Research, Department of the Environment

G. D. Howells, MSc, PhD, Head of Biology, Central Electricity Research Laboratories, CEGB

M. J. Hudson, BSc, MIBiol, Regional Officer, Nature Conservancy Council

M. S. Johnson, BSc, PhD, Lecturer in Botany, University of Liverpool

A. M. Jones, BSc, PhD, Lecturer in Biological Sciences and Director of the Orkney Marine Biology Unit, University of Dundee

N. Lee, BSc, PhD, Senior Lecturer in Economics, University of Manchester

S. Legore, BSc, PhD, Director of Marine Services, Hazleton Environmental Sciences Corp, USA

A. MacLennan, MA, PhD, Assistant Regional Officer, Nature Conservancy Council

P. J. Nelson, MA, MSc, DipTP, MRTPI, Principal, Land Use Consultants

G. D. R. Parry, BSc, PhD, Director, Environmental Advisory Unit, University of Liverpool

N. W. J. Pumphrey, BSc, PhD., Environmental Section, Research and Development Dept, ICI, Mond Division, Northwich, Cheshire

R. D. Roberts, BSc, PhD, Lecturer in Biology and Director, Industrial and Applied Biology Group, University of Essex

T. M. Roberts, BSc, PhD, MIBiol, Research Officer, Biology Section, Central Electricity Research Laboratories, CEGB, Leatherhead

C. Swanwick, BA, MSc, Associate, Land Use Consultants

W. J. Syratt, BSc, PhD, MIBiol, Senior Ecologist, Environmental Control Centre, BP International Limited

J. R. Thompson, BSc, PhD, MIBiol, Research Assistant, Imperial College, University of London. Currently Country Parks and Forestry Officer, Essex County Council

R. Turnbull, Deputy Chief Planner, Scottish Development Department

P. Wathern, BSc, PhD, Lecturer in Botany and Microbiology, University College of Wales, Aberystwyth

C. Wood, BSc, DipTP, MA, MRTPI, Senior Lecturer in Town and Country Planning, University of Manchester.

To
Ann and Eileen
for their
technical and moral
support

CHAPTER ONE

Introduction

Professional planning is commonly defined as an activity which is concerned with the rational allocation or exploitation of resources for man's maximum short- and long-term benefit. The science of ecology is the study of the basic components of these resources (soil, water, air etc.) and their interrelationships with living organisms. Planning and ecology therefore have many common interests and, as such, have long been associated in resource management. This association was, for example, reflected as early as the 19th century in the writings of Patrick Geddes, the progenitor of modern town planning, whose training as a biologist led him to re-interpret the phenomena of urbanization in ecological terms (Geddes, 1886).

Since the time of Geddes the place of ecology has declined in planning circles as other professions and considerations, initially public health and engineering, latterly economic and sociological, have become more central. The reasons for this shifting emphasis are three-fold. First, they reflect changing social attitudes and aspirations in the post-war period. Secondly, many other professions and persuasions (agriculturalists, foresters, industrialists etc.) became organized as pressure groups and helped shape public attitudes and governmental actions. Thirdly, and by far the most important reason, is the prevalence of misconceptions about ecology. Too many people, perhaps misled by the debate surrounding the 'environmental movement', equate ecology with conservation. In reality, conservation is but one component part of ecology. The place of ecology as a profession rather than a 'political persuasion', its involvement in agricultural productivity, forestry practices, air and water pollution control, soil improvement, landscape management to name but a few, has been severely neglected in the recent hubbub of the 'environmental debate'.

In some countries, a realization of the consequences of this declining emphasis on ecology has resulted in various attempts to reverse the process. In the USA, for example, ecological considerations were brought to the forefront with the enactment of the *National Environmental Policy Act* in 1969 and in several other countries by strengthening of environment – related legislation. In contrast, in Britain, ecological considerations have often been treated as minor components in planning issues, to be considered as an afterthought, once other issues had been settled. Where ecology is given attention it has often been for negative reasons; 'conservation' information is often collected by a planning authority when it wishes to oppose a proposed development, and by developers as a defence against such opposition.

This book is written in the belief that ecology has a major, and positive, role to play in planning for good environmental management. Indeed, since ecology is essentially *the science* concerned with the functioning of resources, it can be argued that sound (short- and long-term) planning cannot be achieved without due consideration to ecological information. This book seeks to examine the nature of the interaction between planners and ecologists and the means by which maximum benefit may be obtained from this relationship. This is presented as a discussion of planning procedures and specific ecological methods rather than a discourse on ecological principles appropriate to planners as has been attempted in most other publications in this field (Edington & Edington, 1977; Selman, 1981 inter alia).

The book has developed from a British Ecological Society Workshop (held in 1980) and although many of the case-studies deal specifically with the UK experience there is a section devoted to North American and European environmental planning procedures. The reader needs to bear in mind the intensity of competition for natural resources in the UK which contains a population in excess of 55 million in a total land area of 24×10^6 ha. Consequently, although only 10% of this area is designated as urban, the remainder is subject to intense pressure for recreation, conservation and industrial development as well as agriculture and forestry.

1.1 PLANNING AND ECOLOGY AS COMPONENTS OF ENVIRONMENTAL MANAGEMENT SYSTEMS

Although the major theme in this book is the interface between planning and ecology, many broader issues in environmental management are also discussed. In recent years, the term 'environment' has been widely used in three different senses. Firstly, the 'natural environment' of air, water, soils and organisms which provide the setting in which a plant or animal lives. Secondly, the 'built environment' comprising the areas of urban and industrial developments. Thirdly, the 'social environment' relates to culture, law, economics, music etc. The book deals with the built and natural environments and in particular, with the effect of one on the other.

Management also spans a spectrum of activities but is used here to mean deliberate interference by man with an environmental system (see Holdgate, 1977). This can involve direct management by actions at the implementation level, with spade, fertilizer, livestock, water treatment plant etc. In recent decades, the actions of these direct managers have, however, been influenced increasingly by indirect managers acting at several removes from the implementation stage. This level of management sometimes called 'policy-making' is the area of action in which people make decisions about broad objectives and methods of environmental management: what to use land for and how to do it; what changes to tolerate and what standards to impose and how to achieve them.

In addition to the managers involved in the decision-making and implementation pathways, an additional series of disciplines, is also involved in environmental management systems. These are the various professions, sociologists, biologists, chemists, economists etc., who are responsible for the collection and interpretation of the data required at the decision-making level.

Planners operate in environmental management at two main levels, policy-making and interpretation. The first is concerned with stating goals, options and the broad-zonation of land-use priorities involving the integration of agriculture, forestry, fishery, water catchment, conservation, recreation, urban and industrial development and transport. The second level involves developmental control in which specific proposals are appraised for their conformity with the overall goals and their appropriateness to specific location. As discussed by Booth (Section 2.1), within the framework of environmental management systems, the planning profession essentially plays the role of administrator and interpreter in the reconciliation of conflicting interests and demands. The skills of planners lie in the collation and interpretation of data and application of broad weightings in development control and larger components of balancing in policy plan preparation.

Ecologists on the other hand, are involved at two, and sometimes three levels.

The first level is direct management such as farming, forestry, or conservation with which ecologists have long been associated. The second, and probably least frequent, is in the role of interpreter in the policy-making and regulatory levels. The third, and most common, is the research level where professional ecologists are involved in collection and interpretation of information on organisms and environmental conditions. As discussed in Chapter Three, the types of ecological information most pertinent to environmental management are those which relate to (a) description of spatial and performance patterns of organisms and associated environmental variables, (b) prediction of likely future patterns of organisms and environments under existing or different management options and (c) prescriptions for attaining specific objectives.

The main skills of ecologists lie in the area of information collection by desk and field studies and experimentation for which ecologists have long-established methods. As scientists, ecologists are commonly trained in quantitative skills rather than those involving aggregation and value judgements, which are crucial to the planning profession.

1.2 CONFLICTS IN THE INTERACTIONS BETWEEN ECOLOGY AND PLANNING

Ecologists and planners operate at different levels within environmental management systems and, consequently have developed rather different skills. In the language of Lord Rothschild, the planner at the administrative level represents 'the customer' for the information provided by the ecologist as a 'contractor'. The challenge with which we are faced is that, good environmental management can only come from positive interaction between the two professions. The professions are different, but it is vital that they work together as a team.

In reality, this teamwork has been achieved only rarely in the past, although there are several examples of the excellent benefits obtained when interactions are achieved. The reasons, for this paucity of teamwork are complex, and relate to numerous misunderstandings and conflicts between the professions. In subsequent sections we explore many of these problems and examine how, in specific case examples, they have been overcome. In addition, they are reviewed briefly in this section as a framework for integrating subsequent sections.

There are several conflicts surrounding the statutory procedures within which ecological, and other, information is incorporated in the planning process. In reality, early and comprehensive consideration of the ecological component is essential for ensuring an adequate analysis of effects on the environment. In the UK the environment is not usually considered as a single unit in planning. Normally, ecological factors are considered under a diversity of headings such as landscape quality, ecology of flora and fauna, forestry, agriculture, conservation, recreation, pollution, reclamation etc. The concern is that this fragmented treatment can lead to omission of important consequences of an action or to conflicting policies due to overlap of areas of concern. A review of the more formal structured procedures developed in Europe and North America is given in Chapter Four as are the implications for the UK of the EEC draft directive on 'Environmental Assessment of Projects'.

Many conflicts in environmental management in the UK arise because much of the activity in rural areas is not subject to planning controls and the Central Government Agencies responsible for rural development have rather specific objectives (Chapter Five). For example the objectives of MAFF and the Forestry Commission have simply been to maximize production. This has involved, over recent decades, an intensification and extensification of these activities and has resulted in intense conflicts with both conservation and recreation interests.

This largely reflects an assumption during immediate post-war years, when the first comprehensive planning acts were being framed, that agriculture and forestry would automatically enhance the countryside if allowed to prosper. Governmental control in agriculture is confined to the system of grants and subsidies operated by the Ministry of Agriculture, Fisheries and Food and is determined by central government's desire to increase the level of net self-sufficiency in agricultural produce from 50 to 60% in the short-term, and 75% by the turn of the century (MAFF, 1975). Recently planning and agriculture have been brought closer together by the requirement for approval for grant-aided improvements in National Parks and Areas of Outstanding Natural Beauty (from the Local Planning Authority) and in Sites of Special Scientific Interest (from the Nature Conservancy Council). Similarly forestry, the second major land user of the countryside is unaffected by planning legislation but is required by the *Forestry Acts* of 1967 and 1974 to give attention to the aesthetic and recreational role of forests. In Scotland, the Forestry Commission has invited planners to co-operate with them in developing regional forest policies (SDD, 1975) but comparable moves have not yet evolved in other parts of the country. The Countryside Commission does, however, play a role in stimulating actions to improve the amenity, landscape and recreational facilities in rural areas.

Chapter Six deals with the role of ecologists in strategic planning and examines the special requirements imposed by decision-making at the policy level. Planning controls in the UK are largely concerned with maintaining or improving the quality of the built environment and with resolving competing claims on resources. There is little influence over nature conservation or industrial impacts which are not in direct breach of planning consent condition. This reflects the views prevalent during the formative years of planning legislation in the UK that conservation of wildlife and control of pollution were essentially local issues to be attended to by specialist agencies and covered by separate effluent control legislation. Hence, no comprehensive overview was deemed necessary. However, planners may refuse permission for, or impose conditions upon, development in the interests of amenity. One Statutory Instrument refers to 'detriment of amenity by reason of noise, vibration and smell, fumes, smoke, soot, ash, dust or grit'. There is therefore scope, albeit discretionary, for rigorous planning consent conditions. Indeed planning authorities, as part of their Structure Plan formulations, now supplement the legislative requirements with voluntary actions and consult many of the organizations involved in pollution control, wildlife conservation, agricultural production and other aspects of environmental management. In Scotland the issuance of National Planning Guidelines to protect the coastline during the 'oil boom' in the late 1970s allowed a considerable amount of information to be collected for the 'primary search areas' prior to planning applications. In addition, procedures for Environmental Impact Analysis have been used at an early stage of the strategic planning of water resource development in the North West of England.

Chapter Seven provides a number of case-studies relating to the use of ecological information in project planning. These examples illustrate the positive aspects of environmental assessments both in the context of development control and compliance with effluent control legislation. These examples put into sharp focus the differences in approach to (a) planning major energy developments in Canada (with a formal EIA system) and the UK (with less-structured, more flexible planning requirements), (b) oil-related developments in Scotland by the developer compared to consultants working for the local authority and (c) effluent control procedures from the viewpoint of both industry and the regulatory agency. The papers illustrate the advantages of early consideration of ecological information in development planning, the importance of good base-line surveys and the need to study the impact of existing developments in order to predict the effects of large-scale new developments.

The importance of environmental audits of both EIA as a planning procedure and the accuracy of environmental impact statements is taken up in Chapter 8. The difficulty with the former is the lack of adequate controls in that there can only be a comparison of one development (with formal EIA) and a similar development (without EIA). Nevertheless it is clear from these papers that adequate environmental planning can be carried out using a variety of procedural arrangements.

Assessing the accuracy of predictions made in environmental impact statements has also proven to be extremely difficult. However, monitoring the effects of 'novel developments' must considerably improve the accuracy of future EIAs of similar developments. Indeed, it is becoming clear that the 'snapshot' provided by the initial EIS must be supplemented by subsequent audits to ensure that environmental policies are met in the light of continual changes in industrial practices and environmental legislation. Consideration of environmental effects must therefore be a dynamic process and environmental auditing clearly plays an important role.

1.3 TERMINOLOGY

There is considerable confusion in the literature concerning the use of various terms relating to the incorporation of environmental information in the planning process. Wherever possible the following terms have been used:

Environmental assessment

The base-line monitoring and compilation of data relating to quantification of the important physical and biological characteristics of an area.

Environmental impact analysis

The process of identifying and evaluating the effects on the environment of impact factors arising from a change in use. This may include an environmental assessment, the prediction of changes and an evaluation of the significance of these changes. It does not imply any particular procedure.

Environmental impact assessment

The specific procedural system developed under the American *National Environment Policy Act* (1969) for relating impact factors of a land-use change to the environment. This consists of a specified series of steps including (a) description of proposed development, (b) description of existing and projected environmental conditions, (c) assessment of probable impacts (usually using matrix tables and numerical weighting), (d) compliance with regulations, (e) review of alternatives, (f) preparation of a non-technical summary.

Environmental impact statement

A documentation of the matters considered under the Environmental Impact Assessment.

REFERENCES

Edington, J.M. & Edington, M.A. (1977) *Ecology and Environmental Planning*, Chapman and Hall, London.

Geddes, P. (1886) Quoted in Boardman, (1979) *Town & Country Planning*, **48** (6).

Holdgate, M.W. (1977) *Biologist*, **24**.

Ministry of Agriculture, Fisheries and Food (1975) *Food from our own resources*, HMSO, London.

SDD (1975) *Planning Advice Note* **2**, *Forestry Guidelines*, HMSO, Edinburgh.

Selman, P.H. (1981) *Ecology and Planning*, George Godwin Ltd, London.

CHAPTER TWO

Ecology and Planning

2.1 THE PLANNING COMPONENT

A. G. Booth

2.1.1 Introduction

The United Kingdom was one of the first countries to provide for formal planning procedures through legislative action with the introduction of the *Town and Country Planning Act* of 1947. During its formative years planning as a profession was largely concerned with improving public health particularly in decaying urban centres in post-war Britain. Subsequently the emphasis in planning has shifted through engineering considerations, sociological and economic aspects and latterly environmental quality, reflecting changing concerns amongst society at large. In addition to these changes in the flavour of planning the formal procedures themselves, the framework within which decisions are taken, have also evolved to meet the society's changing requirements.

The central role of the planning profession as an information gathering and evaluation system has remained. Contrary to popular belief planners do not make the final decisions; their role is essentially to ensure that decisions are made by elected representatives from a position of informed knowledge. Planning is therefore a synthesizing activity and the skills of most planners lie in this area. It is here that the interface between planners and ecologists occurs in much the same way that interaction occurs between planners and many other specialist professions. After all planning is concerned as much with maintaining the physical fabric of urban environments in order to promote or maintain employment, residential and social facilities as it is with measuring or predicting the impact of a particular proposal.

In the UK, planning plays a central co-ordinating role in environmental management and resource allocation. It is by definition concerned as much with ecology as it is with economics, social sciences, land utilization and politics. This Chapter reviews the UK approach to planning and examines the context in which ecological information may be used.

2.1.2 Present planning procedures in the UK

There are three distinct planning procedures for the UK as a whole, relating individually to England and Wales, Scotland (which is similar) and Northern Ireland. For reasons of simplicity I shall deal with the first of these in this Chapter.

The 1947 *Town and Country Planning Act* introduced into the UK a formal system of planning based on Development Plans. These 'old-style' Plans were maps, normally at small scale (1:63 360) showing general policies for a County as a whole, with larger scale maps (1:10 560) showing primary land use allocations for individual major settlements within the planning authority area. The derivation of these policies and land-use allocation patterns were based on assessment of a number of considerations including population patterns, urban facilities and infra-structure, employment and recreation facilities, mineral extraction and public utility requirements and protection of the countryside (Table 2.1). A considerable amount of survey work was carried out for the original Essex Development Plan and was updated for its Review in 1964. However, agricultural land classification maps were not available then and only National Nature Reserves were mapped. Areas of Great Landscape Value were defined but on a very subjective basis. Both the policies and land use allocation maps had to be approved by the Minister of Housing and Local Government after public discussion in the form of a public inquiry. Development control decisions were normally taken on the basis of whether a particular proposal was in accordance with its development plan allocation, in some cases aided by any non-statutory policies which may have been formulated, such as village plans.

Despite the machinery for the review of the development plans which existed within the legislation, these 'old-style' development plans were increasingly criticized during the 1960s because of their inflexibility. As a consequence, the Minister of Housing and Local Government appointed a Planning Advisory Group in the early 1960s to review the system of town and country planning and recommended changes to meet the needs of the next 20 years and beyond. The Group reported in 1965 that although the system of control was effective, this control was based on plans which were out of date and technically inadequate. The result of this report was introduced in the *Town and Country Planning Act* 1971. This Act established the present system of Structure Plans and Local Plans and gave the Secretary of State for the Environment powers and duties to co-ordinate and approve the policies of local planning authorities and indeed to initiate policies.

The Structure Plan for a county or part thereof, is a written statement of the county planning authority's broad policies and general proposals that look forward 15 or 20 years, well beyond the period within which most land allocations and site definitions are made. This plan differs from the old-style

Table 2.1 Principles underlying the Essex Development Plan

(a) To limit within Essex the spread of the continuous built-up area of Greater London
(b) To define the Essex sector of the Metropolitan Green Belt and extended green belt around London and to preclude the coalescence of towns therein
(c) To provide for the expansion of towns outside the Metropolitan Green Belt and extended green belt; it is intended that the larger towns shall be expanded to the greatest extent
(d) To have regard to the agricultural importance of the County, to the capital cost of expanding public services, and to the limitations of the transport system and of the sources of water for public supply
(e) To provide sufficient land to accommodate a population of about 1 400 000 by 1981
(f) To provide for a much improved system of main traffic roads
(g) To provide for the decentralization of industry and offices from Greater London where it accords with Government policy
(h) To reserve for schemes under the Town Development Act, 1952, certain areas allocated for residential development at Braintree and at Witham
(i) To correlate the programme of residential, industrial and commercial development with the availability of social and public utility services, especially water supply and drainage, and with the improvement of the main road system and of the railway services
(j) To provide for appropriate development of those villages beyond the Metropolitan Green Belt and extended green belt where adequate social and public services exist or could be encouraged
(k) To provide for the controlled working of sand and gravel, brick-earth, clay and chalk, and to secure suitable after treatment
(l) Subject to sub-paragraphs (c) to (k)
  (i) To preserve the countryside of Essex and to enhance its appearance
  (ii) To protect the coast including the shores of the estuaries and creeks and the hinterland from development that would restrict their recreational value

Development Plan in that it is primarily a written document and is not concerned with the use of particular sites: but may have key diagrams which illustrate the main policies of the county strategy but they are not drawn on an O.S. base.

An important part of the plan preparation is the requirement for extensive public participation exercised at all stages, culminating in an Examination in Public (EIP). This is a unique device in the United Kingdom planning process, intended to be less formal than a public inquiry, the proceedings being designed to put more emphasis on a broad examination of strategic issues while not completely excluding a consideration of some detailed objections. An objector does not have a statutory right to appear at an EIP and can only do so on the invitation of the Secretary of State. This point has caused a certain amount of concern, but the argument in its favour is that the EIP is an examination of strategic issues, and the objectors have a statutory right to appear at subsequent local plan inquiries. Its advantage should reduce the time between submission of the Structure Plan to the Secretary of State and its eventual approval.

Although guidelines on content of the Structure Plan are issued by the Department of the Environment, the detailed content and precise information taken into account during its preparation is left to the discretion of the planning authority. Thus the factors considered and the relative weightings placed on them, vary from county to county reflecting the prevailing characters of the areas and the concerns of local residents. In the case of the Essex County Council Structure Plan particular consideration is given to employment prospects, housing developments to meet local needs and best use of natural resources including mineral deposits, high-grade agricultural land, wildlife reserves and areas of landscape value (Table 2.2). These considerations reflect the relative importance of agriculture and mineral extraction in the Essex environment and the special problems created by rapid housing growth in the 1960s–70s with inadequate facilities and employment. In contrast, the Merseyside Structure Plan places a greater emphasis on environmental pollution and land condition reflecting the smaller size of the planning area, the juxtaposition of industrial and residential areas and the relative awareness of local inhabitants. As a consequence detailed desk and field investigations of air and water pollution, waste, litter, noise and land dereliction were carried out as part of the preparation of the Merseyside Structure Plan (Handley, Section 5.2).

The Structure Plan, therefore, is basically a written statement of general policies and reflects national, regional and sub-regional strategies at a county level. It forms a framework within which a series of local plans formulate detailed policies and actual land allocations. These are now normally prepared by a District Council, but can be prepared at county level. They can cover any part, however small, of a local planning authority's area and can be in one of three forms: District Plans, Action Area and Subject Plans.

District Plans are based on a comprehensive consideration of matters affecting the development and other use of land in a particular area. This may be a whole local planning authority area or part thereof, perhaps a particular town or village. The type of information taken into account will vary according to the character of the area and the context of the plan. Table 2.3 for example lists the considerations included in the development of a District Plan affecting a small village in rural Essex. In the draft proposal for this village an emphasis is placed on the desirability of residential expansion

Table 2.2 Essex County Structure Plan: summary of strategy

(a)	To encourage economic growth, the expansion of existing employment and the introduction of new employment in order to improve the balance between workers and jobs in the county. To take positive measures to attract employment in areas with special problems. To make provision for about 650 hectares of industrial land and 250 000 m^2 of large offices
(b)	To provide for about 90 000 houses (plus intensification and vacancies) to cater for the County's own housing needs and to meet regional demands. To restrict further housing growth to land already committed for development, with exceptions. To gradually reduce the rate of housebuilding
(c)	To concentrate development in the towns and tightly contain villages but to allow sufficient growth in appropriate villages to help maintain rural community life
(d)	To maintain the Metropolitan Green Belt and provide a Southend Local Green Belt
(e)	To make the best use of the County's natural resources by preserving high grade agricultural land; safeguarding wildlife reserves and areas of valued landscape; concentrating recreational development where there is least impact on other resources
(f)	To allow mineral extraction where there is an identified need to obtain the mineral from within the County and where there would be no undue impact on other natural resources or the environment
(g)	Keep town centres as the urban focus and adapt them to meet future shopping demands
(h)	To maintain and make best use of the existing transport system by developing a hierarchy of roads; maintaining a basic level of public transport; increasing the use of traffic management and carefully locating new development
(i)	To keep future development to a level which is technically and financially feasible
(j)	To co-ordinate the provision of services with development
(k)	To provide a sound and workable framework but not to over commit the County so that the Plan may remain flexible and sensitive to the implementation of other plans and policies

to sustain an adequate level of community and transport facilities and the limitations imposed on expansion by current paucity of sewage treatment capacity, of prevention of further encroachment into good quality agricultural land and mineral reserves by concentrating development within a village envelope, and protection of important landscape and conservation features. In contrast, the Local Plan for the Mersey Marshes in Cheshire, an area of intense petro-chemical industrialization, placed a greater emphasis on

Table 2.3 Considerations included in the development of District Plan for a small village in Essex

Population	• twenty year growth pattern • size relative to adjacent settlements
Housing	• predicted future demand for area • availability and location of space within village limits
Employment	• number and type of jobs within village • pattern and scale of community
Transport	• number of car owners, rail and bus facilities, road links
Community facilities	• number of schools, doctors, dentists, shops, etc., in relation to population sizes needed to sustain such facilities
Public utility services	• availability of sewage treatment capacity, gas, electricity, etc.
Natural resources	• distribution of Grade 1 and 2 agricultural land • distribution of areas of great landscape value • distribution of nature conservation sites • distribution of mineral resources
Recreation	• availability of walking, horse riding, sporting facilities, allotments, village hall, etc.
Future requirements	• predicted population growth over 10 years for the local area • local employment prospects • predicted changes in utility services, recreation facilities and transport patterns • demands of increased leisure time

the likely effects of new developments on environmental quality criteria and included information from a number of specially commissioned surveys (Table 2.4) of existing standards of meteorology, air and water pollution, land drainage, waste disposal, noise, hazard and medical characteristics of the area.

Action Area Plans may be developed for an area which has been selected for comprehensive treatment in the near future by means of either development, redevelopment or improvement, or by a combination of these methods. For example the Essex Structure Plan indicated the preparation of an Action Area Plan for that part of the town of Braintree that is affected by the construction of new roads and improvement of others to form an inner relief road. Nationally, very few Action Area Plans have been prepared.

Subject plans are the third type of local plan, and these are designed to enable detailed treatment to be given to particular aspects of planning such as mineral workings, reclamation of derelict land, or countryside recreation. For example the Coast Protection Subject Plan defines the extent of the area to which the Coast Protection Policy in the Essex Structure Plan applies. Guidance on the future location of coastal water recreation facilities is given in a supplementary report; this is not a statutory local plan but 'Supplementary Planning Guidance' which is commended to District Planning Authorities to include in their local plans and also to assist them in the determination of planning applications.

Table 2.4 Environmental conditions surveyed in the Mersey Marshes Area

Meteorology	• 35 year climatological summary • temporal extremes in wind and atmospheric stability • spatial variations in wind and atmospheric stability
Air quality	• identification of potential sources of major pollutants • temporal and spatial variations in SO_2, smoke, acidity, particulates, ammonium, nitrate, chloride, fluoride, sulphate, heavy metals and ethylene
Water pollution	• identification of potential sources of major pollutants • survey of river classifications • summary of existing disposal facilities
Waste disposal	• identification of wastes arising in study area • survey of existing disposal facilities
Noise	• temporal and spatial variation in noise levels
Hazard	• summary of national fatal accident frequency patterns • identification of potential hazard sites in study area
Medical aspects	• assessment of age distributions for local settlements • assessment of death rates • assessment of crude death rates for respiratory disease asthma in local area on national data

In contrast to Structure Plans, Local Plans do not require the approval of the Secretary of State, but nevertheless they do require an extensive programme of public participation and consultation throughout their preparation which in some cases may culminate in an inquiry. They also require a certificate from the County Council to ensure that the policies and proposals are in general conformity with those in the Structure Plan.

The structure of planning in the UK was further fundamentally changed by the *Local Government Act* of 1972, which resulted in local government reorganization in 1974. This Act initiated the new, normally larger, District Councils, and in terms of planning the most important change was the transfer of most development control functions to them from the County Councils. The transition from the old to the new has not always been a smooth one and in many instances there have been conflicts between County Councils and District Councils as to which authority should determine certain applications. The situation has been clarified somewhat by the 1980 *Local Government, Planning and Land Act.* Development proposals are now generally processed at District level although the County Council retains responsibility for some areas of regional or strategic importance, e.g. mineral exploration and waste disposal.

Development control decisions are made on the basis of whether a proposal is in accordance with relevant Structure Plans and Local Plans (although compliance is not mandatory) and 'any other material considerations' and also in the light of written submissions from any party who may be directly affected by the proposal. The planning authority has a statutory obligation to consult various Statutory Bodies (the County Surveyor, Regional Water Authorities, Nature Conservancy Council and Ministry of Agriculture, Food and Fisheries, Fire Brigade and Electricity Board) and may, depending upon the nature of the proposal, also consult with various non-statutory organizations. The final decision, which is made by elected representatives, may take the form of a refusal, permission with certain consent conditions or non-conditional permission. The applicant has the right of appeal to the Secretary of State who may arbitrate either through written submissions or by ordering a public inquiry. The Secretary of State may also 'call-in' applications which he considers to be of regional or national significance, (e.g. marinas on rural coastlines; hypermarkets on the edge of towns; hotels in Green Belt areas or along motorways; major road schemes; airports; reservoir and mining developments) for assessment by written submissions or public inquiry. 'Call-in' may also be exercised when a development is a departure from the approved Development Plan or where significant public interest is aroused. The public inquiry system involves submission of evidence, and cross-examination, before an Inspector who then submits a report with recommendations to the Secretary of State. The decision is made either by the Inspector, or in the more important cases by the Secretary of State. These decisions are final, excepting on a point of law, when there can be an appeal to the High Court.

The legal definition of development in various Acts specifically excludes certain classes of proposals from planning controls. The implications for labour requirements, traffic density, air quality etc. in a change within a single industrial category and in some cases from general to light industry, cannot be formally assessed by the planning authority. Government departments and statutory undertakers are similarly excluded from planning controls although in certain circumstances local and national concerns may influence the Secretary of State to hold a public inquiry for a proposed development. There are two important implications to these exclusions. The first is that the local authority is not in a position to evaluate all the real pressures which arise for development in its area since many of them do not need its permission to proceed. The second is that a number of changes in the use of land and existing buildings with important local consequences, can take place outside the control of the planning authority.

All agricultural practices and erection and use of agricultural buildings smaller than 5000 m^2 are also excluded from the planning framework. Indeed the role of the planning process in the rural environment is limited essentially to the prevention of encroachment of urbanization. This means that a vast array of rural activists may proceed without formal permission even though, as illustrated in Chapter Four, changes in agricultural practices and reafforestation programmes are by far the largest cause of controversy between conservationists and developers.

2.1.3 Planning and impact analysis

The UK planning procedures have evolved over a long period of time and have been modified periodically as a result of experience, the development of new technology and the changing requirements of society. The procedures are rigorous and the various Acts are regarded in general as being very demanding even by post-NEPA American standards. Certainly, the conditions existing in America prior to NEPA were very different from those pertaining to present planning in the UK. Environmental Impact Assessment was introduced into the USA as a result of neglect, perceived by some groups, of balanced social and economic planning brought about by the separation of various facets of planning in several independent legislative and managerial institutions. In contrast, in the UK, although there is a large range of environmental control legislation and agencies (e.g. Table 2.5), the local planning authority plays a central co-ordinating role during policy development and project control.

Current conditions pertaining to UK planning are therefore very different from those of pre-NEPA America. The introduction of formalized procedures for impact analysis such as those embodied in EIA in the USA would therefore be difficult to justify. One of the major achievements of the UK planning procedures has been the development of a flexible approach to planning. Impact analysis, if it could be incorporated into the present framework, would increase this flexibility and may well have an important role to play.

Planners generally use ecological advice in a negative way. For example they use sites of nature conservation interests to restrict development and promote a particular type of management. Whilst this approach is important, for example in Essex where pressures for the development of

Table 2.5 UK laws relating to pollution

Pollution target	*Legislation*	*Enforcement agencies*
Air	*The Alkali etc. Works regulations Act* 1906 *Public Health Acts* 1936, 1961 *Clean Air Acts* 1956, 1968 *Radioactive Substances Act* 1960 *Road Traffic Acts* 1960, 1972 *Road Safety Act* 1967 *Motor Vehicles (Construction and Use) Regulations* 1969 *Control of Pollution Act* 1974 *Health and Safety at Work etc. Act* 1974	HM Alkali and Clean Air Inspectorate Local Authorities Department of the Environment
Freshwater	*The Salmon and Freshwater Fisheries Act* 1923 *The Public Health Acts* 1936, 1961 *The Public Health (Drainage of Trade Premises) Act* 1937 *The Water Acts* 1945, 1948 *Rivers (Prevention of Pollution) Act* 1951, 1961 *Clean Rivers (Estuaries and Tidal Waters) Act* 1960 *Water Resources Act* 1963 *Water Act* 1973 *Control of Pollution Act* 1974	Regional Water Authorities
Marine (+oil pollution)	*Petroleum (Production) Act* 1934 *Oil in Navigable Waters Act* 1955, 1963, 1971 *Continental Shelf Act* 1964 *Sea Fisheries Regulations Act* 1966 *Prevention of Oil Pollution Act* 1971 *Merchant Shipping (Oil Pollution) Act* 1971 *Dumping at Sea Act* 1974 *Control of Pollution Act* 1974	Department of Energy Department of Trade MAFF Local Authorities
Waste disposal	*Public Health Act* 1936 *Civil Amenities Act* 1967 *Litter Act* 1967 *Dangerous Litter Act* 1971 *Disposal of Poisonous Wastes Act* 1972 *Control of Pollution Act* 1974	Local Authorities
Radioactivity	*The Radioactive Substances Act* 1948, 1960 *The Atomic Energy Authority Act* 1954 *The Radiological Protection Act* 1970	Department of Energy Department of the Environment MAFF HM Alkali and Clean Air Inspectorate Radiochemical Inspectorate

marginal land (wetlands and ancient woodlands) and coastal areas are strong, planning is increasingly taking on a more positive role. Introduction of ecological methodologies such as impact analysis may well assist in this positive approach, for example in the selection of pollution control requirements. In addition, ecological principles concerned with the functioning and repair of ecosystems may also have application to many types of developments.

The level within the planning system at which impact analysis and other ecological information is to be used needs careful consideration. Structure Plans reflect national, regional and sub-regional strategies at a county level. It is important that the need for EIA for any major proposals are clearly defined at the highest level, i.e. from national policies downwards, as once these polices become localized the scope for alternative strategies which an EIA may indicate becomes more limited. Nevertheless it is essential that Structure Plans indicate clearly the need for EIA in specific cases, not only in instances which are pre-empted by national or regional policies but, even more important, in policies or strategies in a Structure Plan which are not the result of a localized application of these national or regional policies. This can be done within the framework of existing legislation as the *Town and Country Planning Acts* of 1971 and 1972 require local planning authorities to consider measures for the improvement of the environment and management of traffic. The type of proposal dealt with at Structure Plan level for which EIA might be considered necessary would normally come into one or both of these categories. Examples might include toxic-waste disposal, nuclear power stations, mineral exploration, gas terminals, oil refineries, airports, motorways etc. Similarly at Local Plan level, the requirement for EIA should similarly be repeated for proposals or policies which result from decisions at a higher level and only initiated for proposals which do not come within this category. At Local Plan level the number of cases where this is likely to occur must surely be limited and are likely to include only such proposals as marinas, new settlements and large housing estates and agricultural improvement schemes.

There is a clear justification for EIA becoming an integral part of the planning process, and indeed a form of it can be incorporated into the existing structure without great difficulty, even allowing for possible amendments which may occur in new legislation. Indeed there are numerous cases, many of them dealt with in detail in subsequent sections of the publication, where EIA has been carried out within the existing planning framework. Properly applied, with a definite time limit set for its preparation, there is no reason why it should lead to further delays in decision making, something the Government is committed to eliminate.

2.1.4 Conclusions

Planning is a learning process and must reflect society's concerns about the environment. Planning has evolved from an engineering/public health bias in its early development through to reconstruction after the second world war; followed by an emergence of geographers and other social scientists, economists and sociologists, in the 1960s. The emphasis here was largely concerned with the social and economic effects of land-use planning and economic growth (getting back to the 'people') more recently concerns have been expressed about countryside and the quality of the urban environment. These feelings have been reflected in relevant regional and local plans; present day Structure Plans are conservative in comparison with the Development Plans of the 1960s.

The UK planning procedures have evolved to meet these changing emphases. The present approach is based on a system of flexibility. Ecological approaches, whether in the form of impact analysis or less structured methodologies, have a role to play provided they can be accommodated in such a way as to maintain this flexibility. If they assist in the positive aspects of planning implementation or 'getting things done', then they will serve an important role. However, the final decision is based on broad ranging and diverse considerations and it is politicians who have the ultimate say, not planners or ecologists. They can only try to ensure that all the arguments are presented to decision-makers with ecological factors treated in the same way as social and economic considerations are.

2.2 THE ECOLOGICAL COMPONENT

A. D. Bradshaw

2.2.1 Introduction

Land-use planning is essentially a process of arbitration, of choosing between often mutually exclusive uses of land and of ensuring, either through permission, refusal or consent conditions, that a new land-use proceeds only with an acceptable level of environmental impact.

The process is, however, complex; it involves an assessment of the proposal and the development site and a prediction of the effects that the new development, no matter of what sort, will have on the environment. New developments are essentially physical in basic character, whether they are a new chemical works, quarry, or sewage-treatment plant. So they cause, primarily, physical disturbances such as in water use, landform or release of chemical compounds. But let nobody think that, as a result, we are dealing with a series of essentially simple disturbance situations about which precise predictions can easily be made.

The complexity of impact analysis arises because both developments and the environments they may affect have many different attributes. The complexity of the environment was well outlined by Catlow and Thirwall (1976) who suggested that there are at least 12 recognizable subdivisions of the environment which may be separately and differently affected by a new development. They indicated that these subdivisions were only a first level of division. In nearly all categories especially that of the natural environment considerable further subdivision is indeed necessary.

This analysis of the environment into identifiable components has been followed by the further analysis by Leopold *et al.* (1971) and Clark *et al.* (1976) who recognized that developments themselves can have a variety of different components, each of which can have its own specific effects. As a result they suggested that environmental impacts can only be analysed by realizing that we are dealing with a potentially complex matrix of interactions between components of the development and components of the environment. Such a matrix analysis does not seem at all unreasonable, even if it is complex: in the matrix produced by Clark *et al.* there are 936 cells. Indeed there is no other way environmental impacts can be analysed and land-use planning put on a logical basis even if the process must be kept as simple as possible.

Where in all this then is the potential contribution of the ecologist? The background of an ecologist is biological science, and his primary concern is with living organisms. But professionally his, or (from the beginnings of ecology) her, competence and experience is in understanding the factors that determine how and where things live, how they are affected and interact with their physical environment and with each other. Ecologists in land-use planning are able to assess the potential of land for a variety of uses and to predict and estimate impacts on biological systems and on the organisms contained within them. They are therefore the profession who because of their special training should have the responsibility for the biosphere in land-use planning.

What this means in practice is that they should have direct responsibility for about 115 cells of the impact matrix devised by Clark *et al.* (1976) and have interest in about 66 others, where biological problems may occur (Fig. 2.1). This is a considerable reduction in the number of areas requiring study, but it still leaves some very complex problems to solve, which become apparent if we look at environmental impacts in practice.

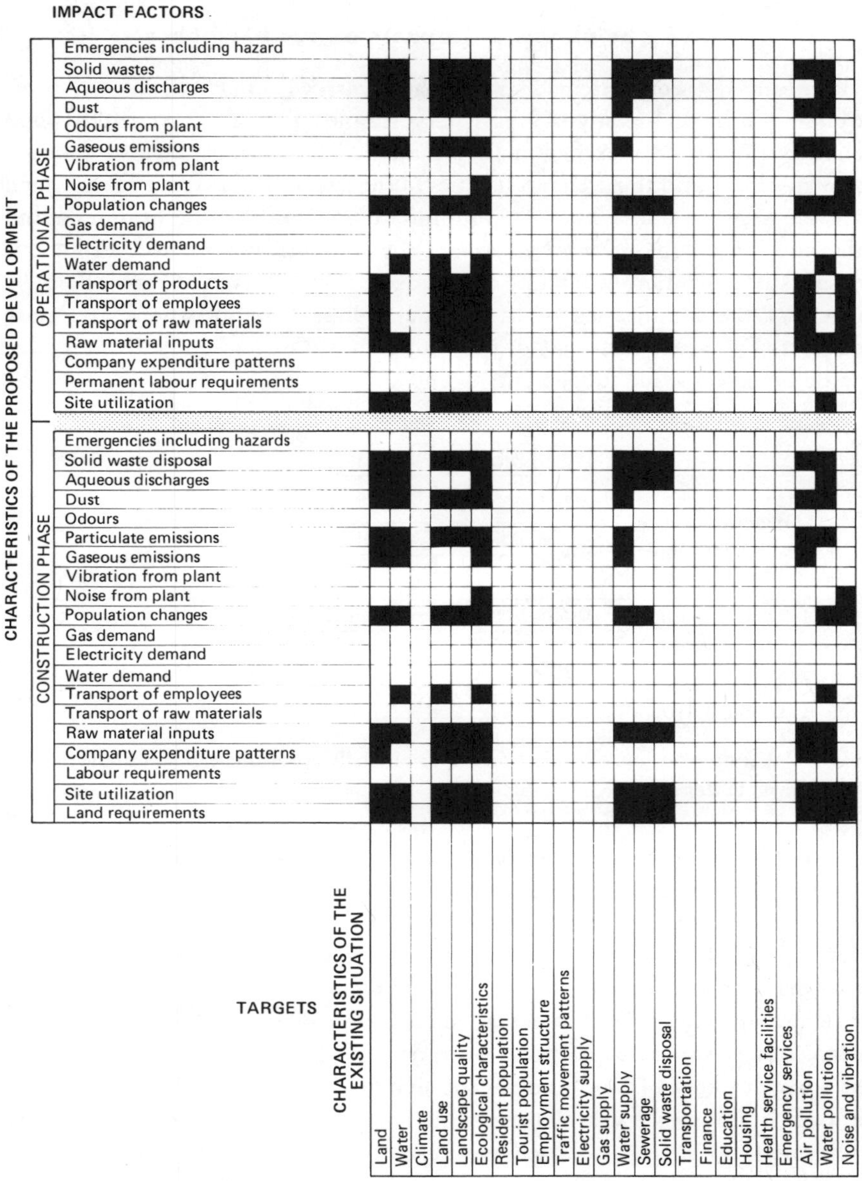

IMPACT FACTORS

CHARACTERISTICS OF THE PROPOSED DEVELOPMENT

OPERATIONAL PHASE

Emergencies including hazard
Solid wastes
Aqueous discharges
Dust
Odours from plant
Gaseous emissions
Vibration from plant
Noise from plant
Population changes
Gas demand
Electricity demand
Water demand
Transport of products
Transport of employees
Transport of raw materials
Raw material inputs
Company expenditure patterns
Permanent labour requirements
Site utilization

CONSTRUCTION PHASE

Emergencies including hazards
Solid waste disposal
Aqueous discharges
Dust
Odours
Particulate emissions
Gaseous emissions
Vibration from plant
Noise from plant
Population changes
Gas demand
Electricity demand
Water demand
Transport of employees
Transport of raw materials
Raw material inputs
Company expenditure patterns
Labour requirements
Site utilization
Land requirements

TARGETS

CHARACTERISTICS OF THE EXISTING SITUATION

Land
Water
Climate
Land use
Landscape quality
Ecological characteristics
Resident population
Tourist population
Employment structure
Traffic movement patterns
Electricity supply
Gas supply
Water supply
Sewerage
Solid waste disposal
Transportation
Finance
Education
Housing
Health service facilities
Emergency services
Air pollution
Water pollution
Noise and vibration

Fig. 2.1 The contribution of the ecologist: those cells of a typical impact matrix (that of Clark *et al.*, 1976) which should be the responsibility of ecologists.

2.2.2 The nature of impacts on the biosphere

When any one component of the impact matrix is analysed the action required for impact analysis seems very simple (Fig. 2.2). The individual character of the development acting as an *impact factor* must be distinguished and specified, and those characteristics of the site and its surroundings which are the *target*, defined. The potential interaction can then be examined and information on the *predicted outcome* of the interaction obtained. From this essentially logical *analysis* an *assessment* (see Figs 2.2 and 2.6) can then be made of the significance of the impact – these two operations are quite distinct. This assessment may or may not be based on logic: it will often have to be based on value judgement.

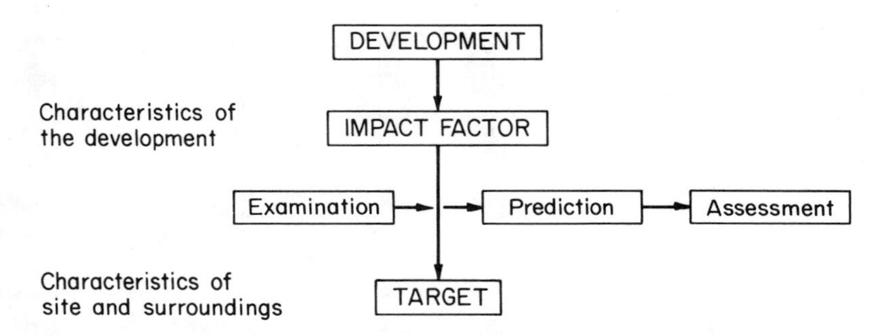

Fig. 2.2 A simple outline of the action required for the assessment of any single impact factor/target interaction.

All this should apply to impacts on the biosphere. To see whether the resulting impact analysis is easy or difficult, and how far the special skills of the ecologist are important, it is instructive to examine a number of past developments which have had biological impacts: there is no better tutor than past experience.

There are many developments which have effects which are simple and predictable. Any past quarrying operation, for instance, has caused the loss of a certain area of land and all the living material within it: it has little impact outside except sometimes visual appearance, dust, disturbance by noise and problems associated with transport. The impact factor and its target can be readily defined, and predictions about the impact are easy to make and can be precise.

More complex developments, such as chemical works, can similarly give rise to simple, though sometimes substantial, impact factors acting on easily recognizable targets. There was, for instance, a chromate works at Little Lever near Bolton which can only be said to have been a most unsavoury operation with very severe effects, yet in essence these were simple. The process deposited large amounts of alkaline chromate waste on a few acres of land, which, because of the alkalinity and the residual chromate, left the land completely sterile and unusable (Gemmell, 1973). The chromate is readily soluble so the drainage water from the heaps maintained high levels of chromium in the receiving River Croal. The section of the river below the works was effectively devoid of aquatic life until it joined the River Irwell (Breeze, 1973). Its impact has therefore been simple and definable. However, it did require careful ecological work by the two authors mentioned to identify the significant impact factors: the toxicity of chromate in particular had not been reported previously.

(a) Complexities in impact factors

In many situations, however, a development can produce an impact factor which is less definable. Thus the presence of an apparently innocuous phosphoric acid plant (now disused) near Liverpool has been shown by analysis of aerial photographs taken at different periods of its history to be associated with the progressive destruction of a 200 year old oak wood which surrounded it. The causal link was not at all clear in the absence of any obvious impact factor. However, it was found that there was considerable emission of phosphoric and hydrochloric acids which only reached ground level for short intervals on rare occasions in particular meteorological conditions: they were therefore very difficult to identify as the cause without continuous monitoring which had not been carried out (Vick and Handley, 1977). Indeed there are many situations where the existence of an impact factor has been recognized after damage has been caused for instance chlorinated hydrocarbons causing decline in birds of prey, and ozone causing chlorotic dwarfs in eastern white pine.

There is, of course, one major example of an apparently well-known impact factor whose significance has recently given rise to much discussion. This is the sulphur emissions often associated in many peoples' minds with electricity generation, but in reality produced from many sources. The very high levels of sulphur dioxide produced at ground level in the past by domestic coal burning and early power stations are well recognized. However, emissions from domestic sources have now been reduced substantially as a result of legislation: and although emissions from power stations have not decreased, their effect on ground level concentrations of SO_2 have been substantially reduced by the development of tall stacks to ensure effective dispersion. The result is that SO_2 levels in polluted areas are now down to about 120 $\mu g\ m^{-3}$ during the winter, which perhaps do not have significant effects. But meteorological conditions cause major fluctuations in SO_2 levels and peaks can still reach 500 $\mu g\ m^{-3}$ or more (Fig. 2.3).

At the same time we now recognize that SO_2 is only one part of the total

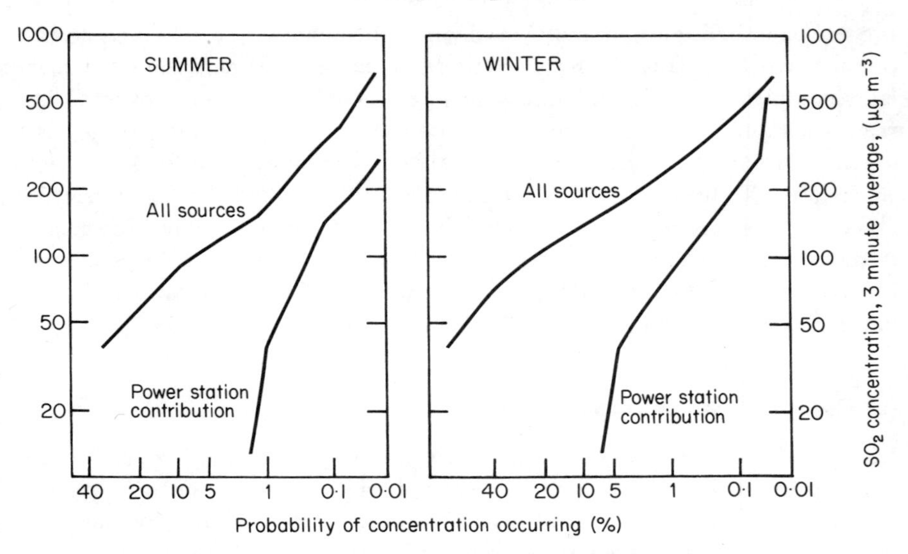

Fig. 2.3 An example of the fluctuations in levels of an impact factor: frequency distribution of SO_2 levels 6 km from a 2000 MW coal-fired power station in S. Yorkshire.

impact factor. What may be of more general importance is the oxidized sulphur which can be distributed very widely across national barriers and is probably contributing to lowering of pH in surface waters and soils (Drablos and Tollan, 1980). In Europe we now have a fairly clear picture of what sulphur goes where, but its significance as an impact factor is not yet clear.

(b) Complexities in targets

The significance of an impact factor can really only be judged when we know its target and the effects that it has on it. Sulphur emissions show the problems of identifying targets and quantifying effects. The past effects of high levels of SO_2 on the growth of trees and lichens are well defined. Apart from having prevented the growth of conifers in high SO_2 areas we have not thought the effects on the biosphere to be very important. But recently it has become clear that the bogmoss, *Sphagnum* sp., is very susceptible to SO_2, and there are historic records of the disappearance of *Sphagnum* from the Pennines at the time of the industrial revolution. This could be dismissed as being important only to ecologists, but for the fact that the widespread erosion of peat in the Pennines began at this time (Tallis, 1964).

At the present time we tend to think that the lower levels of SO_2 do not have significant effects. Lack of visible injury has been held to indicate lack of damage. However, significant reductions in yield of crops, without visible injury, due to SO_2 levels as low as 200 μg m^{-3} are now well established (Bell and Clough, 1973). What we still do not know properly, however, is the impact of short-term fluctuations in concentrations. We have also to take into account the possibility of genetic variability with evolution of tolerance in the target organism (Horsman *et al.*, 1979) as can occur in relation to metal toxicity (Bradshaw, 1976). So about the effects of a major impact factor we are still much in the dark.

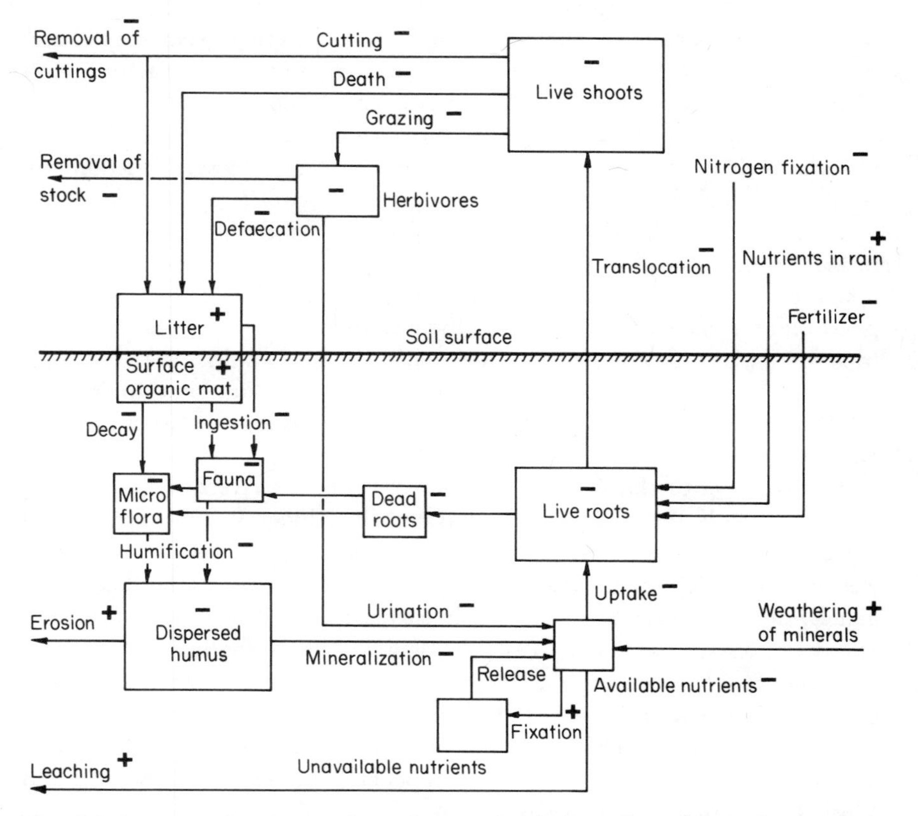

Fig. 2.4 An example of a moderately complex impact factor/target interaction: likely changes in nutrient cycling in an urban grassland as a result of SO_2 pollution: +, increase over normal; −, decrease over normal; boxes indicate approximate normal size of stores.

We also know too little about the effects of sulphur in precipitation – acid rain. Acidity in river water can certainly cause death of fish, but how far it is the direct cause of the disappearance of fish in river of areas such as Scandinavia (Roy. Min. Foreign Affairs, 1971) is debatable (Anon, 1979). The effect of sulphur on soils can be more readily understood: it must cause an increased loss of exchangeable bases and a lowering of soil pH. How rapidly this will occur will depend on the rate of sulphur deposition and the base status of the soil. In agricultural situations any deterioration of soil quality can be prevented by the addition of lime.

However, impact factors rarely act on single targets in the biosphere. They are more likely to act on whole ecosystems. This is very clear if we take a detailed look at what is likely to happen when increased quantities of sulphur fall on the simple ecosystems of a park grassland or for that matter of any unfertilized, unlimed grassland in Britain with a low base status. A very complex series of changes occurs, connected with lowering of surface pH and loss of bases (Vick and Handley, 1975), which mean that ultimately a grass sward of inadequate growth and wear resistance, poor colour, considerable surface mat and poor drainage is produced (Fig. 2.4).

Perhaps the best example of complexities in a target ecosystem leading to complexities in impact, particularly because the ecosystem is being subject to many different impacts, is shown by the present state of the Norfolk Broads, when considered in relation to one recent development – the apparently innocent and beneficial construction of a number of sewage disposal plants which have discharged their effluent into the rivers. A sewage treatment plant gets rid of organic matter which would otherwise have substantial polluting effects mainly by causing deoxygenation. However, it discharges an effluent which although clear contains high levels of nitrogen and phosphorus.

As a result of discharge of sewage effluent into the rivers and, therefore, into the Broads there have been massive algal blooms, disappearance of aquatic macrophytes and drastic declines in the duck and other birds which feed on the macrophytes (George, 1977). All this should have been predicted; but the possibility of such a disastrous change was never considered and the necessary predictive work not done. The disentangling of the significant impact factor causing the deterioration has taken a great deal of work because a large number of different impact factors are operating, mostly related to increased tourism and not associated with the sewage treatment plants, but the output of phosphorus appears to be crucial (Osbourne and Moss, 1977; Phillips *et al.*, 1978).

This example serves to show the problems of impact analysis related to identifying impact factors. But the same example shows the complexity of targets when an ecosystem is involved. There has been an astonishing disappearance of the common reed (*Phragmites australis*) which is crucial as a wildlife habitat. There has also been a substantial increase in the occurrence of the disease, botulism, which has killed large numbers of wild fowl. Both of these changes appear to be related to an increased frequency of anoxic conditions in the water. Many of the biological impacts are therefore sequential, and the whole structure of the environmental impacts on the Broads is exceedingly complex (Fig. 2.5).

A tidal barrier is now proposed at Great Yarmouth to prevent the

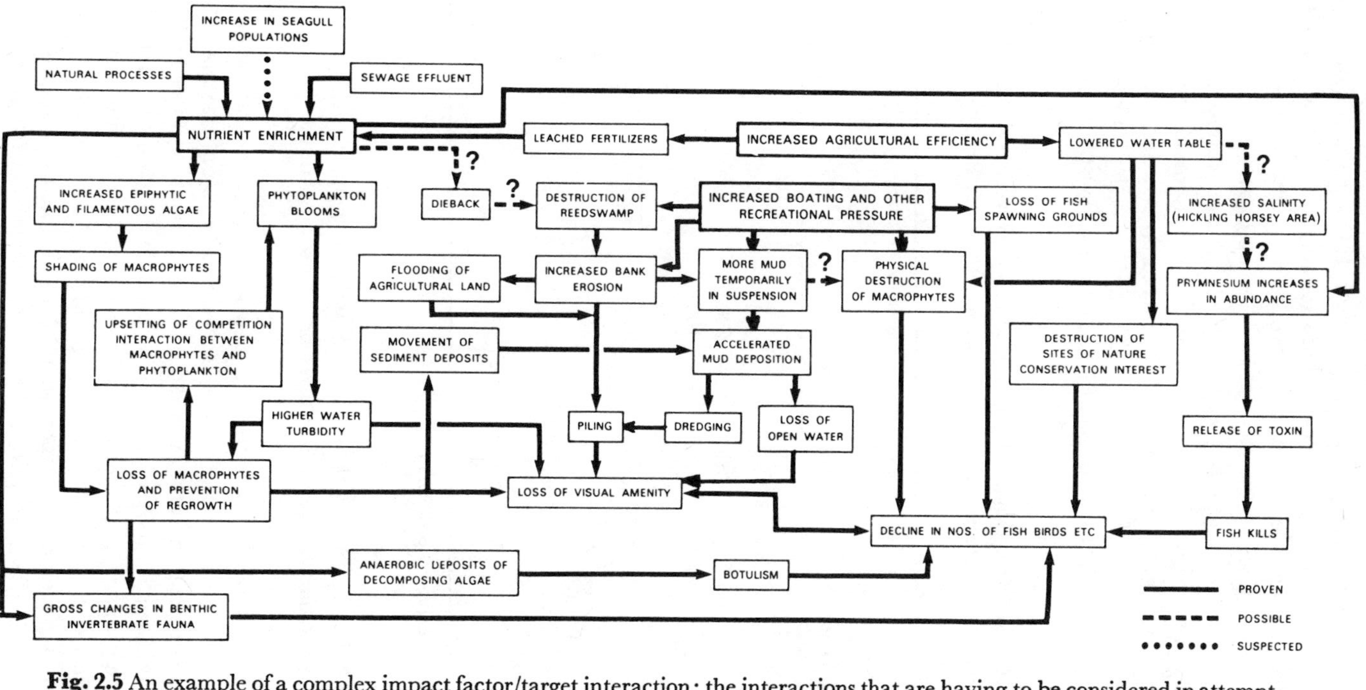

Fig. 2.5 An example of a complex impact factor/target interaction: the interactions that are having to be considered in attempting to understand the present very serious decline in water quality and species abundance in the Norfolk Broads (George, 1977).

occasional winter flooding that now occurs. If we are to predict its environmental impact properly a very elaborate analysis will be necessary. It will affect water movement and therefore salinity and nutrient levels. The security from flooding will increase the amount of arable agriculture (Table 2.6): this will directly cause loss of habitats important for wildfowl. The better drainage will cause loss of ditches which are important refuges for many aquatic plants and animals. The increased aeration of the soil together with increased use of fertilizer will add further to the nutrient input to the system. The targets likely to be affected are very complex: they will only be

Table 2.6 Predicted changes in agriculture in the Norfolk Broads following the construction of a tidal barrier at Great Yarmouth (from Rendel *et al.*, 1977) (land area in hectares)

		Future with no barrier		
Enterprise	*1977*	*Present flood risk*	*Reduced flood risk**	*Future with barrier*
Wheat	2 190	2 800	4 740	6 950
Barley	240	240	530	1 090
Roots	290	290	630	1 230
Peas and vegetables	200	200	420	820
Improved grass	1 940	2 550	4 210	6 530
Sub total – improved area	4 860	6 080	10 530	16 620
Unimproved grass	13 370	12 150	7 700	1 410
Waste created	—	—	—	200
Total land area	18 230	18 230	18 230	18 230

* Assuming local flood protection.

identified by a sequential impact analysis or some equivalent with considerable ecological input. If this is not done, the impact analysis will be inadequate and misleading.

The biosphere is a complex of coexisting species and processes. Any prediction of the impact of a new development can only be made by a very careful ecological analysis. It would be very satisfactory if this implied that a careful ecological analysis will always provide an accurate prediction. Unfortunately this is not true: ecology has developed considerably in the last three decades but the biosphere is complex and ecologists have not yet had sufficient involvement in impact analysis to be able to make predictions with confidence. But as can be seen from this publication the situation is being rectified.

2.2.3 Evaluation of the significance of impact

It is not sufficient to predict an impact: its significance has to be evaluated. The agricultural value of lost land can be quantified in relation to market values or to its land use capability classification (MAFF, 1974). Loss of quality in water to be used for water supply can be assessed in terms of the increased cost of its treatment. However, the value of any natural or semi-natural vegetation is more difficult to quantify. Recently it has become possible to distinguish primary and secondary woodland on the basis of the composition, in terms of species, of the ground flora (Peterken, 1974). When this is used in combination with historic records and the evidence of surface features (Rackham, 1976) it is possible to recognize original woodlands with some precision. Since these original woodlands are now becoming rare, less than 20% of the remaining deciduous woodland, the destruction of any more is equivalent to the destruction of an original work of art such as a Chippendale chair. On this basis a recent proposal for the development of a quarry which would have destroyed Darenth Wood in Kent was not allowed, despite the undeniable economic losses incurred.

In much of Britain, however, vegetation is not original, and other criteria for its value must be found which include an evaluation of its associated animals. These are discussed by Goode (Chapter Five). The very important account of Sites of Special Conservation Value recently published by the Nature Conservancy Council (Ratcliffe, 1977) goes some way to ensuring that the sites which are outstanding on a number of different criteria are recognized and so can be preserved at all costs. It is backed up by the system of Sites of Special Scientific Interest (SSSIs) which are intended to ensure that a sufficient quantity of appropriate environments remain which will ensure the continued existence of our wildlife. However, the criteria for these are based rather too much on floristic diversity, when other characteristics such as rarity, intrinsic interest and wildlife potential must also be taken into account.

Unfortunately we can readily get bogged down in discussion on criteria and it is too easy with SSSIs, for which there is no statutory requirement for their preservation, to decide that their value is unimportant in the face of some important proposed development. The trouble is that *nature conservation value* is difficult to define and is inevitably mainly subjective. It is true that we should be careful not to lose individual species because they may be useful to us. But in the end it is whether or not we wish to live in a biological desert which is the ultimate criterion. It is all too clear that erosion of our wildlife resources is going on at a frightening rate (Moore, 1962; Bradshaw, 1977). What is worrying is that this is the outcome of a slow piecemeal series of changes such as those discussed by Handley (Chapter 6). Ecologists have, therefore, a very important part to play in defining the extent of the resource, documenting its erosion and producing a rational set of criteria for conservation at all levels which can be used in impact assessment. This is in fact an important activity of the Nature Conservancy Council (NCC) at the present time (NCC, 1975).

2.2.4 Impact modification

It is easy to take a simplistic and rather narrow view of impact assessment – that it is something to be done once when a development proposal has been made, so that the proposal can be judged and either accepted or rejected. This seems very likely if environmental impact analysis is solely the responsibility of planning authorities as is suggested by Clark *et al.* (1976).

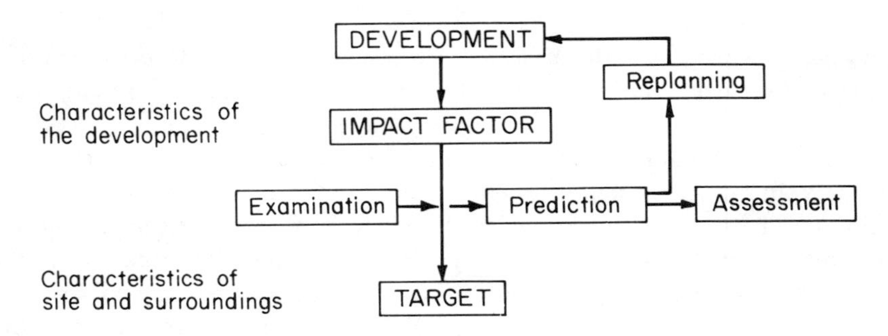

Fig. 2.6 An outline of the action required for the constructive use of impact analysis: the analysis should involve positive feedback to ensure reduction of impact and cost, and maximization of benefit (*cf.* Fig. 2.2).

But it is surely sensible that the predictions from an analysis should be used at an early stage by the developer to prompt him to examine alternative ways to carry out the development. The programme of analysis for any individual component of the matrix which was suggested earlier requires a feedback loop (Fig. 2.6). If you ask an ecologist to predict the impact of a particular development you will find that he will nearly always indicate that his prediction would be different if certain modifications were made to the development. Indeed it seems a fundamental aspect of impact analysis that different options should be considered.

In this case the analysis should be carried out by the developer, and then for impacts on the biosphere the ecologist should work with the developer. This would give him an advantage since he would be able to make suggestions at an early stage in the development and carry out any necessary experimental work in good time.

(a) Reduction of impact

There are many examples where the ecologist can provide substantial reductions in impact. In Norwegian hydroelectricity schemes, large and therefore obtrusive quantities of rock are produced by the excavation of tunnels. By use of pocket planting, trees can be established direct on the waste: if the fines from the excavation process are removed and spread last, they can, with the use of appropriate fertilizers, be an effective medium for grass growth (Hillestad, 1973). Both techniques are inexpensive and get rid of the environmental impact almost totally. Low cost ameliorative treatments of this sort are possible in many situations (Bradshaw and Chadwick, 1980).

In coal mining, whether by surface or deep workings, a major problem is the pyrite contained in the associated strata. This oxidizes to cause extreme acidity – down to pH 2·5 – which makes plant establishment on the wastes extremely difficult and can lead also to serious problems of acid mine drainage. Treatment by high levels of liming is possible. However, the different strata adjacent to a coal seam nearly always vary in pyrite content. If an impact analysis is carried out as the development is being planned it may be possible to identify the strata containing high levels of pyrite and ensure that these get buried under more kindly materials (Grim and Hill, 1974).

(b) Reduction of costs

One major problem that developers foresee arising from the imposition of detailed environmental assessment is that their costs will rise because their developments will be limited or they will be required to undertake extensive impact control measures. Yet any impact analysis where alternative methods of development are being considered can include examination of ways in which environmental standards can be met at reduced cost. Again these must be considered at the outset of planning a development and will often only become apparent from a full impact assessment.

In operations that involve large-scale earth movement it is becoming normal to envisage topsoil replacement. In many situations this is the most effective or the only way in which the productivity of the land can be restored after mining. However, there are situations for example in the Bowen Basin Coalfield in Australia or in kaolin mining areas in Georgia, where investigations have shown that the subsoil, when appropriately treated can form a soil which is as productive or even more productive than the original. The savings in haulage costs can markedly alter the profitability of such mining operations.

In Britain there are several situations where to comply with planning requirements topsoil has been purchased. But it has been of such poor quality and has been used over such high quality subsoil that its use has not only cost large sums of money but it has given a poor end-product and even degraded the ultimate value of the land (Bloomfield *et al.*, 1981).

(c) Maximization of benefit

There are many situations where developments have radical effects on the environment. This is particularly true in mining, quarrying and other developments involving land disturbance. It is all too easy in these cases to propose that the land should be returned to its former use. Yet after a proper impact analysis alternative uses may well suggest themselves. Where developments of necessity cause radical changes in land form – this is particularly true of quarries – a new land use must be sought.

New land uses range from agriculture to industry or housing, but they can include forestry, nature conservation, or recreation. The possibility of the change and the benefits which arise can sometimes be obvious: but they may often not be apparent without careful work.

Gravel pits in low-lying alluvium usually flood when working has finished. We have now become aware that with ecologically-based landscape treatment the open water complexes can be made into immensely attractive recreation areas. But these areas with forethought can also be made into nature reserves to reverse the present appalling losses in wetland habitats: the Cotswold Water Park, and the Sevenoaks Gravel Pit Nature Reserve (Harrison, 1974), are excellent examples of promise and achievement. In the United States after careful ecological treatment an area of disused gravel working was sold for high-class housing for more than its original cost.

Quarries, particularly, are sites of considerable potential: yet so far in Britain this has not been realized. We now know that the development of a natural vegetation can be achieved rapidly if appropriate techniques are employed. As a result quarries can be turned into wilderness areas of great recreation and wildlife potential. There is no reason why quarries in National Parks cannot be assets rather than liabilities, acting as recreation and wilderness areas taking pressure off the more fragile areas of original landscape (Bradshaw, 1979). Old naturally colonized quarries such as Millersdale in Derbyshire, and Grays in Essex show the potential.

Even extreme situations have potential. Effective afforestation techniques of colliery spoil heaps, as in Germany in particular, can turn objectionable eyesores into areas of considerable forestry and recreational value. Disturbed land offers immense possibilities (Bradshaw and Chadwick, 1980). So it is extremely important that the possible benefits that can be derived from a development are considered at a very early stage so that any necessary manipulative work can be carried out at low cost during the life of the development, rather than afterwards at great expense.

2.2.5 Biological monitoring

It would be foolish to presume that the impact of a development would always turn out in practice as predicted from an impact analysis. Characteristics of the development acting as impact factors might deviate from original specification, or might have effects different from those predicted. Deviations of emissions etc. can be checked by careful physical monitoring, but failures to predict effects correctly can only be discovered by biological monitoring.

Biological systems are always in a state of dynamic equilibrium and are sensitive to environmental fluctuations. As a result there is constant change in any parameter of the system which is measured. Any short- or long-term effects of an impact factor which are being monitored must be distinguished from these natural fluctuations, which themselves can be short or long term.

It is always possible that an impact factor has effects on a totally different target from those originally envisaged, such as in the case of the common reed in the Norfolk Broads. Perhaps the best example of this, however, which has extremely serious consequences for wildlife was the totally unpredicted effects of chlorinated hydrocarbons being used as pesticides on egg shell thickness (Ratcliffe, 1967). This caused a substantial decline in the populations of a large number of much-valued raptor species of birds such as the peregrine falcon.

It is very difficult to devise a foolproof monitoring system, but there is no doubt that any development must be preceded by a properly designed *base-line study* which records with precision all the biological attributes of the environment which might be affected and takes into account fluctuations of natural origin. Only a single detailed survey may be necessary at one point in time. But if natural fluctuations in any of the targets within this environment are expected, the base-line study must include surveillance over an appropriate time period. For this the many existing surveillance schemes, for example of common birds, carried out by voluntary bodies can be very important. Biological monitoring is not easy but considerable strides have been made recently in its science and practice (NERC, 1976).

The main value of monitoring is in providing a time dimension to the analyses and therefore the assessment of a particular development. Its information will of course be of value in the feed-back loop of Fig. 2.6 since its results can be used to suggest modifications to the development while it is in progress, so that performance can live up to promise.

But it can be of crucial value for subsequent developments in the same or similar industry or in the same or similar environments. We have already seen that because the biosphere is so complex we are not yet able to predict its response to impacts with precision. Monitoring provides an important way in which we can accumulate experience. But the results of such monitoring exercises must, within the normal limitation which may have to be set by industry, be published, otherwise the experience cannot be passed on. In public reaction to environmental impact the performance of any industry is judged as a whole and not as its separate enterprises.

2.2.6 Conclusions

So the contribution of the ecologist to land-use planning is substantial. The biological environment is not only our supplier of food: it is also our spiritual heritage. It seems unlikely that anyone will ever wish to live in an environment of arable fields surrounded by concrete, where no birds sing.

There are several elements to the ecologist's contribution (Table 2.7). Not

Table 2.7 Work required of ecologists in impact analysis

Work	Details
1. Identification of major impact factors affecting biosphere	• nature • level
2. Identification of primary targets in biosphere	• nature of species, soil, water etc. • nature of ecosystems • quality of these
3. Prediction of primary impacts on biosphere	• on species, soil, water etc. • on ecosystems
4. Prediction of secondary impacts on biosphere	• on species, soil, water etc. • on ecostems
5. Experiments to confirm predicted impacts (if necessary)	• on particular targets • on particular ecosystems
6. Recommendations to modify development	• to minimize impact • to maximize benefits • to minimize costs
7. Biological monitoring	• base line studies • observations on known targets • observations on possible targets dissemination of findings

Table 2.8 Skills required of ecologists involved in impact analysis

1. Knowledge of impact factors
2. Knowledge of relevant species and ecosystems
3. Knowledge of past case histories
4. Ability to predict impact in relation to special characteristics of particular development
5. Ability to carry out realistic experiments on impact problems
6. Knowledge of development sufficient to be able to propose sensible modifications to reduce environmental impact

all the work is, as we have seen, easy because the environment is complex and predictions correspondingly difficult. It requires skills which are the special province of the ecologist (Table 2.8). But the need to work with developers and planners, as well as the need to have accumulated experience of the ways in which environmental impacts manifest themselves, means that those ecologists who wish to contribute to environmental management must leave their ivory towers and be prepared to use their academic knowledge for practical ends and to understand and work with industry and the professions involved. We shall only have the environment we work for.

REFERENCES

Anon (1979) *Effects of* SO_2 *and its Derivatives on Health and Ecology*, IERE Report Vol 2, CEGB, Leatherhead.

Bell, J.N.B. & Clough, W.S. (1973) *Nature (London)*, **241**, 47–9.

Bloomfield, H.E., Handley, J.F. & Bradshaw, A.D. (1981) *Landscape Design*, **135**, 32–4.

Bradshaw, A.D. (1976) in *Experimental Studies on the Biological Effects of Environmental Pollutants* (ed. T.A. Mansfield), Cambridge University Press, London, pp. 135–139.

Bradshaw, A.D. (1977) *Proc. R. Soc. Lond.* B, **197**, 77–96.

Bradshaw, A.D. (1979) *J. Roy. Town Planning Inst.*, **65**, 85–8.

Bradshaw, A.D. & Chadwick, M.J. (1980) *The Restoration of Land*, Blackwell, Oxford.

Breeze, V. (1973) *J. Appl. Ecol.*, **10**, 513–25.

Catlow, J. & Thirwall, C.G. (1976) *Environmental Impact Analysis*, Department of the Environment, HMSO, London.

Clark, B.D., Chapman, K., Bisset, R. & Wathern, P. (1976) *Assessment of Major Industrial Applications: A Manual*, Department of the Environment, HMSO, London.

Drablos, D. & Tollan, A. (1980) *Ecological Impact of Acid Precipitation*, Proc. Int. Conf., Sandefjord, Norway, SNSF, Oslo.

Gemmell, R.P. (1973) *Environ. Pollution*, **5**, 181–97.

George, M. (1977) *Trans. Norfolk and Norwich Naturalists Soc.*, **24**, 41–53.

Grim, E.C. & Hill, R.D. (1974) *US Env. Protection Agency*, Cincinnati, Ohio.

Harrison, J. (1974) *The Sevenoaks Gravel Pit Reserve*, WAGBI, Chester.

Hillestad, K.O. (1973) *Springstein Tipp og Landskap*, Norges Vassdrafs – og Elektrisitets vesen, Oslo.

Horsman, D.C., Roberts, T.M. & Bradshaw, A.D. (1979) *J. Exp. Bot.*, **30**, 495–501.

Leopold, L.B., Clarke, F.E., Hanshaw, B.B. & Balsley, J.R. (1971) *Geol. Survey Circular* 645, US Dept. of Interior, Washington.

Ministry of Agriculture, Fisheries and Food (1974) *Technical Bull.*, **30**, HMSO, London.

Moore, N.W. (1962) *J. Ecol.* **50**, 369–92.

Nature Conservancy Council (1975) *First Report*, HMSO, London.

Natural Environmental Research Council (1976) *Publications Series B*, **18**, London.

Osbourne, P.L. & Moss, B. (1977) *Freshwater Biol.* **7**, 213–33.

Peterken, G.F. (1974) *Biological Conservation* **6**, 239–45.

Phillips, G.L., Eminson, D.F. & Moss, B. (1978) *Aquatic Bot.*, 4, 103–26.

Rackham, O. (1976) *Trees and Woodland in the British Landscape*, Dent, London.

Ratcliffe, D.A. (1967) *Nature (London)*, **215**, 208–10.

Ratcliffe, D.A. (ed.) (1977) *A Nature Conservation Review*, 2 Vols, Cambridge University Press, Cambridge.

Rendel, Palmer & Tritton (1977) *Yare Basin Flood Control Study*, Anglian Water Authority, Norfolk & Suffolk River Division, Norwich.

Royal Ministry for Foreign Affairs (1971) *Air Pollution Across National Boundaries. The Impact on the Environment of Sulfur in Air and Precipitation*, Kungl. Botryckeviet, Stockholm.

Tallis, J.H. (1964) *J. Ecol.*, **52**, 345–59.

Vick, C.M. & Handley, J.F. (1975) *J. Inst. Parks Recreation Admin.*, **40**, 39–48.

Vick, C.M. & Handley, J.F. (1977) *Environmental Health*, May 1977 115–17.

CHAPTER THREE

Ecological Methodology

OVERVIEW

It is clear from the preceding section, that ecologists have important roles to play in the planning process. It is equally clear that interactions between planners and ecologists are required at the two levels at which planners operate – the policy or plan-making level and the implementation or development control stage. At the *policy level*, where planning is concerned with stating goals, options and the broad zonation of land use priorities ecologists contribute in two ways. First they provide baseline information – a status quo description of air and water quality, land pollution, landscape and wildlife and through ecological analysis they add some prediction of how these patterns may change in the future and how resilient they may be to perturbation. At the *developement control level* of planning, ecological knowledge contributes to an evaluation of the impact of a development on its environment. Site surveys provide a baseline and help identify impact targets; experimentation and modelling studies provide a basis for indicating the responses of the site to the impact and help to identify and evaluate measures to mitigate these effects. The ecologist, or indeed the planner, is not concerned in the final decision-making step but rather their role is to ensure as far as possible through their skills in information gathering, synthesis and evaluation that the final decision is taken in the light of a full appreciation of the ecological consequences of each alternative.

These activities require that ecologists should acquire a number of basic methodologies and skills. Bradshaw (Section 2.2) suggests a number of essential attributes which can be aggregated into three groupings:

(a) Descriptive skills required for baseline survey.
(b) Explanatory and predictive skills required for impact prediction and the development of mitigation measures.
(c) Evaluation and communicative skills required to transmit ecological information into the planning procedure.

Traditionally, ecologists and indeed most scientists have tended to be descriptive and explanatory. More recently attention has been paid to the development of predictive and communicative skills although these are still less well developed than the more traditional attributes. This section examines ecological methodologies for descriptive and predictive exercises and illustrates how available techniques impinge on the role of the ecologist in planning processes. Evaluation and communicative skills are perhaps the least developed attributes of ecologists and are considered throughout this book and discussed more fully in the conclusions (Chapter 9).

The paper by Parry *et al.* (Section 3.2) deals with baseline surveys in the context of metalliferous mining proposals. Here the major impact was considered to be habitat loss and the paper considers the techniques for collecting ecological information as a basis for (1) the design of mining operations to protect, where possible, the most important habitats and (2) the development of appropriate restoration techniques.

Descriptive survey work is probably the best developed attribute of most ecologists. Most descriptive information was produced as a result of general survey work in the 18th and 19th centuries. Amateur naturalists observed wildlife and kept records describing the distribution and behaviour of many species. Within the last 20 years or so information-recording and retrieval has become more sophisticated and new techniques such as the computer mapping of flowering plant distribution have evolved.

In cases where the major impact of a development is likely to be habitat loss the initial objective must be to record and evaluate what is present in the area. The main problem here is one of simplification and abstraction (that is, how to reduce both the scale and the complexity of the system to proportions which can be handled with the resources and time available). Simplification and abstraction in ecological surveys is usually achieved by a reduction in sampling area through quantitative sampling techniques and a reduction in complexity through the aggregation of units with a small number of dominants as unit indicators. It is obvious that the four basic vegetation types included in Ordnance Survey maps – woodland, wetland, farmland and moorland – contain complex variations. It is less obvious, but equally evident that heather moor, oak woodland and *Agrostis*-fescue grassland also contain much within-type variation. Variability and heterogeneity can be reduced or lost as a result of simplification and abstraction and it is essential that subjective recognition of sampling units with distinctive criteria are agreed prior to field studies.

The paper by Edwards (Section 3.3) is concerned with the prediction of potential consequences on the nature conservation and fisheries values of aquatic systems as a result of a proposed reservoir development. In this case, the potential impacts are more numerous, diverse and complex than the loss of habitat discussed by Parry *et al.* Similarly in view of the greater geographic extent and habitat diversity of the affected areas and the fluctuating nature of aquatic environments, the status of organisms and the environment and the relationships between them will also be more complex and less readily assessed. The paper by Edwards illustrates how in such circumstances the likely consequences of a development can only be predicted from a basis of detailed and rigorous surveys and explanatory studies.

Explanatory studies may be technically difficult and time-consuming. This is due in large part to the complexity of ecological systems. This complexity arises because of the great number of interactions between organisms and between organisms and the environment. Explanatory studies are intended to explain why these organisms live in certain places and in association with other characteristic organisms as a basis for predicting their resilience or sensitivity to an environmental perturbation.

As illustrated by Edwards (Section 3.3) the ecologist's ability to predict the consequences of an action on a quantitative basis may require very detailed explanatory studies. The danger here is one of maintaining a balance between essential information and data over-kill. The tendency during the early days of EIA in the USA was to err on the side of over-kill, but with experience and the development of a 'scoping process' a balance now appears to have been achieved (Legore, Section 4.2). Where previous experience has been obtained from a particular type of proposal the lessons learnt from quantitative studies should reduce the data base required by future proposals. Where novel developments are involved, detailed explanatory studies may be essential and it is imperative that the ecologist becomes involved in the planning process at the earliest possible stage.

The paper by Wathern (Section 3.4) is concerned with ecological methodologies for the evaluation and communication of ecological information and the development of quantitative impact predictions. The quantitative predictive component of planning is probably the most difficult task for ecologists. This is partially due to the relatively recent requirement for predictive studies combined with the problem of complexities. Early impact analyses relied upon the use of matrices for predictions. However, these involve a largely subjective estimate of ecological consequences and are more usefully thought of as evaluation and communicative methodologies. EIA as practised in the USA and Canada involves a quantitative approach to impact prediction in marked contrast to most impact analysis in this country where matrices and other charts have been used in the guise of predictive methodologies. More recently mathematical and simulation modelling techniques are being developed for impact predictions. The potential advantage is that modelling systems may retain a degree of flexibility and a capacity for assessment of functionally isolated components of ecosystem. This advantage is derived from the capacity of models to accommodate complex relationships between environmental variables, time-dependent relationships; to state the nature of relationships and to explore the implications of uncertainty. They should have the capacity both to facilitate consideration of all potential impacts whilst at the same time identifying the critical impacts in areas of uncertainty. However, the difficulty with mathematical and simulation models appears to relate to the application of models of physical or chemical phenomena to biological systems.

3.2 ECOLOGICAL SURVEYS FOR METALLIFEROUS MINING PROPOSALS

G. D. R. Parry, M. S. Johnson and R. M. Bell

3.2.1 Introduction

By definition, establishment of a comprehensive database in the form of an environmental baseline study is an integral part of the planning process. A significant component of baseline studies is the development of an inventory of the biological landscape so that objective assessments can be made about the ecological significance of development proposals. Such information is commonly used in the planning process in a negative way, that is, permission may be refused for a development if it impinges on the conservation value of an area. Baseline studies also form an integral part of Impact Analysis and, as such, offer scope for a positive approach to planning. Impact Analysis provides the foundation for discussions between developers and control agencies involved in resource management. From this starting point programmes can be evolved to attempt to preserve features of wildlife interest whilst permitting development activity. This can greatly reduce the likelihood of conflict at the later stages of a planning inquiry.

This paper examines the positive role of ecological survey by reference to restoration proposals for metalliferous mine workings. Reclamation of mine workings has hitherto been based on amelioration of inherited problems whereas the opportunities for an imaginative approach to reclamation have been restricted by the nature of the waste materials. The situation is now changing because of the recognized need to align existing ecological values with the mining system, and the importance of planning rehabilitation to a defined after-use consistent with the original character of the site and surrounding countryside. New mining techniques and forward planning may now permit accommodation of the proposed development within the environment in a manner which contributes to the conservation of ecological interest. Development proposals should include detailed plans for phased restoration, where possible, and a programme of final restoration to a defined after-use which may include replacement or reproduction of selected habitats and communities. Restoration proposals should always be prepared with a positive attitude including, for example, an examination of the possibilities of long-term ecological gains which may develop from sensitive and carefully prepared rehabilitation plans (Harrison, 1974).

3.2.2 Methods and aims of ecological surveys

The essence of survey is the abstraction of data, and ecological surveys are based on the simplification of the variety in nature. The commonest units involved in studies of terrestrial ecosystems are vegetation types because these can be represented on the greatly reduced scales needed for planning purposes. The characterization of vegetation units by one or two dominants (e.g. oak woodland, heather moorland) involves a reduction in the variability that is typical in nature.

There are two broadscale approaches to ecological survey. The first is based on objective recording of variation to generate the units for interpretation at the planning level and later stages. The second, much simpler, approach assumes that vegetation can be classified into a series of types, or units, which together cover the whole study area and form recognizable boundaries. These boundaries can be delineated by ground truth surveys and by the use of air photography, in particular false colour infra-red imagery. This latter system provides little insight into the biological variability of an area (i.e. oak woodlands in Wales and East Anglia are very different from one another) and the location of rare species may be overlooked. Generalizations of this nature would normally require additional objective studies to fulfill the requirements of an Environmental Baseline Study.

Holdgate (1976) has identified four systems of ecological survey:

(a) Within-site studies designed to describe variation within ecological systems within defined areas.
(b) Between-site surveys designed to describe variation between comparable ecological systems from site to site (e.g. oak woodlands).
(c) Species surveys intended to qualify variation within a species over its habitat range.
(d) Surveillance studies to describe variation with time, commonly in response to factors such as land-use and pollution.

The preparation of an Environmental Baseline Study involves at least two of these four systems. Within-site surveys are needed to describe the ecological components of a proposed development area. Such studies have been undertaken over very extensive areas (e.g. county or district) where 'unit' identification has been executed essentially subjectively. Only in recent years have objective methods of within-site survey seen practical applications.

Once details of habitat, flora and fauna variation have been recorded objectively the data can be used to map broad patterns of distribution of environmental features and to identify components of special interest. A second type of survey is then required to establish the exact significance of these special interest components in local and national terms. Surveillance studies can also be useful in environmental monitoring because surveys can be used to examine the progress and effectiveness of management strategies.

3.2.3 Ecological value of derelict mine workings

It is undoubtedly ironic that in some cases the drastic modifications to the physical and chemical environments produced by mining activity have provided novel habitats that have acquired plant and animal communities of ecological, scientific and conservation value. The creation of cliff faces, shafts and adits has provided important refuges for a wide range of breeding birds (Table 3.1), whilst the often toxic spoil tips have developed a unique flora comprising metal-tolerant ecotypes of a limited but characteristic range of species (Table 3.2). Other relatively innocuous spoils within metal mine complexes support locally and regionally uncommon plant species, which are often absent from other potentially suitable habitats in the locality (Table 3.3).

Reclamation of derelict mine sites may be highly desirable from many viewpoints but it must be recognized that unselective or insensitive schemes will be destructive to conservation interests in certain instances (Table 3.4). However, effective measures for controlling metal dispersal are unquestionably necessary in some situations. The practical problems of waste removal and disposal elsewhere mean that the principal system is based on re-contouring and vegetative stabilization. The latter may involve amelioration by surface blinding, soiling and fertilizing before seeding, or direct sowing of metal tolerant grasses such as *Festuca rubra* var. 'Merlin' (Johnson and Bradshaw, 1979).

The pressures on mineral operators have increased in recent years. This is particularly apparent in the Peak District National Park where the mining industry has responded to changing attitudes by developing a systematic programme of land reclamation encompassing both disused opencast workings and areas of tailings disposal. The latter, which extend from 8 to 30 ha in surface area, are viewed as the most serious problem, partly for amenity reasons but also for the environmental susceptibility of extensive deposits of fluoride and heavy metal contaminated waste subject to windblow, leachate dispersal and, to a minor degree, water erosion. Revegetation is essential because vegetative stabilization and landscaping provide the opportunity to minimize the inherently intrusive nature of the operations.

Although metalliferous tailings are one of the most intractable forms of derelict land from a revegetation viewpoint, changes in the sources and composition of crude ore in line with modern recovery objectives, combined with improvements in processing technology, have resulted in a tailings product far more amenable to vegetation establishment than those produced in the past. A detailed yet economic programme of restoration and

Table 3.1 Common breeding avifauna of metalliferous workings

*Species**	*Habitat*				
	Disused buildings	*Mineshafts and open-cast sites*	*Adits, leats and culverts*	*Open ground*	*Marshland and damp meadows*
Barn owl (*Tyto alba*)	×	○			
Blue tit (*Parus caeruleus*)	×				
Chough† (*Pyrrhocorax pyrrhocorax*)	○	×			
Coot‡ (*Fulica atra*)					×
Curlew† (*Numenius arquata*)				×	
Dipper (*Cinclus cinclus*)			×		
Great tit (*Parus major*)	×				
Grey wagtail (*Motacilla cinerea*)			×		
Jackdaw (*Corvus monedula*)	○	×			
Kestrel (*Falco tinnunculus*)	○	×			
Lapwing (*Vanellus vanellus*)				×	
Little owl‡ (*Athene noctua*)		×			
Meadow pipit† (*Anthus pratensis*)				×	
Merlin† (*Falco columbarius*)				×	
Moorhen (*Gallinula chloropus*)					×
Pied wagtail (*Motacilla alba*)	○		×		
Raven† (*Corvus corax*)		×			
Red grouse† (*Lagopus lagopus*)				×	
Redstart (*Phoenicurus phoenicurus*)	×				
Ring ouzel† (*Turdus torquatus*)		×			
Robin (*Erithacus rubecula*)	×				
Skylark (*Alauda arvensis*)				×	
Snipe† (*Gallinago gallinago*)				○	×
Spotted flycatcher‡ (*Muscicapa striata*)	×				
Starling (*Sturnus vulgaris*)	○	×			
Stock dove (*Columba oenas*)		×			
Swallow (*Hirundo rustica*)	×				
Swift (*Apus apus*)	×	○			
Wheatear† (*Oenanthe oenanthe*)	×	○			
Whinchat† (*Saxicola rubetra*)				×	○
Willow tit† (*Parus montanus*)					×
Wren (*Troglodytes troglodytes*)	×				

* Principal (×) and subsidiary (○) breeding habitats
† Restricted to workings in central/west Wales
‡ Restricted to workings in north-east Wales.

Table 3.2 Common plant species of toxic lead–zinc mine wastes in Wales

Calcareous substrates (pH 6·9–7·7)	*Acidic substrates* (ph 3·4–5·5)
Alpine penny-cress (*Thlaspi alpestre*)	Common bent grass (*Agrostis tenuis*)
Common bent grass (*Agrostis tenuis*)	Sheep's fescue (*Festuca ovina*)
Common mouse-ear chickweed (*Cerastium holosteoides*)	Sheep's sorrel (*Rumex acetosella*)
Creeping bent grass (*Agrostis stolonifera*)	Sweet vernal grass (*Anthoxanthum odoratum*)
Creeping fescue (*Festuca rubra*)	Wavy hair grass (*Deschampsia flexuosa*)
Harebell (*Campanula rotundifolia*)	Yorkshire fog (*Holcus lanatus*)
Ribwort (*Plantago lanceolata*)	
Sorrel (*Rumex acetosa*)	
Sweet vernal grass (*Anthoxanthum odoratum*)	
Tufted hair grass (*Deschampsia cespitosa*)	
Vernal sandwort (*Minuartia verna*)	
Wild thyme (*Thymus drucei*)	
Yorkshire fog (*Holcus lanatus*)	

management based initially on background research (Johnson, 1977) and then developed jointly by scientists, engineers and landscape architects (Johnson *et al.*, 1976) has greatly reduced the conflict between mining and countryside conservation. Not surprisingly, land-use restrictions remain in terms of agriculture and other sensitive management systems but considerable benefit is being derived from the policy to foster the development of wildlife and ecological interest (Cullimore and Johnson, 1978).

3.2.4 Applications to alluvial tin deposits: a case study

There is increasing recognition that industrial development does not automatically result in biological poverty and that certain old mining sites have acquired wildlife interest, albeit through a series of fortunate coincidences rather than positive planning towards this endpoint (Holliday *et al.*, 1979). In south-west England there are several sites in this category which are now viewed as potentially valuable sources of alluvial tin derived from periglacial weathering of primary cassiterite (tin oxide) together with solifluction, fluvial transport and deposition. Few of these areas have been designated under the key site system for nature conservation in Britain (Ratcliffe, 1977), nor do they correspond to local nature reserves or other scheduled sites indicative of ecological value. However, this cannot be taken to mean an absence of biological interest. Furthermore, it is generally accepted that

Table 3.3 Major botanical features of some toxic and innocuous metalliferous mine-wastes in Britain

Metal(s), gangue mineral, and spoil (pH)	*Location*	*Surface area (ha)*	*Species*	*Author*
Pb/Zn: calcite (7·4)	Cumbria	15	*Minuartia verna*, *Parnassia palustris*, *Viola lutea*	Ratcliffe (1974)
Cu: calcite (6·9)	Gwynedd	1·5	*Armeria maritima*	—
Pb: calcite (7·0)	Derbyshire	0·3	*Cochlearia officinalis*, *Minuartia verna*, *Saxifraga hypnoides*	Smith (1973)
Pb/Zn: calcite (7·1)	Northumberland	4	*Epipactis helleborine*, *E. phyllanthes* *E. leptochila*	Richards & Swann (1976)
Pb/Zn*: calcite (7·7)	Derbyshire	16	*Epipactis atrorubens*, *Thlaspi alpestre*	Wathern (1976)
Pb/Zn: calcite (7·2–7·6)	Derbyshire	0·7	*Dianthus deltoides*, *Draba incana*, *Echium vulgare*, *Gentianella amarella*	Clapham (1969)
Pb: quartz (5·0)	Dyfed	5	*Asplenium septentrionale*, *A. septentrionale* × *trichomanes*	Johnson *et al.* (1978)
Pb: quartz (5·4)	Dyfed	9	*Silene maritima* × *vulgaris*	Johnson *et al.* (1978)
Pb/Zn: quartz (4·7–5·3)	Dyfed	5	*Dactylorhiza maculata*, *Wahlenbergia hederacea*	Johnson *et al.* (1978)
Pb/Zn: calcite (7·6)	Clwyd	20	*Coeloglossum viride*, *Epipactis atrorubens*, *Orchis mascula*	Johnson *et al.* (1978)

* Associated with deposits of fluorspar (CaF_2).

safeguarding of key areas alone is insufficient to protect Britain's flora and fauna, and that it is essential to conserve much of the national capital of wildlife and habitat outside scheduled sites.

The ecological interest of alluvial tin areas comes mainly from their wilderness character and the complex mosaic of plant and animal communities resulting, in many cases directly, from early tin-streaming operations. Many sites include substantial areas of wetland, the protection of which is now regarded as a priority because of losses through recent drainage

Table 3.4 Plant species locally endangered by current site disturbance or proposed reclamation works

Mine site	*Activity**	*Species*
Halkyn (SJ 205706)	Land reclamation (A)	*Minuartia verna, Thlaspi alpestre*
Trelogan (SJ 123805)	Disposal of demolition rubble (A)	*Minuartia verna*
Minera (SJ 275515)	Land reclamation (P)	*Coeloglossum viride, Dactylorchis purpurella, Epipactis atrorubens, Minuartia verna*
Rhosesmor (SJ 212683)	Refuse tipping (A) Land reclamation (P)	*Cystopteris fragilis* *Minuartia verna, Thlaspi alpestre, Vicia orobus*
Goginan (SN 688817)	Removal of material for hardcore and road surfacing (A) Land reclamation (P)	*Silene maritima* × *S. vulgaris*

* A = active; P = proposed.

and reclamation schemes. Clearly, however, detailed biological data are required before the ecological status of such areas can be ascertained and the significance of new mining proposals evaluated. This baseline information is also essential for the development of wildlife protection measures and rehabilitation plans in the event of mining activity. To comply with these objectives, certain basic data must be obtained:

(a) Area of high conservation value at local, regional and national levels must be identified for possible exclusion from mining operations. The preservation of these areas requires a management policy to ensure their security during mining, and subsequently so that they serve as foci for recolonization.

(b) Species of conservation value in habitats otherwise of little importance must be catalogued and mapped. Special measures may be required to ensure their continued survival in the locality (e.g. transplanting).
(c) Habitats which, if permanently lost, would detract from the overall interest of the area should be identified and mapped. This information can serve as a basis for re-creation or replacement as an integral part of rehabilitation works.
(d) Areas of minimal biological importance should be identified, particularly those where a genuine opportunity exists to create new habitats to foster redevelopment of wildlife interest after mining.

Unfortunately, the recording of even basic biological data presents a series of potential problems such that attention is often focused on the vegetational component of ecosystems as an index of overall ecological value (Ratcliffe, 1977). This is commonly justified on the grounds that vegetation represents an integrated expression of numerous environmental factors and is the major determinant of animal populations. It is self-evident, however, that attention must as far as possible be paid to all components of an ecosystem if objective biological assessments are to be made. Statements about 'quality' and 'significance' present an even more intractable series of problems, even though criteria have been developed in order to judge conservation values. Techniques of assessment have evolved from studies of Dorset heathlands (Moore, 1962), primary woodlands (Peterken, 1974) and limestone pavements (Ward and Evans, 1976) and there are numerous more general methods of evaluation (e.g. Tubbs and Blackwood, 1971; Goldsmith, 1975). However, most techniques are either too habitat- and community-specific or too general to allow objective interpretation and it is often necessary to modify even widely-used criteria, such as those advocated by the Nature Conservancy Council (Ratcliffe, 1977), in order to fit artificial, man-made sites. This is particularly true of former tin-streaming areas which consist of a complex network of habitats and communities with no clear pattern apart from a tenuous relationship with early mining operations.

3.2.5 Biological survey and evaluation of alluvial tin deposits

The need to modify current survey methods is well illustrated by a recent study of an alluvial tin prospect at Porkellis Moor, Nr Helston, Cornwall (O.S. Grid Ref: SW 689324). This site comprises two distinct units, the larger south moor covering 105 ha of the basins of two river valleys enclosed essentially by the 125 m contour. The geology, climate, surrounding land-use pattern, rehabilitation plans and detailed biological surveys are described elsewhere (Consolidated Gold Fields and University of Liverpool, 1978) but the methodology and summary results are described below to illustrate one

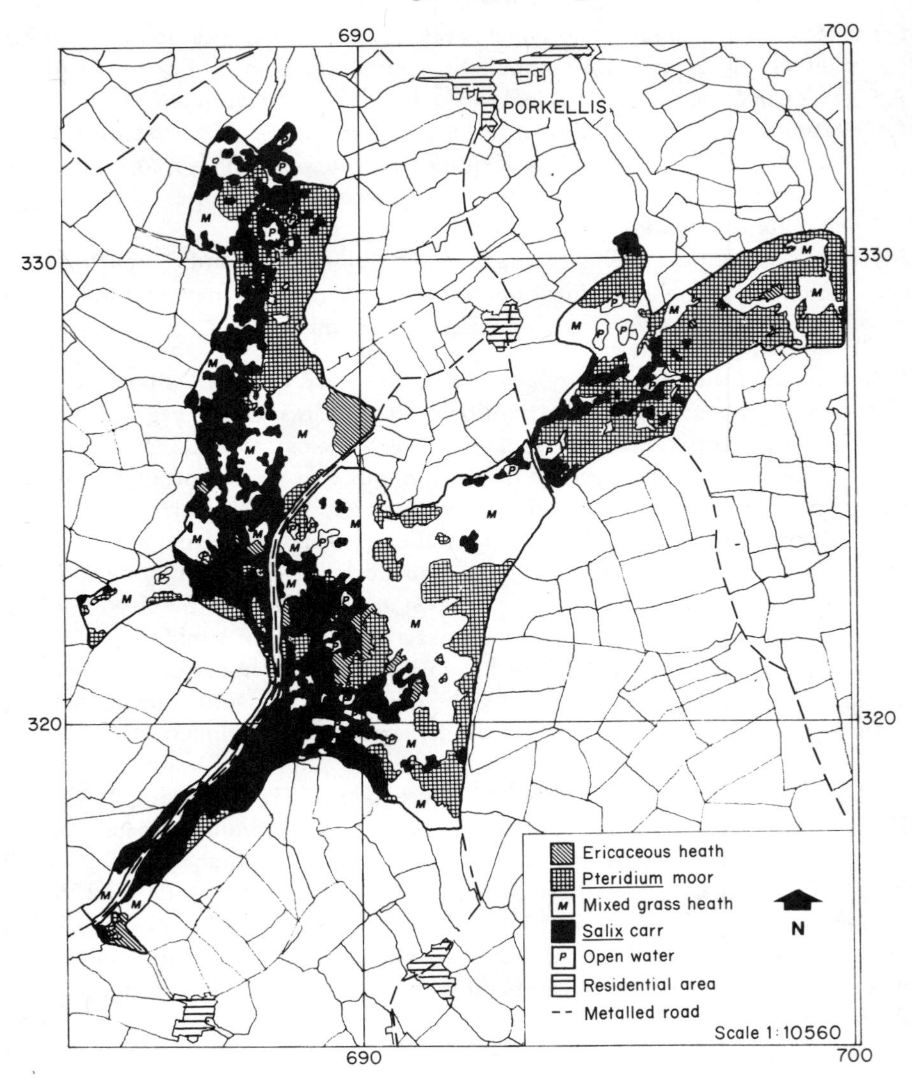

Fig. 3.1 The semi-natural vegetation of South Porkellis Moor.

approach which can be used to provide an overall ecological assessment of such areas in relation to redevelopment proposals.

(a) Vegetation

The semi-natural plant communities were studied by general survey methods supplemented by the use of 1:2500 air photographs. The principal plant communities (e.g. heather moorland, mixed grass heath, *Salix* carr. etc.)

Table 3.5 Main vegetation types, plant communities and important associated species on Porkellis Moor

Vegetation type	*Plant community/dominant species*	*Species of botanical interest*
1. Ericaceous heath	*Calluna vulgaris*; *Erica cinerea*; *E. tetralix*	*Platanthera bifolia*; *Dactylorhiza maculata*
2. *Pteridium* moor	*Pteridium aquilinum*	
3. Mixed heath	Dry (a) *Agrostis tenuis*/ *Holcus mollis*	*Platanthera bifolia*
	(b) *Agrostis setacea*/ (*Erica* spp.)	
	Wet (a) *Juncus* or *Equisetum* marsh	*Dactylorhiza praetermissa*; *Menyanthes trifoliata*; *Potentilla palustris*; *Scutellaria galericulata*; *Veronica scutellata*; *Wahlenbergia hederacea*
	(b) *Molinia caerulea*/ *Eriophorum vaginatum* = acidophilous mire	*Carex curta*; *Carex rostrata*; *Myrica gale*; *Pinguicula lusitanica*; *Osmunda regalis*
4. *Salix* carr	Immature, Mature } *Salix atrocinerea*	*Cordalis claviculata*; *Osmunda regalis*; *Ranunculus tripartitus*; *Usnea articulata*
5. Open Water	Ponds	*Apium graveolens*; *Apium inundatum*; *Littorella uniflora*; *Pilularia globulifera*; *Utricularia minor*; *Utricularia neglecta*
6. Agricultural land and Cornish hedges	Dry grass heath	

and main vegetation types (e.g. scrub, grassland) were mapped, in outline, from the air photographs and then surveyed in detail in order to compile species lists. The latter were produced from line and belt transects positioned to traverse all recognized sub-divisions of the vegetation with extensive replication. Community distribution was mapped along representative transects which were then surveyed to determine topographical variation and other possible determinants of vegetation patterns (e.g. water table, soil texture, nutrient status). Supplementary studies were carried out to identify sub-divisions of each community, outstanding examples worthy of conservation and the locations of uncommon species. This databank was used to compile a map of the spatial distribution of the principal vegetation types (Fig. 3.1). Further maps were prepared to describe (a) community distribution according to topographical and soil conditions, (b) location of areas of ecological importance and, (c) siting of species of local, regional and national importance.

The main vegetation types, plant communities and significant plant species are summarized in Table 3.5, which features a pattern broadly typical of the lowland heath formation that is common and widespread in south-west England. The site is dominated by successional, pre-climax communities and although no botanical features of paramount importance were identified, species of significance because of their localized distribution were recorded, particularly in the open water, wet heath and *Salix* carr communities. Based on the detailed results of the vegetation survey it became clear that on botanical grounds these habitats/communities should form the basis of conservation and rehabilitation plans in the event of mining (Consolidated Gold Fields and University of Liverpool, 1978).

The main vegetation units (Fig. 3.1) provided the basis for investigations and mapping of the native fauna in order to identify habitats and communities to be excluded from mining or requiring faithful replacement in the final land use plan. This was justified on the basis that the flora is the only major component of the ecosystem which can be manipulated directly towards a specific end-point, and re-establishment of vegetation is the initial step in re-creating the total ecosystem.

(b) Odonata

The diversity of Britain's invertebrate fauna limits any systems of ecological evaluation because of the inordinate amount of time and manpower required to carry out a comprehensive survey. Some means of selective screening is invariably necessary to identify invertebrate groups for which objective assessments can be made. This is usefully based on previous studies in the locality, county or region, and in consultation with local conservation groups and organizations together with the Nature Conservancy Council. In this case the order Odonata (dragonflies and damselflies) was distinguished because of the special importance of the wetland components of alluvial tin areas (Turk and Turk, 1976). This group was studied by a modified version of the transect index method for examining the abundance and diversity of butterflies (Pollard *et al.*, 1975) and by sweep netting. This revealed that Porkellis Moor contains 15 of Britain's 43 native Odonata, and that similar

Table 3.6 List of Odonata recorded on Porkellis Moor and other alluvial tin deposits in Cornwall and their relative abundance at local, regional and national levels

	Porkellis Moor	*Local**	*Regional†*	*National‡*
ANISOPTERA (Dragonflies)				
Aeshna cyanea	C	C	C	C_s
Aeshna juncea	C	C	C	C
Anax imperator	O	C	F	F_s
Cordulegaster boltonii	VC	VC	C	C_s
Orthetrum coerulescens	+	F	F	F_s
Libellula depressa	F	F	F	F_s
Libellula fulva	+	R	R	R
Libellula quadrimaculata	O	C	C	C
Sympetrum striolatum	C	C	C	C
ZYCOPTERA (Damselflies)				
Agrion splendens	F	F	F	C_s
Agrion virgo	C	C	C	F_s
Ceriagrion tenellum	C	LF	O	O
Coenagrion puella	F	C	C	C
Enallagma cyathigerum	+	C	C	C
Erythromma najas	R	VR	VR	LF_s
Ischnura elegans	F	C	C	C
Lestes sponsa	C	C	C	C
Pyrrhosoma nymphula	O	C	C	C

KEY
* Refers to the county of Cornwall
† Refers to Cornwall, Devon, Somerset and Dorset
‡ Refers to Great Britain and Ireland
+ Species not recorded on Porkellis Moor but present on other alluvial tin areas
C common
F frequent
O occasional
R rare
V very
L locally
Subscript s, refers to species with a predominantly southern distribution

sites in the locality not subject to mining interest contained this complement together with a further three species (Table 3.6). The listing includes several uncommon species, notably *Ceriagrion tenellum* and *Erythromma najas*; the latter is normally associated with east and south-west England but is absent from Cornwall in recent distribution maps (Chelmick, 1979). The remarkable diversity of Odonata is ascribed to the complex of aquatic habitats suited to breeding (e.g. rivers, streams, ponds, marshes and bog pools).

(c) Avifauna

Recent distribution maps show that south-west Cornwall is lacking in many bird species common throughout the remainder of southern England (Sharrock, 1976). Tin-streaming sites are known to support breeding populations of a wide range of common bird species and are of some importance as passage feeding grounds and overwintering sites. A summary list for Porkellis Moor, based on a breeding bird census and other observations

Table 3.7 The distribution, residential and breeding status of the more important birds on Porkellis Moor

Species		*Habitat*					
		1	2	3	4	5	6
Sparrowhawk	*Accipiter nisus*	*B	*	*		*	
Buzzard	*Buteo buteo*	*		*	*	*B	
Hen harrier	*Circus cyaneus*		*		*		
Montagu's harrier	*Circus pygargus*		*S				
Merlin	*Falco columbarius*		*				
Water rail	*Rallus aquaticus*	*					W*B
Curlew	*Numenius arquata*			*B			
Barn owl	*Tyto alba*	*B		*B			
Little owl	*Athene noctua*					*B	
Short-eared owl	*Asio flammeus*	*	*				
Nightjar	*Caprimulgus europaeus*		*B				
Great spotted woodpecker	*Dendrocopus major*			*			
Nuthatch	*Sitta europaea*					*B	
Treecreeper	*Certhia familiaris*	*	*	*		*B	

Habitat types:
1 *Salix* Carr; 2 mixed heath; 3 agricultural; 4 *Pteridium* moor; 5 woodland; 6 open water
* present in habitat type; B proven breeding; W winter resident; S summer resident

(Table 3.7) indicates the distribution, residential and breeding status of the more important species in relation to the major habitats. At the national and regional levels Porkellis Moor cannot be regarded as an important breeding site though the presence of water rail (*Rallus aquaticus*), nightjar (*Caprimulgus europaeus*) and curlew (*Numenius arquata*) provide features of local importance.

(d) Other groups

Subsidiary investigations were made of the amphibian, reptilian and mammalian faunas of the study area. These revealed a significant local feature, namely the diversity and abundance of bats resident in deep mine workings, buildings and shafts on higher ground beyond the proposed development zone but dependent for their feeding on features of the latter area. Five species were recorded, including whiskered (*Myotis mystacinus*), Natterer's (*M. nattereri*), and Daubenton's (*M. daubentoni*), all of which are common in south-west England and occur within and adjacent to other alluvial tin deposits in the immediate locality (< 54 km). Monthly surveys of the indigenous Lepidoptera (butterflies and moths) indicated that at least 50% of the total number of butterfly species reported in Cornwall over the last 50 years occur in the study area, including the uncommon marsh fritillary (*Euphydryas aurinia*) and holly blue (*Celastrina argiolus*), and the local brown hairstreak (*Thecla betulae*).

(e) Summary of conservation values

All the vegetation types and plant communities of Porkellis Moor are common in south-west England and well represented in similar, but generally smaller, alluvial sites throughout Cornwall. Overall there are few plant and animal species of intrinsic importance in terms of national rarity but there are several features of regional, and particularly local, conservation value. Much of the biological interest centres on the open water areas and successional stages through wet heath to *Salix* carr. These wetlands are clearly important for Odonata and the international decline of wetlands together with the fragmentation of heathlands in Britain emphasizes the need to conserve such features both within and outside sites scheduled by the Nature Conservancy Council (Nature Conservancy Council, 1977). On these criteria it was considered important to develop both conservation and rehabilitation strategies, but on the basis of integrating these proposals with any future mining venture rather than to prevent such an operation occurring.

(f) Mining, conservation and rehabilitation

Any future mining of Porkellis Moor is likely to be based on a system of dredging flooded mining paddocks brought into active use progressively and used subsequently for tailings disposal. This gives the potential for gradual and continuous rehabilitation, and combined with the shallow nature of the alluvium and tailings replacement system, the prospect of a very much reduced impact on landform compared to normal deep mine and open-cast workings.

Despite the inevitable disruption if mining operations proceed it will be possible to maintain the existing conservation value. A progressive system of restoration combined with the preservation of viable and outstanding representatives of the existing communities should be adopted, analagous to that well established for the mineral sand mining industry of Australia (Brooks, 1976). Although the top-soil can be described as pedologically juvenile and therefore not so important for its physical and chemical characteristics, it should be stripped and faithfully replaced to enable recolonization by vegetative fragments and by means of the native seed bank. Certain species which do not readily establish from seed but are essential to rehabilitation plans (e.g. *Salix* spp.) can be rapidly re-established from the indigenous stock by vegetative propagation. The islands of undisturbed indigenous vegetation can act as foci for recolonization and as reservoirs for the transplanting of sensitive and localized species.

Mining operations will inevitably disturb the breeding territories of the native Odonata. The uninterrupted presence of neutral to acid ponds, marshes and bogs is essential to the maintenance of viable breeding populations. This may be achieved by the exclusion of small but specific areas from the mining programme. These will act as nuclei for the eventual recolonization of the abandoned mining paddocks and supporting habitats.

Provided that suitable conservation measures are integrated with a continuous and effective restoration programme, the impact on wildlife and landscape values could be minimal. Indeed, there is considerable scope with an imaginative restoration strategy genuinely to enhance conservation interests in the long term. This policy could make a positive contribution to our ever-declining wetland resources and if applied elsewhere assist greatly in relating the apparently conflicting objectives of nature conservation and mining.

3.2.6 Conclusions

Environmental impact assessment is in many ways not a new concept but rather represents a formal framework for incorporating many of the existing facets of development control. However, broadening the content of impact analysis to include ecological impact as a major component of the exercise requires development of techniques for adequate data base preparation.

Methods of ecological survey and evaluation have developed to a point where objective decisions can now be made on the quality and significance of industrial sites subject to development proposals. By definition such studies are complex requiring not only a biological inventory but the capacity to interpret such data in order to prepare practicable and constructive management frameworks. By basing these on integration of the proposal with conservation requirements where possible, impact analysis based on ecological principles, provides a positive contribution towards minimizing the conflicts between the needs of the developer and the widely accepted need for environmental protection.

Acknowledgements

The authors gratefully acknowledge the co-operation and financial support of the many organizations who have assisted in these studies. Particular acknowledgement is made to the assistance of the Welsh Development Agency, Welsh Office, Nature Conservancy Council, Laporte Industries Ltd, and Consolidated Gold Fields.

3.3. PREDICTING THE ENVIRONMENTAL IMPACT OF A MAJOR RESERVOIR DEVELOPMENT

R. W. Edwards

3.3.1 Introduction

This paper seeks to examine some of the more obvious potential ecological consequences to the River Wye of the enlargement of the existing reservoir, Craig Goch, in its headwaters in mid-Wales and the use of this reservoir for river regulation. It is not about the application of formal procedures of impact analysis to this water resource development, neither does it claim to be comprehensive in its identification and quantitative prediction of environmental perturbations. It does, however, illustrate the importance of careful collection of good base-line data prior to attempting any prediction of the impact of new developments.

The studies summarized here were undertaken principally by the UWIST Craig Goch Research Group, together with post-graduate students of the Department of Applied Biology during the period 1975–9. The UWIST Craig Goch Research Group was established principally to provide base-line ecological and water-quality data which might subsequently be used to assess the impact of water resource schemes in the upper catchment of the River Wye and to predict the likely consequences of developing a major regulating reservoir, Craig Goch, in the Elan sub-catchment on the ecology and water quality of the River Wye and, if necessary, to suggest ways – both in design and operation – of obviating any adverse effect.

Attempts at predicting the consequences of the Craig Goch schemes to the River Wye seem particularly worthwhile as it is a river with valuable salmonid and coarse fisheries, with angler expenditure at around £12·5 million at 1979 prices (Gee and Edwards, 1980) and is a river of considerable nature conservation interest (Ratcliffe, 1977) regarded as worthy of being declared a Grade 1 SSSI. Furthermore, the water to be abstracted from the lower reaches of the river near Monmouth, and stored in Llandegfedd Reservoir, a pumped-storage reservoir in the Usk valley, could change the nutrient status of the reservoir which already suffers algal problems: clearly changes in that status could have consequences to the provision of treatment facilities.

Despite the importance of biological resources of the River Wye and the key role of this river in the water strategy of South Wales, the Wye River Authority, which was responsible for the river management before 1973, did not undertake or commission biological surveys of rivers within its catchment and chemical data were principally restricted to standard indices of pollution (e.g. BOD, NH_3–N, NO_3–N, PO_4–P). However, after 1970 several aspects of water quality were monitored in relation to a water-supply intake at Monmouth. Although no direct surveillance of fish stocks was attempted, annual redd counts of salmon were undertaken and the returns of salmon catches, both from the commercial net fishery in the estuary and from recreational fisheries in the river, were reliable.

3.3.2 The catchment and water resources scheme

The River Wye, 250 km long and draining a predominantly rural catchment of 4183 km^2, rises at Plynlimon (677 m.O.D.) and flows to the Severn Estuary (Fig. 3.2). The major geological division within the catchment is between the Ordovician and Silurian sediments (41% of catchment area) generally above 200 m.O.D., and the Old Red Sandstone Series (55% of catchment area) predominantly below 120 m.O.D. The population density in the upper 42% of the catchment, about 0·15 ha^{-1}, compares with 0·76 ha^{-1} in the remainder of the catchment. The principal agricultural uses of the catchment are permanent grass (37%), tillage (20%), rough grazing (15%), temporary grass (14%) and forestry (9%) (Wye River Authority, 1972).

The average discharges of the major tributaries and stations on the main river (Fig. 3.2) are given in Table 3.8 (Wye River Authority, 1972). The average discharge of the R. Elan, about 2·0 $m^3 s^{-1}$, is only about 50% of the catchment yield, the remainder being exported to the West Midlands. The most significant tributaries are the Ithon, Irfon and Lugg which contribute on average about 43, 36 and 20% respectively to the total flow of the R. Wye below each confluence.

In the review of the water resources in England and Wales (Water Resources Board, 1973) it was recommended that Craig Goch, the uppermost of an existing series of direct-supply reservoirs on the Elan tributary of the Wye (Fig. 3.2) be enlarged and used to regulate both the rivers Severn and Wye at times when natural flows in these rivers were too low to satisfy projected increases in demand on downstream abstractions. Initially it was anticipated that the water yield of such a reservoir might be about $1{\cdot}2 \times 10^6$ $m^3 d^{-1}$ and its storage volume and surface area 534×10^6 m^3 and 1400 ha respectively. In addition to the supplementation of low flows of the rivers

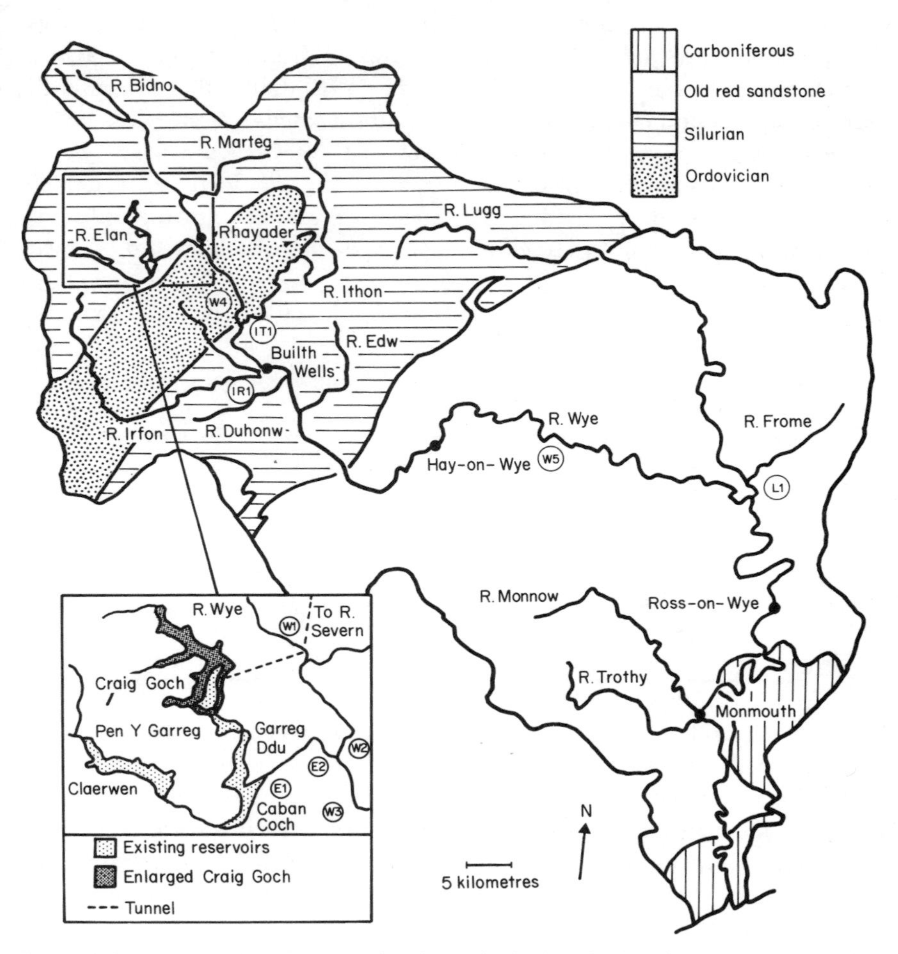

Fig. 3.2 Catchment of the R. Wye showing principal geological features and sampling stations. The insert shows the existing Elan Valley reservoirs and the proposed extension to the Craig Goch reservoir.

Severn and Wye (in the latter river up to 6·1 $m^3 s^{-1}$), the scheme involved the transfer of substantial volumes of water between catchments. The studies described in this paper for the Wye, and similar studies carried out in the Severn catchment by the Severn–Trent Water Authority, were conceived when this major resource development was considered necessary.

Since 1973 there has been a pronounced downturn in the projected water demand within planning horizons which has led to a contraction of the proposed Craig Goch scheme, such that the gathering grounds are now limited to the reservoir's natural catchment and modest sources in the upper Ystwyth which may be transferred by gravity. The storage volume and area have been reduced to 238×10^6 m^3 and 810 ha, water being released at times of low natural flow into the upper Severn or its tributary, the Dulas, by tunnel, and into the upper Wye. Thus the Wye, unlike the Severn, will receive water almost entirely from its own natural catchment and predictions of the ecological consequences to the Wye of the Craig Goch scheme may now be confined to considerations of flow regulation and so avoid the extremely difficult problems of predicting the effects of intercatchment transfers, problems which emanate largely from the stochastic character of such effects.

If alternative ground-water resources are exploited in Shropshire for use

Table 3.8 Average flows of stations on the Wye and its major tributaries (for position of stations see Fig. 3.2)

Station	*Average flow* (m^3 s^{-1})
W_2	7·0
W_4	47·9
E_2	2·9
It_1	7·5
Ir_1	9·9
L_1	10·6

in the West Midlands, the enlargement of Craig Goch may be postponed indefinitely. A reduced regulation scheme has already been approved for the Wye which uses existing reservoir storage in the Elan Valley and redeploys current compensation flows to the R. Elan from existing reservoirs. Such continuous compensation flows will be reduced from 1·34 to 0·8 m^3 s^{-1}, and save water for discharge when natural flows are considered inadequate to sustain abstracted uses. The maximum regulation discharge, 1·9 m^3 s^{-1}, will be released when the natural flow at Monmouth in the lower catchment is below 14 m^3 s^{-1} plus any projected abstractions.

Although long-term average monthly flows even in the summer do not fall below 25 m^3 s^{-1} near Monmouth, there are many days when flows are below 14 m^3 s^{-1} (Fig. 3.3) at which regulation releases would occur. It is important to establish not only frequency with which river flows would be modified, but also the extent of the modification. Figure 3.4, in which cumulative frequency plots of daily flow are shown for 1975–1977 for the months when regulation discharges might normally be expected (May to October), indicates that even in 'normal' summer months flows of about 10 m^3 s^{-1} might be expected. At these flows regulation releases could comprise, even in the downstream reaches of the river, up to about 60% of total flows with the large Craig Goch scheme (max. release 6·1 m^3 s^{-1}) and 20% with the modest Elan Redeployment Scheme (max. release 1·9 m^3 s^{-1}).

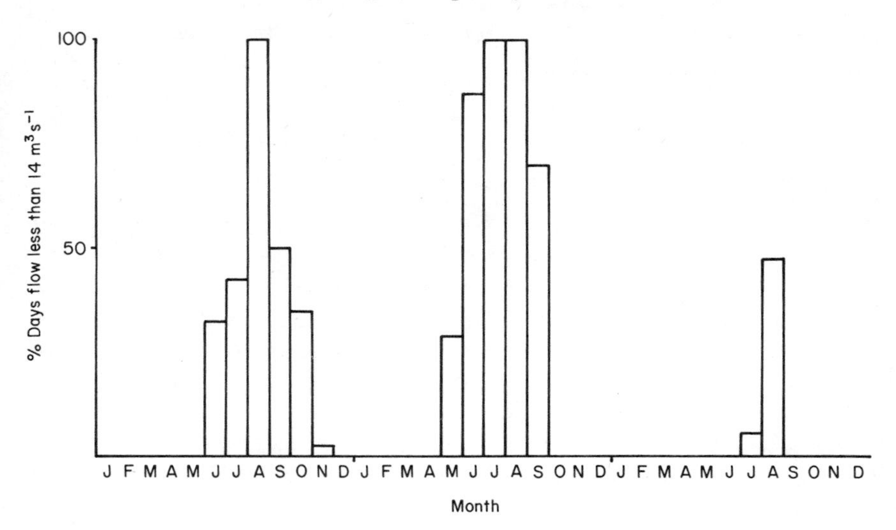

Fig. 3.3 Seasonal periodicity in flows below 14 $m^3 s^{-1}$ at Monmouth in 1975–1977, and requiring supplementation for further abstraction.

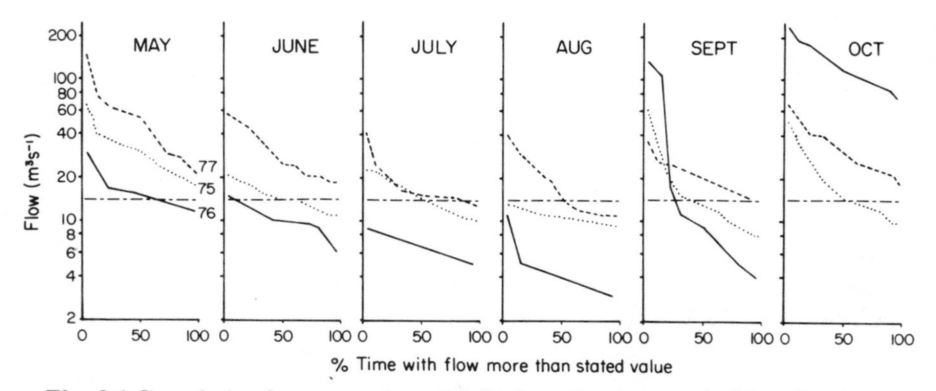

Fig. 3.4 Cumulative frequency plots of daily flows for the months May–October.

Throughout this paper calculations are made assuming maximum regulation discharges from the two schemes, denoted as CG and ER, are 6 and 2 $m^3 s^{-1}$.

3.3.3 Effects of flow regulation

Effects of regulation (Edwards and Crisp, 1982) derive not only from flow modification, particularly low-flow augmentation, but also from changes which occur in the prior storage of water in reservoirs. Thus in considering the consequences of regulation to a catchment it is useful to distinguish between extensive and local effects and between effects resulting from modification to flow and those resulting from reservoir storage. This broad classification will be adopted in this paper rather than the alternative approach of examining each resource in turn.

(a) Extensive effects

(i) Flow modification

Changes in flow may exert their effects through changes in velocity and residence time, water depth and channel width. Flow-tracer studies in the R. Wye suggest that in the lower catchment (which represents about 80% of the river length) flow changes exert their effects principally through velocity whereas in the upper catchment both velocity and cross-sectional area (depth × width) are equally affected (Table 3.9).

Table 3.9 Relationships between velocity (V) and flow (F) in three reaches of the Wye assuming an equation $\log_{10} V = a \log_{10} F + b$ (V being in km h^{1} and F in m^{3} s^{-1})

Reach	*Length (km)*	*a*	*b*	*r**
Upper reach (Nannerth–Newbridge)	19	0·45	−0·35	0·94
Middle reach (Newbridge–Llyswen)	27	0·62	−0·68	0·93
Lower reach (Lyswen–Monmouth)	144	0·02	−1·16	0·94

* r = correlation coefficient.

Using the equation in Table 3.9 and the likely flows at sites down the river when the flow at Monmouth is 14 m^{3} s^{-1}, the threshold for regulation discharges, it is possible to assess the impact of regulation releases on both velocity and residence time (Fig. 3.5). In the lower reach where major tributaries such as the Ithon, Ifron and Lugg, have reduced the proportionate effect of regulation on total flow the ER and CG schemes increase the velocity by only 11 and 39% and decrease the total residence time within the river from the Elan confluence by 20 and 40%. In contrast, in the upper reach, where regulation releases have a large proportionate effect on flow, then the ER and CG schemes increase the velocity by 26 and 76% and decrease residence time by 11 and 20% respectively.

Velocity increases, although small, could possibly induce the erosion of inorganic sediments within a limited size range, roughly equivalent to sands, and prevent the deposition of coarser material (Hjulstrom, 1935) (Fig. 3.6). However, of much greater potential ecological significance at these comparatively low velocities are the effects of increases on the behaviour of organic particles, which are almost neutrally buoyant. Such increases could reduce the deposition and increase the removal of organic debris and benthic organisms.

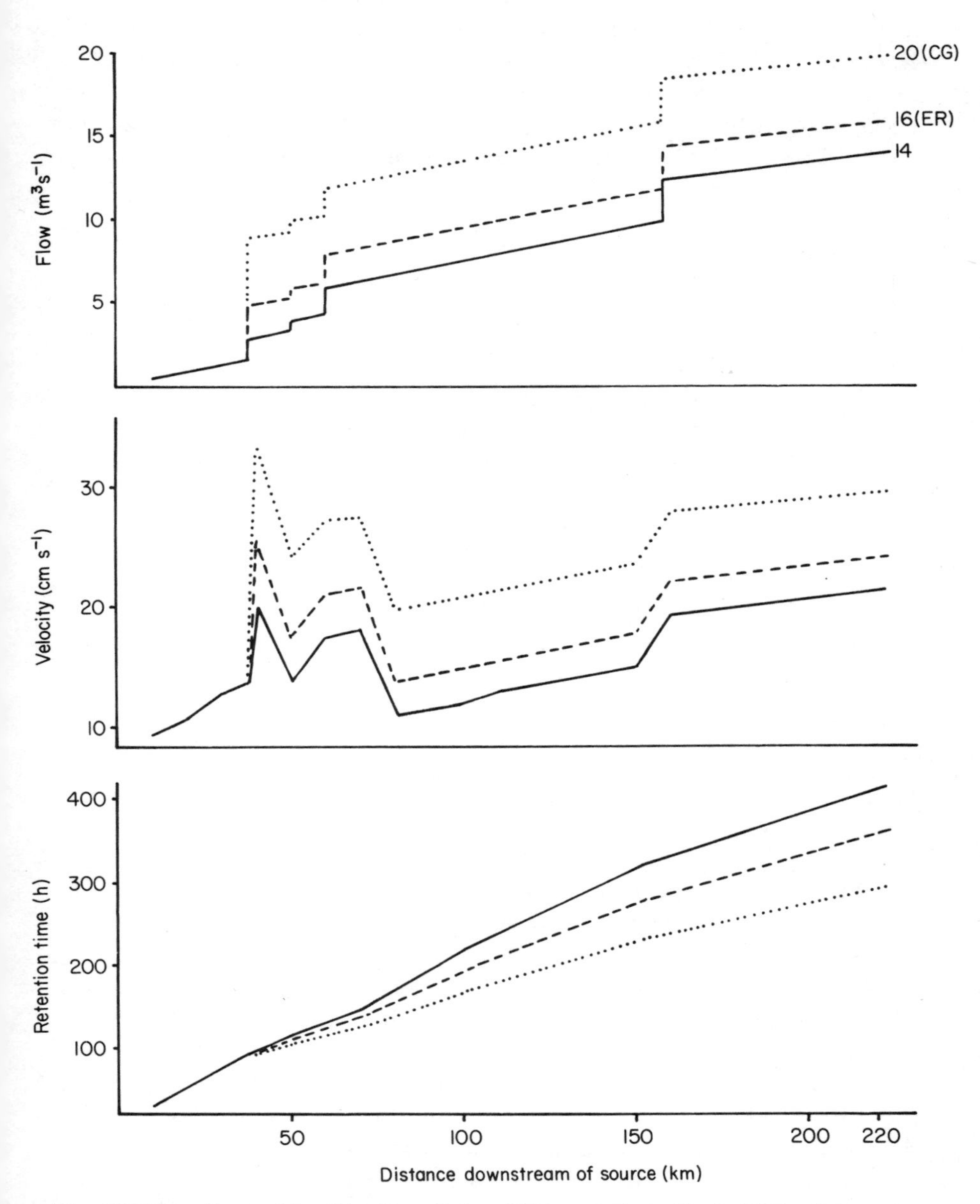

Fig. 3.5 The effects of the Elan Regulation (ER) and Craig Goch (CG) schemes on flow, velocity and residence time at an initial flow at Monmouth (Redbrook) of 14 $m^3 s^{-1}$. Regulation water is assumed to be discharged through the Elan

Far more research has been undertaken on the behaviour of drifting benthic invertebrates than other benthic organisms, such as algae, or organic debris but even with benthic invertebrates focus has been predominantly on daily and seasonal rhythms rather than flow-related effects. Brooker and Hemsworth (1978) observed changes in the drift of benthic invertebrates in the upper Wye before and during an experimental release of water from Caban Coch which increased the flow downstream of the Elan–Wye confluence from 2 to 5 $m^3 s^{-1}$ and the velocity from 17 to 26 cm s^{-1}. Fortunately

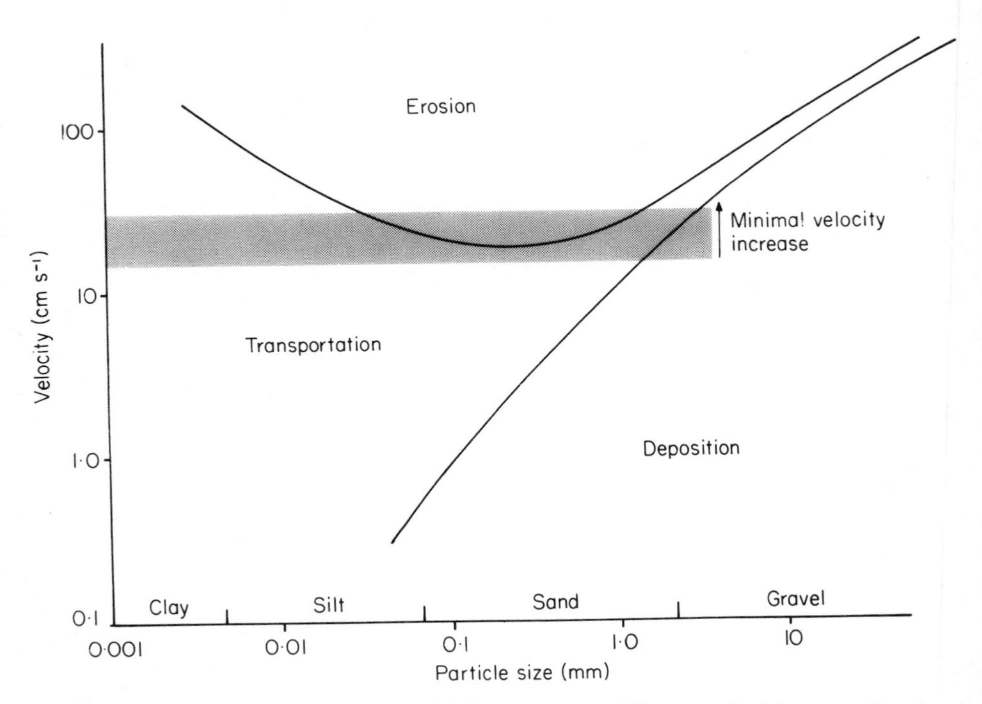

Fig. 3.6 Behaviour of particles of different size at different velocities according to Hjulstrom (1935). The effect of regulation in the Wye on minimum velocity is shown.

there were no substantial changes in water temperature during the release to confuse interpretation of flow effects. During the first day of the release there was an increase in both the total number and actual density of drifting invertebrates, but during the second day both number and density decreased, suggesting that invertebrates were adjusting to the modified flow regime: drift number and density remained relatively constant at an upstream control site throughout the period (Fig. 3.7). Species responded to the increased flow very differently, the numbers of larvae of *Rheotanytarsus* drifting increased almost immediately from 0·8 to $22{\cdot}3 \times 10^3$ h^{-1} whereas the increased drift of the mayfly *Ephemerella ignita* was delayed until the night of the release when

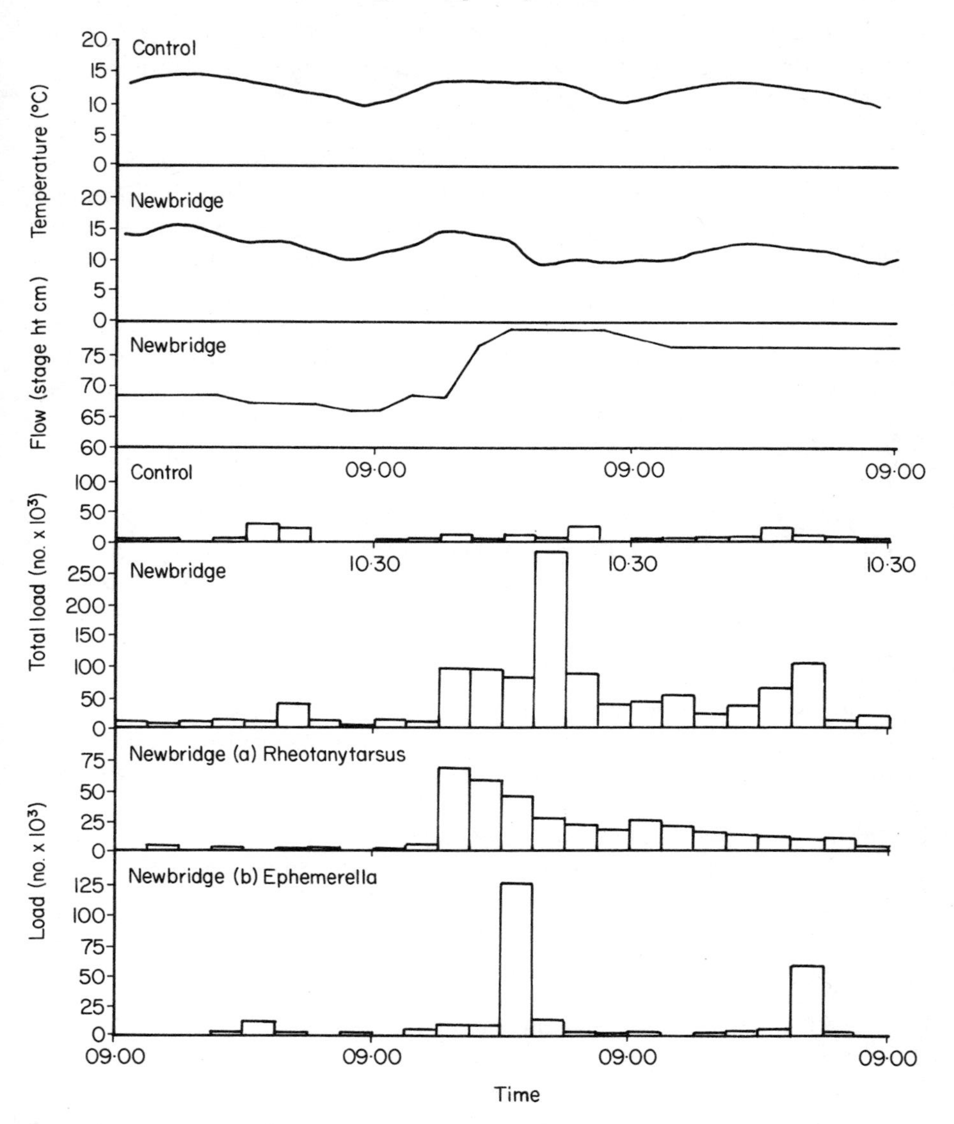

Fig. 3.7 Invertebrate drift and temperature upstream of the Elan confluence (control) compared with downstream (Newbridge) during a release of reservoir water via the Elan.

numbers were over 10 times those of the previous night (Fig. 3.7). From these and other studies elsewhere it seems likely that at the onset of regulation releases increasing numbers of organisms will drift and, at increased velocities, drift further: where releases are sustained for several days the loss rate is likely to decline, although not to pre-release values. The response to sustained flow increases could be modified by a vertical re-distribution of the fauna within the substrate in response to an improvement of oxygen status with increased inter-gravel flow.

Drifting organisms are important in the diet of fish species, particularly for salmonids during the summer (Elliott, 1967) and any increase in availability might be regarded as beneficial to fisheries, provided the loss-rate from the benthos did not lead to inadequate population replenishment, either from enhanced predation or downstream displacement. Hemsworth and Brooker (1979) assessed the downstream population displacement for cohorts of several benthic invertebrates in the R. Wye: most were displaced less than 10 km during the aquatic stages of life-cycles. With increased water velocities causing both increased loss-rates from the benthos and extension of distances transported, some further downstream population displacement might be expected – particularly as regulation occurs during the summer at times of maximal drift densities. Nevertheless with normal dispersive mechanisms, particularly of species with an aerial-flight stage, major shifts in the distribution of the benthos seem most unlikely.

Whilst the same basic considerations of the effect of velocity on benthic loss rate apply to attached algae, the situation is made more complex by the growth of many species both in the 'planktonic' and attached phases (Furet, 1979). Attempts have been made to interpret the spatial and temporal variations in plankton counts of various groups of algae, particularly diatoms, recorded by Furet (1979). However, unrealistically short doubling-times, occasionally as low as 3 h, for suspended diatoms indicate that much of the production is benthic, the cells being scoured off and transported downstream. No significant correlation has yet been established between apparent doubling times and flow either for centric or pennate diatoms, the former having planktonic, as well as benthic, growth.

Water velocity, as well as affecting particle transport, also influences transfer processes of dissolved gases and oxygen is probably of greatest significance in the lower reaches. In 1976 when flows at Monmouth decreased to about 6 $m^3 s^{-1}$ at the end of June, the annual growth of water-buttercup (*Ranunculus penicillatus*) flowered and began to decay. The oxygen concentration decreased to less than 10% air saturation which, coupled with very high water temperatures (>25 °C) caused substantial mortalities of salmon (Brooker *et al.*, 1977). The low water had affected several aspects of the oxygen economy of the river.

Water velocity influences surface re-aeration and, coupled with river depth, is sometimes used to calculate the mass transfer coefficient of oxygen (Owens *et al.*, 1972). It also affects both the respiratory and photosynthetic rates of macrophytes (Westlake, 1967), and although these effects are only detectable well below ($<0{\cdot}1$ cm s^{-1}) the average velocities occurring in the R. Wye even under drought conditions (>10 cm s^{-1}), water flow through weed-beds is restricted and may well be within the range at which velocity effects occur at low river discharges. Of major significance to the oxygen economy is the oxidation of dead plant material, both the site of decomposition and the dilution of the respiratory load. These two factors are influenced in turn by river flow, velocity determining the extent to which decay is *in situ* or dispersed downstream and flow determining the dilution afforded the organic matter and its decay products.

Some of these flow-related effects on oxygen economy, such as surface aeration, are easy to quantify (Owens *et al.*, 1972). It seems likely for example that the mass transfer coefficient (f) of 5·6 cm h^{-1} at a flow of 14 $m^3 s^{-1}$ in the lower reaches would be increased to 6·1 and 6·8 by regulation flows of 2 and 6 $m^3 s^{-1}$ respectively. But such increases would have an insignificant effect on the oxygen economy when compared with flow-related changes in plant metabolism which cannot satisfactorily be predicted. Nevertheless such flow supplementation will improve the oxygen status of the river during the critical period of plant decay by dispersing plant material.

Brooker *et al.* (1978) have demonstrated that plant growth itself is related to average daily flow during the period April–June in the lower Wye, evidence coming from their own detailed studies of biomass during the period 1975–7 and from an analysis of Wye River Authority reports over a longer period. It seems unlikely, however, that regulation could appreciably increase average flows during the three-month growth period, April–June during most years.

The effects of flow, operating principally through changes in velocity, on the upstream migration and behaviour of Atlantic salmon have been reviewed by Banks (1969) and Alabaster (1970). The effects of regulation on salmon movements and on angling in the Wye are not easy to predict, but are not likely to be substantial, for the following reasons:

(a) At Monmouth the supplementary flows are abstracted and therefore, in the lowest reaches and in the estuary, regulation releases may only be detected by changes in water quality.
(b) Supplementation is not likely to be pulsed, so simulating a freshet, but merely balancing abstraction needs at flows below 14 $m^3 s^{-1}$ at Monmouth.
(c) Water released from storage may not mimic water quality changes of freshets which are claimed to induce salmon movement.

However, in the upper reaches where regulation discharges can represent a large proportional increase in summer flows and where operating procedures may impose some flow pulsing, salmon movements and capture may be increased by reservoir releases despite their differences from natural freshets with respect to water quality. The use of sonic tagging techniques in the Wye should resolve such questions in the near future.

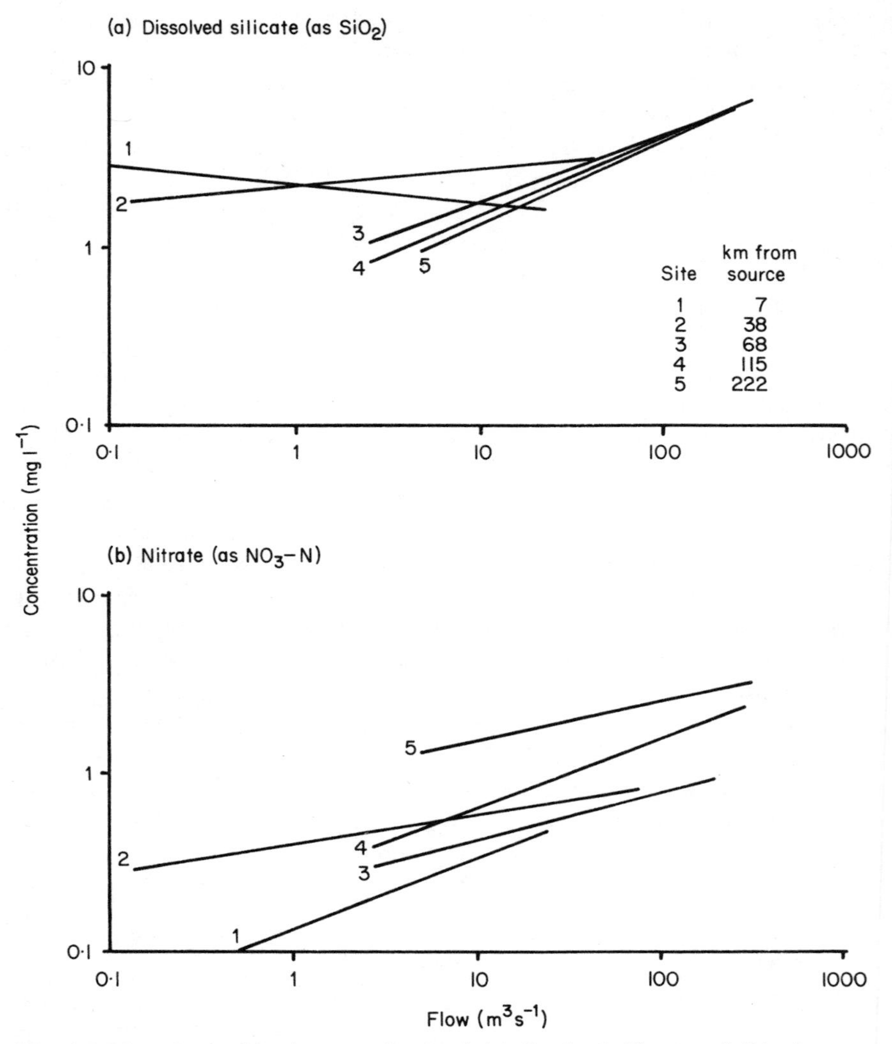

Fig. 3.8 The relationships between flow and (a) dissolved silicate and (b) nitrate at stations down the R. Wye.

Regulation discharges substantially reduce the residence time of water within the river (Fig. 3.5), which reduces the growth time of phytoplankton within the river, and also reduces the opportunity for removal of nutrients by both benthic plants and phytoplankton. This increases residual concentrations in the lower reaches where water is abstracted for subsequent impoundment. The distribution of dissolved silica (Fig. 3.8a) (important in the formation of diatom frustules) seems to indicate the importance of residence time although such indication is confused by seasonal factors. In the head waters, where little change has occurred to the quality of incoming drainage and run-off, the highest concentration occurs at low flows but progressively downstream is a reduction in silicate concentration at low flows (= high retention times). In contrast, at high flows the dissolved silicate concentrations increase downstream, this increase probably reflecting differences in the soil properties of the lower catchment. A downstream reduction in the concentration of dissolved silicate is not evident during the occasional periods of low winter flow when diatom growth is low. It would seem, therefore, that with regulation discharges during the summer higher concentrations of dissolved silicate may occur in the lower catchment although this could well be balanced by the lower initial concentration of the regulation water itself, being derived from the upper catchment and being stored before use. A further difficulty in making accurate predictions is the extent to which diatom growth in the middle and lower reaches of the Wye is currently limited by low dissolved silicate concentrations. Edwards *et al.* (1978) calculated that over 90% of the dissolved silicate input is removed at certain times of year, leaving concentrations below 0·5 mg l^{-1}, at which diatom growth is likely to be restricted (Wetzel, 1975).

In the case of nitrate too, reduced residence times provided by regulation discharges during the summer could increase residual concentrations in the lower reaches. It now seems certain that benthic denitrification is the dominant process determining nitrate losses in rivers, particularly at high summer temperatures, and not plant assimilation (Owens *et al.*, 1972). However, reduced losses of nitrate must be balanced against reduced inputs because regulation discharges, derived from the upper catchment, are likely to contain low nitrate concentrations (Fig. 3.8b) compared with drainage waters from the fertile lower catchment. Any changing nitrate status in the R. Wye related to regulation seems unlikely to affect ecological processes in the river although its effect on plankton growth during subsequent impoundment is less certain.

(ii) Storage

Regulation discharges from upland reservoirs, such as Craig Goch may change water quality in the lower reaches either because of the higher proportion of upland water, generally derived from catchments with base-poor soils and rocks, extensive ranching or forestry and low human populations, or because of temporal changes resulting from reservoir storage.

Several determinands act conservatively and their behaviour may be modelled fairly accurately, that of others which change phase or chemical form or are utilized by biological systems is more difficult to predict. The behaviour of phosphate, generally regarded as notoriously non-conservative in natural waters, may be modelled in the Wye with reasonable accuracy.

Generally, the empirically-derived relationship between river flow and the concentration or mass flow of chemical determinands at a site in the lower catchment before regulation takes place is of little predictive value for, in regulation, the proportions of the total flow derived from different subcatchments are changed by releases from storage. Nevertheless the derivation and combination of separate subcatchment relationships may be somewhat more useful. Such empirically-derived relationships are frequently described using the equation:

$$\log_{10} C = b \log_{10} F + a \quad (3.1)$$

where C = concentration (mg l^{-1}); F = mean daily flow (m s^{-1}), a and b being constants.

In the Wye, with the determinands alkalinity, total dissolved solids, sodium, calcium and magnesium, this simple equation generally explains less than 50% of the variance (Oborne *et al.*, 1980).

An alternative approach in which the separated components of flow (base-flow, run-off, effluents) are hydrologically, and then chemically, modelled has been proposed by Birtles (1977) and Birtles and Brown (1978) who have accounted for up to 83% of the variance of some determinands in the R. Severn and R. Avon. Using this approach, Oborne (1981) has compared the actual and calculated concentrations for chloride, phosphate, total hardness and alkalinity at Monmouth during the period 1976–7 after using data acquired within the period 1970–6 for model calibration. Within the test period the model explained between 59 and 88% of the variance for these determinands. The concentration–flow relationship (Equation 3.1), in contrast, explained only between 7 and 36% of their variance (Sambrook, 1976) during an earlier period.

Using this flow-separation model (Oborne, 1981), the effects of regulation on the water quality at Monmouth have been calculated for the 12 month periods September 1975–August 1976, when flows in the lower catchment were about 38% of the long-term average, and September 1976–August 1977, when flows were about 118% of the long-term average. Table 3.10 shows the results of using the model to calculate the likely effect of such regulation on alkalinity, chloride, phosphate and total hardness. With all these determinands, increasing regulation has the principal effect of reducing mean concentrations and ranges, the most dramatic effects clearly being shown in dry years when regulating releases are most frequent.

Effects of changes in quality may result not only from the degree of quality changes but also from the period for which such changes are sustained. The five-day sustained concentration: mean ratio during the dry period varies between 0·5 and 4·0 with natural flows and between 0·8 and 1·4 with CG regulation (Fig. 3.9). The effects of regulation therefore include a tendency towards homogeneity of water quality in those determinands

Table 3.10 Simulated mean concentrations (mg l^{-1}) and ranges (10–90 percentile) of alkalinity, chloride, phosphate-P and total hardness under natural flow conditions (N), Elan Redeployment (ER) and Craig Goch (CG) regulation conditions during contrasting 12 month periods (see text)

	Dry period			*Wet period*		
	N	ER	CG	N	ER	CG
Alkalinity						
mean	101	89	77	91	88	84
range	66–141	66–112	65–92	60–130	60–123	60–106
Chloride						
mean	17·4	16·1	14·7	16·0	15·8	15·3
range	13·5–23·5	13·5–20·2	13·5–16·5	13·5–20·0	13·5–19·8	13·5–18·3
Phosphate-P						
mean	0·140	0·119	0·097	0·086	0·083	0·079
range	0·069–0·263	0·069–0·203	0·069–0·139	0·064–0·133	0·064–0·120	0·064–0·103
Total hardness						
mean	123	110	97	115	112	107
range	87–169	88–135	84–112	86–154	86–151	86–131

which behave conservatively and which normally increase in concentration at low flows.

The ecological consequences within the R. Wye of water quality changes, whilst difficult to predict, are not likely to be dramatic and may be solely related to changes in calcium content, principally in the upper reaches where concentrations are naturally low. Edwards *et al.* (1978) suggested that calcium may well influence the distribution of certain molluscs and crustaceans in these upper reaches. Although crayfish (*Austropotamobius pallipes*)

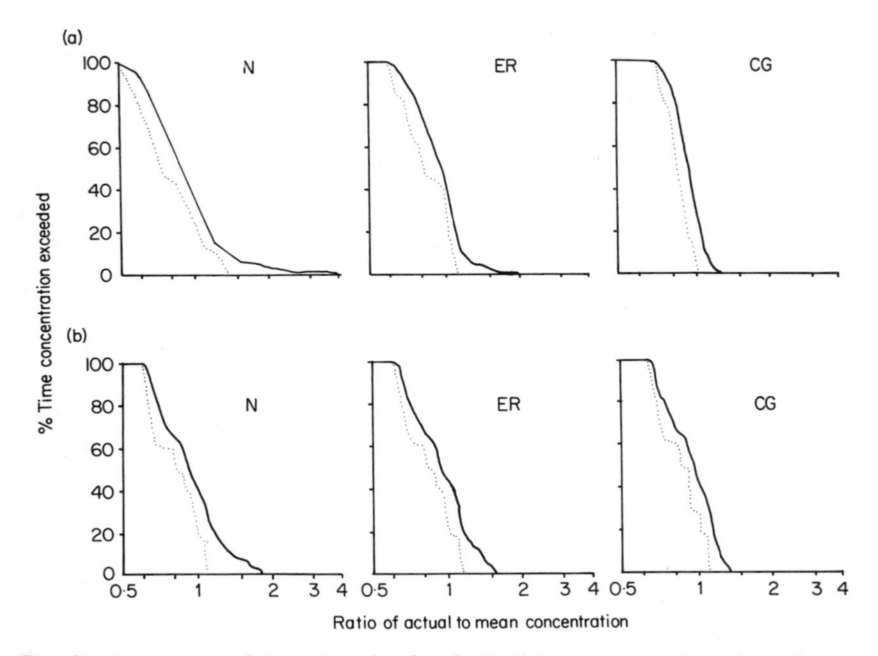

Fig. 3.9 Percentage of time that simulated alkalinity concentrations (actual: mean concentration) are exceeded for more than 5 or 50 days in the (a) dry and (b) wet periods (N = natural flow; ER = Elan Redeployment and CG = Craig Goch). 5 and 50 day periods are shown by continuous and dotted lines respectively.

are not normally found in the main river, their distribution in tributaries seems closely associated with calcium concentrations, absence being general where concentrations are below about 10 mg l^{-1} (Lilley *et al.*, 1979). With regulation the reduction in calcium concentrations in the lower reaches, from 45 to 40 (ER) or 33 mg l^{-1} (CG) at Monmouth, could be of more significance to supplies abstracted for industrial use. Regulation, in softening the water, brings it nearer to the quality of water abstracted from the R. Usk and stored within Llandegfedd Reservoir (22 mg l^{-1}).

(b) Local effects

(*i*) *Flow modification*

Although, in the lower reaches of the R. Wye, increasing flow is reflected predominantly in increasing velocity, in the upper reaches cross-sectional areas are also increased (Table 3.9). In these upper reaches the proportional effects of regulation on flow are large and at Newbridge-on-Wye, about 10 km below the Elan confluence, flow supplementation is likely to occur at a natural river flow of about 3 $m^3 s^{-1}$ (Fig. 3.5) and to increase cross-sectional area from 18·8 to 23·8 (ER) and 32·1 M^2 (CG) and velocity from 16 to 21 (ER) and 28 cm s^{-1} (CG). Such increases in cross-sectional area will increase both water depth and the permanently wetted area of river bed.

Considering wetted area, although increasing river width seems likely to increase the total production of mosses and attached micro-flora which will, in turn, enhance the production of certain invertebrate 'grazers and scrapers' (Cummins, 1975) and the abundance of species particularly associated with such plant growths (Armitage, 1976), energy flow in rivers is largely sustained from allochthonous sources (Mann, 1975), particularly where there is shading from trees. Furthermore, benthic invertebrates frequently occur at considerable depths in river beds (Williams and Hynes, 1974) some species probably moving vertically in response to discharge changes, and in the upper Wye the studies by Morris and Brooker (1979) suggest that about 50% occur at depths greater than 10 cm – some species being exclusive to, and others absent from, superficial sediments. Thus whilst there may be substantial increases in the permanently wetted area, commensurate increases in invertebrate production may not occur. Similarly, Gee *et al.* (1978) have suggested that the mortality of each cohort of salmon fry in the upper Wye after 1st June is not influenced by changes in the wetted area of sites, mortality being determined principally by initial population density.

Depth increases occurring with flow supplementation will, in upper reaches with clear riffle-run-pool sequences, decrease the area of riffles and extend that of pools. In the upper Wye such a shift is not likely to be substantial. Whilst some workers have shown that invertebrate abundance in riffles may sometimes exceed that of pools (Egglishaw and Mackay, 1967) others have found little qualitative or quantitative difference in their faunas (Logan, 1980) and Armitage (1976) has even demonstrated that pools may support a larger invertebrate biomass than riffles. With respect to fish, Jones (1975) has described the distribution of fish in riffle-run-pool sequences, revealing a partial segregation with 0+ salmon, bullheads and lampreys predominating in riffles; 1+ salmon, 0+ trout and stoneloach predominating in both riffles and runs; 1+ trout predominating in runs and pools; and 2+ and older trout, minnows and gudgeon predominating in pools. The significance of such habitat changes, resulting from flow supplementation during the summer, to the production of salmon smolts or large trout is not easy to assess. The production of salmon smolts must be considered primarily on a catchment basis and, in these terms, any local habitat shift is likely to be of minor importance: the enhancement of trout production, on the other hand, has a more local relevance and, below the Elan–Wye confluence, the maintenance of increased water depth during the summer could be beneficial to the stock of larger trout. Depth also influences the upstream migration of salmonids and, possibly, the sites of preferred spawning, but during the autumn when such spawning occurs at high flows, abstraction of river flows for reservoir replenishment is a more likely contingency than flow supplementation.

Bank vegetation has a characteristic vertical zonation and might be affected by depth stabilization during the summer but it seems from the work of Merry (personal communication) in the Wye that the banks of upland sites are dominated by bryophytes and grasses, the former being generally tolerant of periodic submergence and drying and not having very precise vertical zones. In contrast, at lowland sites where depth changes resulting from regulation seem likely to be minimal (Table 3.9) there are several plant species having more specific vertical distributions, e.g. *Gnaphalium uliginosum* (Fig. 3.10). Generally, therefore, substantial changes in the distribution of bank vegetation are not expected.

The impact of increases in water velocity on particle transport has already been discussed: it seems likely that the relationship between the distribution of some invertebrates, particularly filter-feeders, and velocity operates principally through such particle transport. Velocities below 10 cm s^{-1} are frequently inadequate to sustain the food supply of filter-feeders such as larvae of net-spinning caddis-flies (Edington, 1965, 1968) and velocity increases resulting from regulation (Fig. 3.5) could well extend the localized distributions of populations of these and other filter-feeders. The effects of regulation on velocity are compounded by the release of reservoir water which frequently contains high concentrations of planktonic organisms.

Concerning salmonid fish, the immediate effects of velocity on site-selection for spawning, demonstrated by several authors (Fraser, 1975; Peterson, 1978), and on the subsequent embryo development, are not relevant in the present case as such spawning occurs during the late autumn when flows are not modified and development and fry emergence are completed before the summer. There remains the reservation that substrate modification occurring during periods of summer regulation could influence the suitability of sediments for subsequent salmonid spawning and embryo development.

The response of juvenile salmonids and other fishes to water velocity has been recently reviewed (Arnold, 1974). For Atlantic salmon, the species of major importance in the upper Wye, Kalleberg (1958) found that at

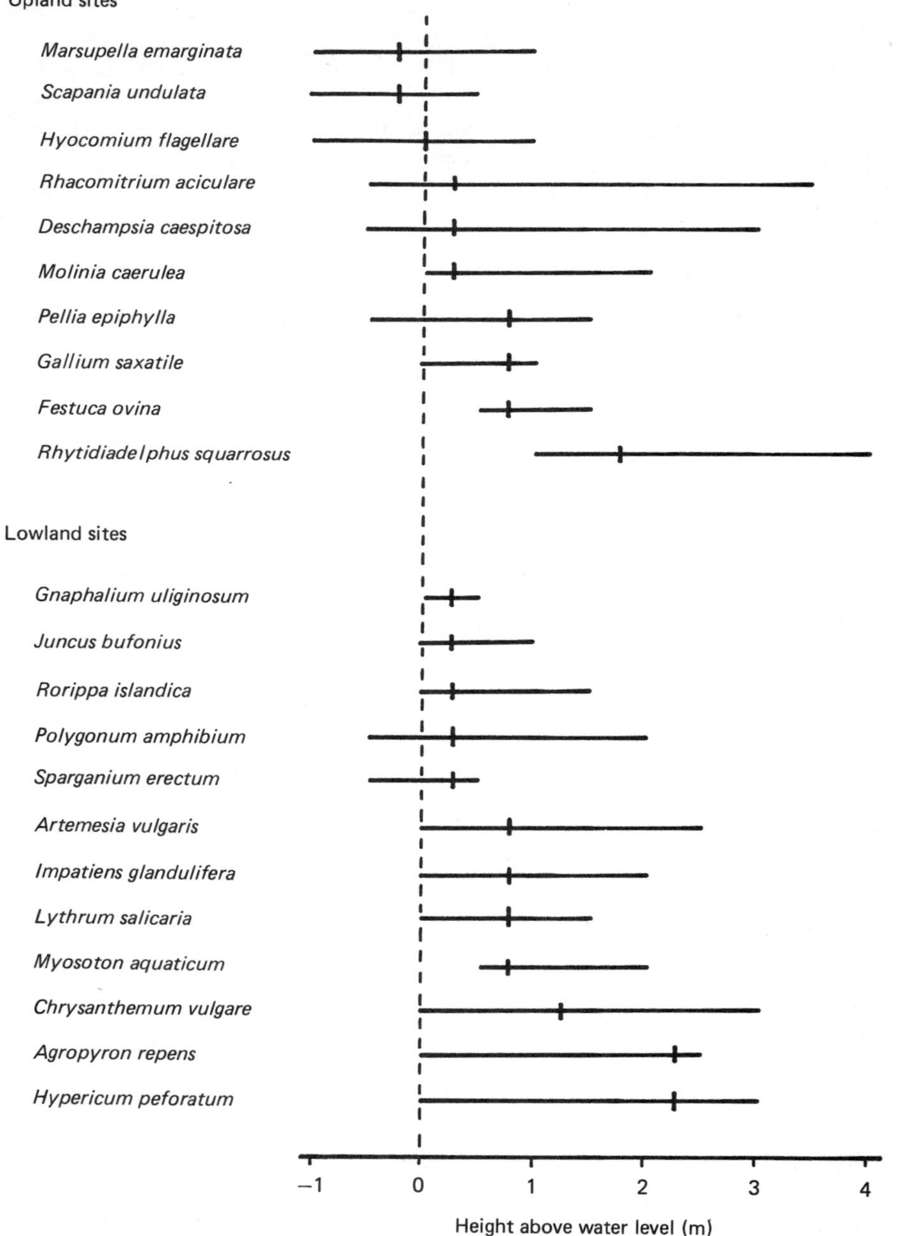

Fig. 3.10 Vertical distribution of some riparian plant species on the Wye in relation to normal summer water level: the vertical bar indicates height (above water level) of maximum abundance.

increasing velocity juveniles leave open water to find shelter at the streambed (5–10 cm s^{-1}) and to establish territories (18–29 cm s^{-1}): these velocities at which behavioural changes occur are within the broad range influenced by both regulation schemes on the Wye (Fig. 3.5) although velocities at which juvenile salmon are purported to achieve maximal densities (50–65 cm s^{-1}) (Symons and Heland, 1978) are considerably above this range. Whilst regulation might therefore be expected to increase the survival of juvenile salmon, there is no evidence that factors other than the initial fry density are of major significance (Gee *et al.*, 1978).

(ii) Storage

Figure 3.11 shows the mean monthly temperatures at sites in the Elan and the Wye above and below the Elan confluence (see Fig. 3.2) during 1978, a reasonably typical year. It is evident with the release of hypolimnetic waters from Caban Coch by current operational procedures, temperatures in the R. Elan are substantially below those in the neighbouring Wye, particularly from May until August. Nevertheless within the R. Elan there is an appreciable increase in temperature from sites E1 to E2, and below the confluence (W3) the effect of the Elan is not readily detectable. Even with substantially larger reservoir releases associated with regulation, temperature effects are likely to remain relatively local, the river rapidly equilibrating to the heat sources and sinks related to meteorological, topographical and geological factors. The localization of temperature effects is also influenced by the proximity of two major tributaries, the Ithon and Irfon, only 12 and 20 km respectively downstream of the Elan confluence.

Despite the likelihood that temperature effects of regulation releases are likely to be confined to the reaches of the Wye immediately downstream, no adequate predictive model has yet been developed and tested. Nevertheless several authors have related growth rates and development periods of invertebrates to water temperature (Nebeker, 1973; Mackey, 1977; Humpesch, 1980) and life-cycle changes and reduced production could occur during summers when regulation discharges are frequent (see Fig. 3.3).

The growth of salmonids is similarly related to temperature but temperature optima, about 14 °C for brown trout (Elliott, 1975) and 16–18 °C for rainbow trout (Brown, 1971), are lower than summer maxima. Using the growth equation for brown trout on maximum feeding proposed by Elliott (1975) it is evident from Fig. 3.12 that growth at most sites is very similar except for the site on the R. Elan immediately downstream of the reservoir release (E1) and the most upstream site on the Wye (W1). The lower Elan site (E2), although generally cooler (see Fig. 3.11b), supports growth rates similar to those generally found in the Wye because of its relatively homothermous nature. The effects of temperature modification, resulting from regulation, on the growth rate of trout, and probably juvenile salmon are likely to be local but could be beneficial in the enlarged Craig Goch scheme where direct releases are proposed into the head waters of the Wye near W1 (see Fig. 3.2): at times of such summer regulation, with low natural flows, reservoir releases will reduce temperature bringing it near to the growth optimum.

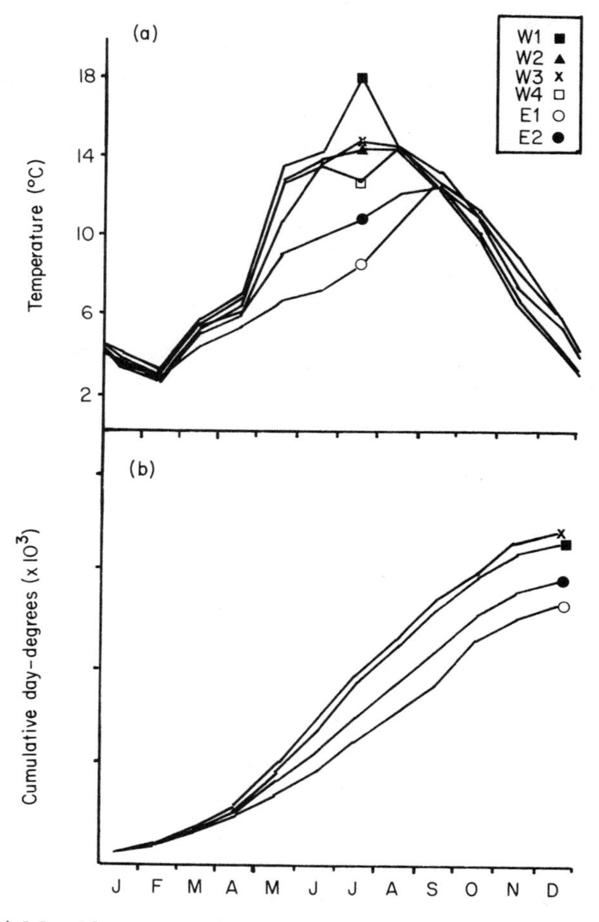

Fig. 3.11 (a) Monthly mean water temperatures at sites in the upper Wye and Elan for 1978 (see Fig. 3.2 for location of sites) and (b) cumulative number of day-degrees for 1978 at sites in the upper Wye and Elan.

It is known that temperature also influences the induction of smolt migrations, movements being suppressed below about 10–12°C (Solomon, 1978). In the upper Wye such temperatures are generally reached by the beginning of May (Fig. 3.11a), well before the period when cooler reservoir waters might be released for regulation purposes (Fig. 3.3).

Hypolimnetic waters in Caban Coch are substantially de-oxygenated during the summer months. Should a similarly deoxygenated hypolimnion develop in the enlarged Craig Goch reservoir then natural re-aeration of water discharged to the river seems likely to localize the deoxygenated zone.

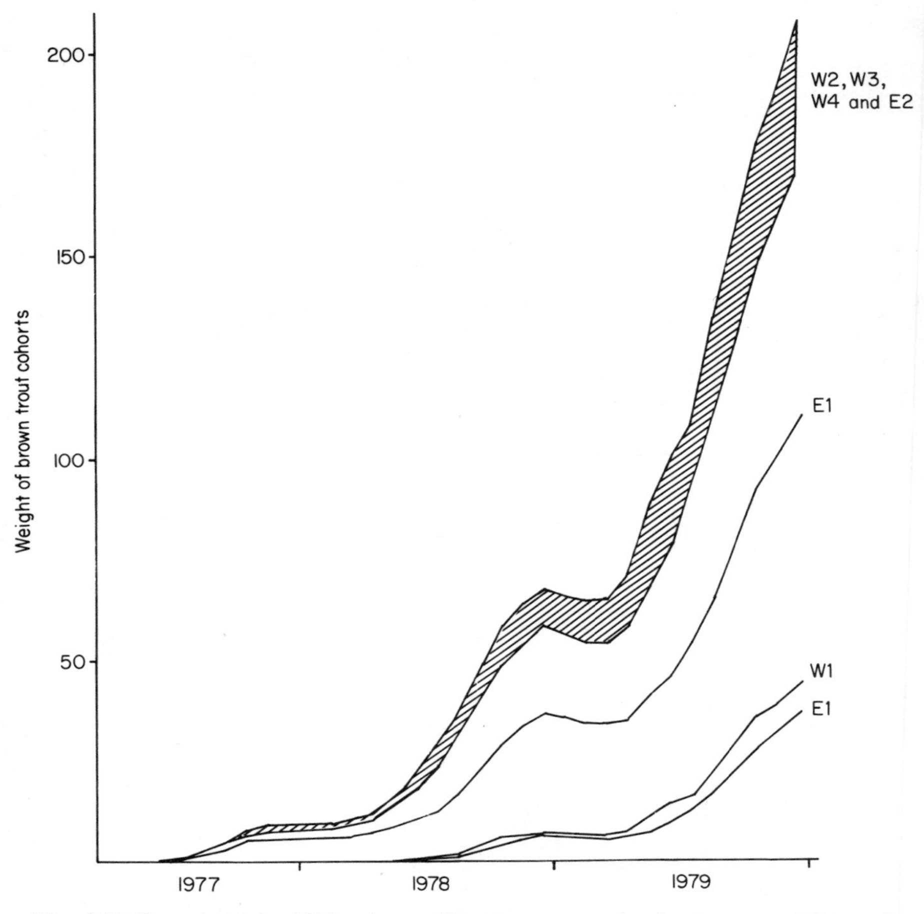

Fig. 3.12 Growth of the 1977 cohort of brown trout at sites in the upper Wye and Elan calculated from the model for maximum ration developed by Elliott (1975). No temperature data are available for W1 in 1977 so growth of the 1978 cohort at W1 and E1 are compared.

In rivers of generally similar velocity and depth, mass transfer coefficients for oxygen have been sufficiently high (f 30–40 cm h^{-1}) to raise the dissolved oxygen concentration from 5 to 8 mg l^{-1} in less than 500 m.

Whilst upland reservoirs are important as sediment traps, particularly for peat (Armitage, 1977; Crisp, 1977), they produce other organic particles, (phyto- and zoo-plankton) which, on release downstream, can lead to the development of high densities of filter-feeders (Armitage, 1978). From studies both in the Elan (Hopper, 1978) and the Upper Tees at Cow Green reservoir (Armitage and Capper, 1976), where most zoo-plankton was lost from the river-water within a few kilometres, it seems likely that faunal changes are likely to be local.

Particles which contain 40% organic material, 6% iron and 4% manganese are discharged from Caban Coch and accrete on the bed of the Elan at a rate of 0·1–0·2 g m^{-2} d^{-1} up to maximal amounts of 400 g m^{-2}. These deposits probably contribute to the distinctively different fauna of the river which is characteristic of rivers polluted with ferric hydroxide. With the ER scheme which redistributes flows only to a small degree (2 m s^{-1} compared with the current compensation flow of 1·3 m^3 s^{-1}) it is unlikely that there would be any appreciable change in the deposition of such solids unless water were withdrawn from the epilimnion during the summer. With the CG scheme it is possible, should the enlarged Craig Goch reservoir become partially deoxygenated, that similar deposits might develop during the summer downstream of the tunnel discharge (Fig. 3.2). Although these deposits might lead to the eradication of certain sensitive groups, it is unlikely that a 'replacement fauna', similar to that in the Elan, would develop.

High concentrations of iron and managanese occur not only in solids accreting on the river bed of the Elan but also in solution. The mean total concentrations in the Elan (E2) are about 500 μg l^{-1} (Fe) and 100 μg l^{-1} (Mn), of which 75 and 50% respectively are associated with particulate matter. Regulation by the CG scheme might be expected temporarily to elevate downstream total concentrations of iron and manganese to 220 and 45 μg l^{-1} respectively (compared with the maximum acceptable concentrations for potable supply of 300 and 50 μg l^{-1}) but with the conditions prevailing in the Wye during the summer it seems likely that much will be removed from solution in passage downstream.

3.3.4 Conclusions

Figures 3.13 and 3.14 summarize the more likely general and local effects induced by flow modification and storage associated with the regulation of the Wye during the summer months by the enlargement of the Craig Goch: changes associated with the redeployment of existing compensation water from the existing reservoirs via the Elan are likely to be less discernible. Considering extensive effects (Fig. 3.13), fisheries are likely to be improved by the avoidance of low oxygen concentrations in the lower reaches at times of plant decay which have previously caused severe salmon mortality and by the enhancement of the drift of benthic invertebrates which form a significant component of salmonid diet during the summer. Flow regulation is also

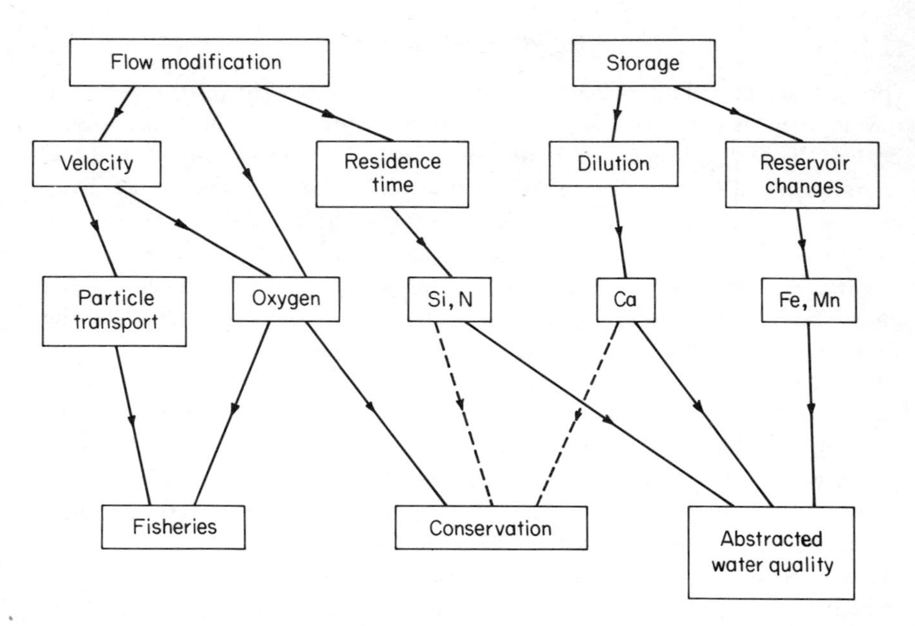

Fig. 3.13 Possible extensive effects of regulation on aquatic resources of the River Wye. More doubtful effects are shown by hatched lines (see also Fig. 3.14).

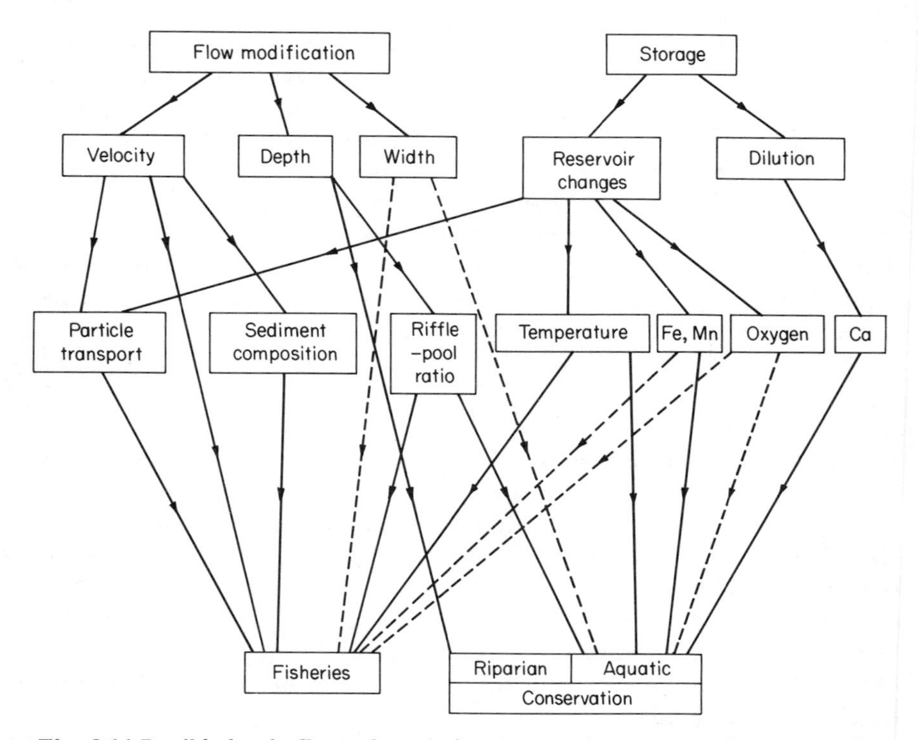

Fig. 3.14 Possible local effects of regulation on aquatic resources of the River Wye.

likely to protect certain components of the benthic invertebrate fauna which may also be sensitive to low oxygen concentrations. Of the quality aspects of abstracted water which may change, a degree of softening is the most predictable. Other possible changes, such as those to silica and combined nitrogen concentrations, have not been adequately modelled depending as they do on the growth of microorganisms as well as the predictable changes in residence time and influent concentrations. Iron and manganese concentrations are likely to increase unless the quality of source water can be improved, perhaps by abstraction of epilimnetic rather than hypolimnetic waters, but such increases are not likely to damage water quality, in part because of phase shifts.

Local effects of regulation will probably be more varied (Fig. 3.14) but are only of concern to *in situ* resources, there being no important abstraction immediately downstream of the regulation releases. With respect to fisheries, the major effects are likely to be associated with increases in water velocity which will influence the upstream movement of salmon and with local modifications to the temperature and oxygen regimes. These latter effects resulting from the release of hypolimnetic water could be reduced or avoided by the provision of multi-level abstraction facilities and the withdrawal of epilimnetic water. The discharge of zoo-plankton and phyto-plankton will enhance the food supply of filter-feeders but the restriction of the period, and inter-year variation of regulation may reduce the opportunity for population response. There may, however, be some benefit in the local food supply of fish although it now seems likely that salmonid growth is principally determined by temperature (Edwards *et al.*, 1979). It seems probable that the principal effect of regulation on the distribution of the invertebrate fauna will result from a reduction in calcium concentrations and, at the low concentrations prevailing, the upstream limit of distribution of some molluscs (e.g. *Ancylus fluviatilis*) and crustaceans may change. Whether such distribution changes related to modifications in water quality will be readily detectable seems dubious, for the longitudinal succession of invertebrate associations down the river already varies considerably from year to year (Brooker and Morris, 1980) in response to other environmental factors.

Acknowledgements

The author wishes to thank Dr M. P. Brooker for reading the script, Mr P. Logan for drawing the figures and undertaking tedious calculations, the Llysdinan Trust without whose help the UWIST Craig Goch group could never have been established and, finally, to the group itself (Dr M. P. Brooker, Dr A. S. Gee, Mr R. J. Hemsworth, Dr N. J. Milner, Mr D. L. Morris and Dr A. C. Oborne) which undertook the investigations and made possible this synthesis.

3.4 ECOLOGICAL MODELLING IN IMPACT ANALYSIS

P. Wathern

3.4.1 Introduction

The ability to identify, predict and assess the consequences of changes in land use is an essential component of environmental management. The framework within which these activities take place and the results of the appraisal made available to decision-makers is referred to as Impact Analysis. This paper contains a review of methods developed to facilitate the overall appraisal of potential impacts, but specific techniques for predicting individual impacts within this methodological framework are not discussed. Many methods have been developed, particularly in the US, but most have been used only rarely, while others have been designed to aid only one aspect of the assessment, for example impact prediction. No attempt will be made to consider these partial and infrequently used methods, as they have been catalogued elsewhere (Clark *et al.*, 1980). They should not be dismissed out of hand, however, as their use may be justified in certain circumstances. In this paper, four approaches to impact analysis, selected because they illustrate the main types of method, are described in detail. These approaches are:

(a) Matrices
(b) Networks
(c) Quantitative methods
(d) Simulation modelling

Each approach has strengths and weaknesses and it is important in selecting a particular method to match its favourable attributes with the characteristics of the proposal under consideration. One difficulty in evaluating the utility of different impact analysis methods is the dearth of material detailing their use. This problem is compounded in the UK where so few environmental impact statements (EIS) have been produced; so much so that it is difficult to discuss the value of methods strictly within a UK context.

3.4.2 Matrices

Matrices provide an ideal format for relating the proposed development to the local environment. Clark *et al.* (1976), for example, have advocated the use of a matrix as a check list to identify interactions which may lead to environmental impacts. They recognized that both developments and environments have many different attributes which may interact in many ways and have suggested that environmental impacts can only be analysed as a complex matrix of interactions. They have proposed a two-dimensional check list where development components on the vertical axis can be related to features of the pre-development environment as the horizontal axis. It must be stressed, however, that completing a matrix of potential impacts is merely the first stage of assessment as the likely significant effects of development must be given detailed consideration subsequently. The other methods, however, use matrices as more than a simple two-dimensional check list for identifying potential impacts.

(a) The Leopold matrix

The method developed by Leopold *et al.* (1971) represents one of the first attempts to relate development actions and components of the environment using a matrix format. The horizontal axis of the matrix consists of 100 'project actions', attributes of development which may induce environmental impacts, while the vertical axis includes 88 components of the environment which may be affected (Fig. 3.15). For each of the 8800 possible interactions between project actions and environment components, those preparing an impact assessment must decide whether an impact is likely to occur. Although there is a large number of possible interactions, Leopold *et al.* suggest that only 25–50 are likely to be important for a major development. The magnitude and significance of each potential impact is recorded in the appropriate cell of the matrix, the upper half of the cell representing magnitude, the lower half significance. A written report containing a description of the potential impacts is also prepared. Leopold *et al.* indicate that this discussion should focus not only on those project actions which have high individual values for magnitude or significance, but also on those which affect a number of environmental components.

The Leopold matrix has been criticized mainly on its inability to reveal indirect effects of development (Environment Canada, 1974). Only direct effects of individual project actions on separate environmental components are assessed. Although environmental components are related to project actions in the method, they are not related to one another, so that the complex interactions between components of the ecosystem which lead to indirect impacts are not assessed. In practice, this may not be a serious constraint if those undertaking an assessment have sufficient expertise to be aware of the ramification of inter-relationships within an ecosystem. The main criticism lies in the failure to consider indirect impacts systematically.

Leopold *et al.* have attempted to produce a matrix which can be used for a variety of developments. Consequently, it is unwieldy. This encyclopaedic consideration of impacts has been fostered in the US by a system which encourages the production of an EIS simply to comply with the provisions of the *National Environmental Policy Act* of 1969 (NEPA), rather than investigating effects to improve environmental planning (Munn, 1979). The consideration

Instructions

1. Identify all actions (located across the top of the matrix) that are part of the proposed project
2. Under each of the proposed actions, place a slash at the intersection with each item on the side of the matrix if an impact is possible
3. Having completed the matrix, in the upper left-hand corner of each box with a slash, place a number from 1 to 10 which indicates the MAGNITUDE of the possible impact; 10 represents the greatest magnitude of impact and 1, the least, (no zeroes). Before each number place + (if the impact would be beneficial). In the lower right-hand corner of the box place a number from 1 to 10 which indicates the IMPORTANCE of the possible impact (e.g. regional vs. local); 10 represents the greatest importance and 1, the least (no zeroes)
4. The text which accompanies the matrix should be a discussion of the significant impacts, those columns and rows with large numbers of boxes marked and individual boxes with the larger numbers

Sample matrix

	a	b	c	d	e
a		2/1			8/5
b		7/2	8/8	3/1	9/7

			A. Modification of regime													B. Land transformation and construction																			C. Resource extraction						
		Proposed actions	a. Exotic flora or fauna introduction	b. Biological controls	c. Modification of habitat	d. Alteration of ground cover	e. Alteration of ground water hydrology	f. Alteration of drainage	g. River control and flow modification	h. Canalization	i. Irrigation	j. Weather modification	k. Burning	l. Surface or paving	m. Noise and vibration	a. Urbanization	b. Industrial sites and buildings	c. Airports	d. Highways and bridges	e. Roads and trails	f. Railroads	g. Cables and lifts	h. Transmission lines, pipelines and corridors	i. Barriers including fencing	j. Channel dredging and straightening	k. Channel revetments	l. Canals	m. Dams and impoundments	n. Piers, seawalls, marinas and sea terminals	o. Offshore structures	p. Recreational structures	q. Blasting and drilling	r. Cut and fill	s. Tunnels and underground structures	a. Blasting and drilling	b. Surface excavation	c. Subsurface excavation and retorting	d. Well drilling and fluid removal	e. Dredging	f. Clear cutting and other lumbering	g. Commercial fishing and hunting
CHEMICAL CHARACTERISTICS	1 Earth	a. Mineral resources																																							
		b. Construction material																																							
		c. Soils																																							
		d. Land form																																							
		e. Force fields and background radiation																																							
		f. Unique physical features																																							
	2 Water	a. Surface																																							
		b. Ocean																																							
		c. Underground																																							
		d. Quality																																							
		e. Temperature																																							
		f. Recharge																																							
		g. Snow, ice and permafrost																																							

Fig. 3.15 A Section of the Leopold matrix (US Geological Survey).

of all impacts is seen as one way of avoiding litigation, and the encyclopaedic approach should ensure that all impacts are reviewed. However, it is possible to produce a reduced matrix by selecting a subset of environmental components and project actions for a particular development. The apparent comprehensiveness of the matrix might restrict the consideration of other aspects for a particular proposal. Alternatively, it may be necessary to expand one of the broad Leopold categories in order to investigate the full ramifications of development. For example, within the category 'birds', resident woodland passerines may benefit from certain types of afforestation, while breeding upland waders and raptors may be adversely affected.

The inclusion of numerical values of magnitude and significance for all potential impacts obscures the credence which may be placed in the assessment of different impacts. There are three main drawbacks. First, no indication is given of whether the data on which numerical values of magnitude and significance are based, are qualitative or quantitative. Secondly, the probability of a project impact occurring is not specified. Finally, the validity of individual projections cannot be assessed because details of the techniques used to predict impacts are excluded. It is unlikely that a decision-maker will read the written report in detail when accompanied by an apparently admirable summary matrix. There is also a danger that decision-makers will attempt to add numerical values in order to produce a composite value for the environmental impacts of the development. This would lead to erroneous results as the individual impacts are on an ordinal scale and aggregations are invalid (Clark *et al.*, 1979).

(b) The Component Interaction matrix

The inability to detect indirect impacts systematically is a major limitation of the Leopold matrix. Indirect, or higher-order, interactions are those which depend upon the presence of an intermediary. To overcome this constraint in using matrices, Environment Canada (1974) proposed the use of simple matrix multiplication to reveal higher-order interactions within an ecosystem based on all known first order (direct) linkages. Both axes of the matrix consist of identical listings of environmental components and each possible interaction between components is scored if the row component is dependent upon a column component. Squaring the matrix reveals second order, two link dependencies. Repeated multiplication of the matrix will give successively third, fourth and fifth order dependencies. In the development proposal considered by Environment Canada, matrix manipulation was terminated at this stage, because it was considered that all significant linkages had been detected with the inclusion of fifth order dependencies.

The use of a binary system to indicate dependency can be criticized because it assumes that all direct dependencies are of similar magnitude.

	1. Live Plants	2. Roots	3. Standing dead	4. Litter	5. Soil	6. Sheep	7. Sheep dung	8. Dung invertebrates	9. Slugs	10. Earthworms	11. Leachates
1. Live plants											1
2. Roots	1										
3. Standing dead	1										
4. Litter			1				1	1	1		
5. Soil		1		1		1	1	1	1	1	
6. Sheep	1		1								
7. Sheep dung						1					
8. Dung invertebrates							1				
9. Slugs	1						1				
10. Earthworms				1							
11. Leachates	1										

Fig. 3.16 Component Interaction Matrix showing first order interactions (qualitative data).

	1. Live Plants	2. Roots	3. Standing dead	4. Litter	5. Soil	6. Sheep	7. Sheep dung	8. Dung invertebrates	9. Slugs	10. Earthworms	11. Leachates
1. Live plants	1								1		
2. Roots											1
3. Standing dead											1
4. Litter	2					1	2				
5. Soil	3		2	1		1	3	1	1		
6. Sheep	1										1
7. Sheep dung	1		1								
8. Dung invertebrates						1					
9. Slugs						1					
10. Earthworms			1				1	1	1		
11. Leachates											1

Fig. 3.17 Squared Component Interaction Matrix (figures denote the number of second-order interactions between components).

Within an ecosystem individual linkages are likely to be of varying importance. Consequently, Lennon (1979) applied both qualitative and quantitative data to an upland grassland ecosystem threatened by impoundment. One of the major effects of development would be the removal of sheep grazing from inundated lowland 'in-bye' land. The use of this area is essential to maintain the current level of grazing on the surrounding unimproved upland grassland. Relaxation of grazing would probably lead to considerable changes in these upland grasslands. Matrix manipulation can be used to investigate the importance of sheep in the ecosystem.

	1. Live Plants	2. Roots	3. Standing dead	4. Litter	5. Soil	6. Sheep	7. Sheep dung	8. Dung invertebrates	9. Slugs	10. Earthworms	11. Leachates
1. Live plants	2										1
2. Roots	1										2
3. Standing dead	1										2
4. Litter	2		1			2	1	1	1		3
5. Soil	2	1	2	1		1	1	1	1	1	3
6. Sheep	1		1								2
7. Sheep dung	2		2			1					3
8. Dung invertebrates	3		3			2	1				4
9. Slugs	1		3			2	1				4
10. Earthworms	3		2	1		3	2	2	2		4
11. Leachates	1										2

Fig. 3.18 Minimum Link Matrix.

Perkins *et al.* (1978) have produced data on energy flow within a comparable upland grassland system, which seems applicable to this situation. These data show direct dependencies and indicate the magnitude of each. Each possible interaction between environmental components was scored 1 if there was a direct dependency (Fig. 3.16). The matrix was squared to reveal second-order interactions (Fig. 3.17). Matrix manipulation was continued, with no additional indirect dependencies beyond the fourth-power matrix. A minimum link matrix was constructed from the direct and indirect dependencies, where the numbers in the cells indicate the shortest chain between environmental components (Fig. 3.18). Direct energy flow between environmental components was substituted for the binary qualitative

	1. Live Plants	2. Roots	3. Standing dead	4. Litter	5. Soil	6. Sheep	7. Sheep dung	8. Dung invertebrates	9. Slugs	10. Earthworms	11. Leachates
1. Live plants											0.169
2. Roots	1.0										
3. Standing dead	1.0										
4. Litter			0.597				0.086		0.017		
5. Soil		0.334		0.645		0.026	0.010		0.002	0.092	
6. Sheep	0.802		0.198								
7. Sheep dung						1.0					
8. Dung invertebrates							1.0				
9. Slugs	0.812						0.188				
10. Earthworms				1.0							
11. Leachates	1.0										

Fig. 3.19 Component Interaction Matrix showing magnitude of first-order interactions.

	1. Live Plants	2. Roots	3. Standing dead	4. Litter	5. Soil	6. Sheep	7. Sheep dung	8. Dung invertebrates	9. Slugs	10. Earthworms	11. Leachates
1. Live plants	0.198										0.203
2. Roots	0.198										0.298
3. Standing dead	0.198										0.298
4. Litter	0.815		0.615			0.089	0.089		0.017		0.135
5. Soil	0.926	0.334	0.460	0.737		0.102	0.076		0.015	0.092	0.157
6. Sheep	0.192		0.198								0.198
7. Sheep dung	1.169		0.198			1.0					0.192
8. Dung invertebrates	1.136		0.198			1.0	1.0				0.169
9. Slugs	1.086		0.037			0.188	0.188				0.055
10. Earthworms	0.802		0.615	1.0		0.089	0.089		0.017		0.115
11. Leachates	1.198										0.198

Fig. 3.20 Matrix showing accumulated dependence up to sixth-order interactions.

data within the matrix and the matrix was raised successively to the sixth power (Figs 3.19 and 3.20). This level of termination was an arbitrary decision, but it can be justified because energy flows at this stage were small, less than 3% of the total energy flow into individual environmental components.

Energy data have been used in impact analysis on previous occasions, but generally in energy flow diagrams, for example, Gilliland and Risser (1977). Including quantitative data in a component interaction matrix greatly extends the utility of the approach, because it allows an assessment of the magnitude of inter-relationships. Other parameters indicating the extent of dependency, such as nitrogen, phosphorus or carbon budgets could be used as alternatives to energy flow.

There are, however, several disadvantages of this approach. The use of quantitative parameters means that any form of dependency which cannot be expressed numerically in terms of that parameter can not be taken into account. Thus, for example, relationships where plants provide cover but are not consumed by a particular animal cannot be assessed. There are practical problems of implementation. The appraisal of the upland ecosystem in terms of its energetics in this study was possible only because of the availability of the Perkins' data. Without such data it would probably not be possible to collect quantitative data within the time scale of most development proposals. Thus, use of this extension of the component interaction matrix generally will be fortuitous and dependent upon the availability of appropriate data.

The main value of the approach is in inter-relating the existing environmental components. The tacit assumption that a statement of the existing energetics infers the likely consequences of modifying the ecosystem may not be justified. Indirect impacts implied from matrix manipulation may not occur because of the redirection of energy flow and the establishment of new equilibria.

3.4.3 Networks

Sorensen (1971) proposed the use of a network approach for revealing the indirect effects of development in the coastal zone. The approach depends upon the ability to follow the effects of development through changes in environmental parameters (Fig. 3.21). Potential causes of change associated with a proposed use are identified using a matrix format. These activities result in certain primary condition changes which are manifest in the form of specific environmental effects. The environmental effects, impacts, may result directly from primary condition changes or indirectly through some induced secondary or tertiary condition change. A condition change may result in several different types of impact each of which can be followed

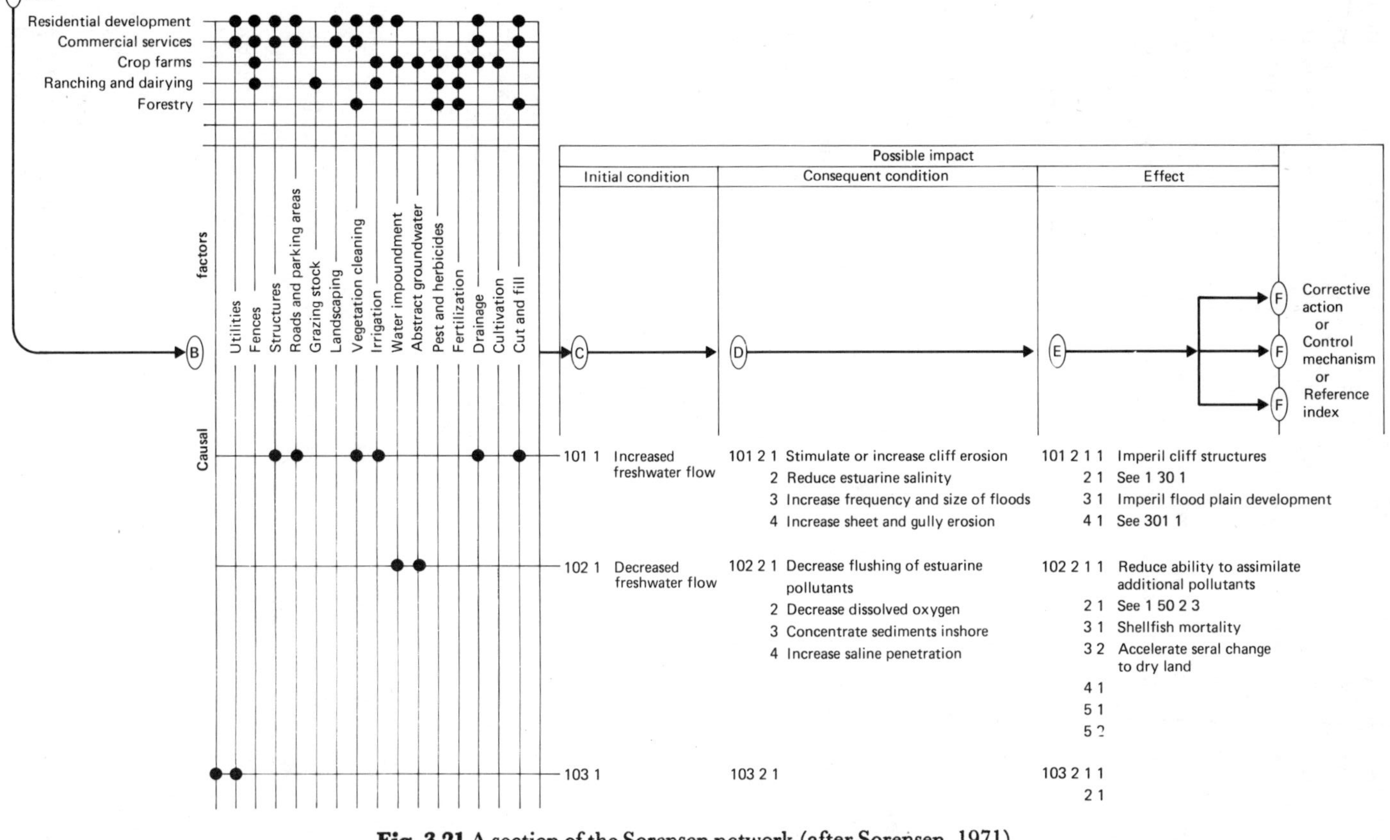

Fig. 3.21 A section of the Sorensen network (after Sorensen, 1971).

through the network. Sorensen also considers that the method should lead to the identification of remedial measures and monitoring schemes.

Although the Sorensen network can be used to follow the sequence of change induced by development, it does not establish the magnitude of inter-relationships between components or the extent of change. It requires considerable knowledge of the environment under consideration to construct the network and its main advantage is the ease with which the ramification of proposed development can be demonstrated. Thus its main value is as an aid to presenting information.

3.4.4 Quantitative methods

One of the problems faced by decision-makers is the necessity to relate data on a number of different types of impact, some beneficial some adverse, before arriving at a decision. At each stage the decision-maker must choose between alternatives, either between a number of different development proposals or between development and maintaining the *status quo*. Several methods have been developed which attempt to express all impacts quantitatively in a form that then can be aggregated. These methods are described in this section.

(a) Totality indices

Totality indices were used by Odum *et al.* (1975) to assess the impact of eight alternative highway proposals. Fifty-six environmental parameters were identified. The maximum impact of any alternative on an environmental parameter was given a value of 1 and the impacts of the other alternatives on this parameter were scaled accordingly. This procedure was repeated for each parameter. Impacts were multiplied by weights of relative importance on a scale 0 to 10 where 10 represents the greatest impact. The effect of each alternative on each parameter was determined by multiplying the scaled value by the weighting factor. Impacts were further classified into two groups depending upon whether they were short-term or long-term impacts. The latter were further multiplied by a factor of 10. Adverse impacts were given negative values and beneficial impacts positive values. Composite impact scores, totality indices, were determined for each alternative by aggregating individual impacts.

The method could lead to erroneous results if gross errors resulting from the subjectively derived values, such as estimation of scales impact values and weights, were incorporated. In an attempt to set confidence limits on the totality indices, the data were recomputed assuming up to 50% error in subjectively derived values.

(b) Environmental impact indices

Stover (1972) proposed a procedure for the preparation of Environmental Impact Statements. The most significant feature of this approach is the system for providing a numerical value indicating the impact of development on 50 predefined components of the environment. Numerical values of significance are ascribed subjectively to both short-term, initial impacts and long-term, future impacts on a seven point scale (Table 3.11). Initial and future impacts are generally associated with the construction and operational phases of development respectively. Future impacts are considered to be more important and attract higher values than short-term impacts. The duration of an impact directly affects its significance. Consequently, the numerical rating of each future impact is multiplied by its duration expressed in years, while initial impacts are given no additional weighting. The environmental impact index for each impact is obtained by adding the weighted scores for initial and future impacts.

Table 3.11 Weighting system for long-term and short-term impacts (after Stover, 1972)

	Initial impact	*Future impact*
Extremely beneficial	+5	+10
Very beneficial	+3	+6
Beneficial	+1	+2
No effect	0	0
Detrimental	−1	−2
Very detrimental	−3	−6
Extremely detrimental	−5	−10

Alternatives can be compared directly when all impacts are standardized to the same numerical base. The maximum value of each impact is reduced to 1, irrespective of which alternative proposal is responsible for it, and the numerical values for the other alternatives scaled accordingly. A composite score is obtained by adding the scaled values for each alternative, the preferred option having the lowest composite score for environmental impacts.

(c) Environmental evaluation system

A major criticism of the quantification procedures proposed by Odum *et al.* (1975) and Stover (1972) is that scaling an impact to a common base implies a linear relationship between the magnitude of an impact and its significance. This type of relationship may not apply for all impacts. Dee *et al.* (1973) have proposed an approach, termed the environmental evaluation system, which accommodates non-linear relationships. It is argued that a measured environmental parameter can be converted into environmental quality scores on a scale 0–1 where 1 represents maximum environmental quality. All parameters can be treated in a similar way to normalize scores. Graphs of value functions showing the relationship between environmental parameters and environmental quality have been produced for all environmental

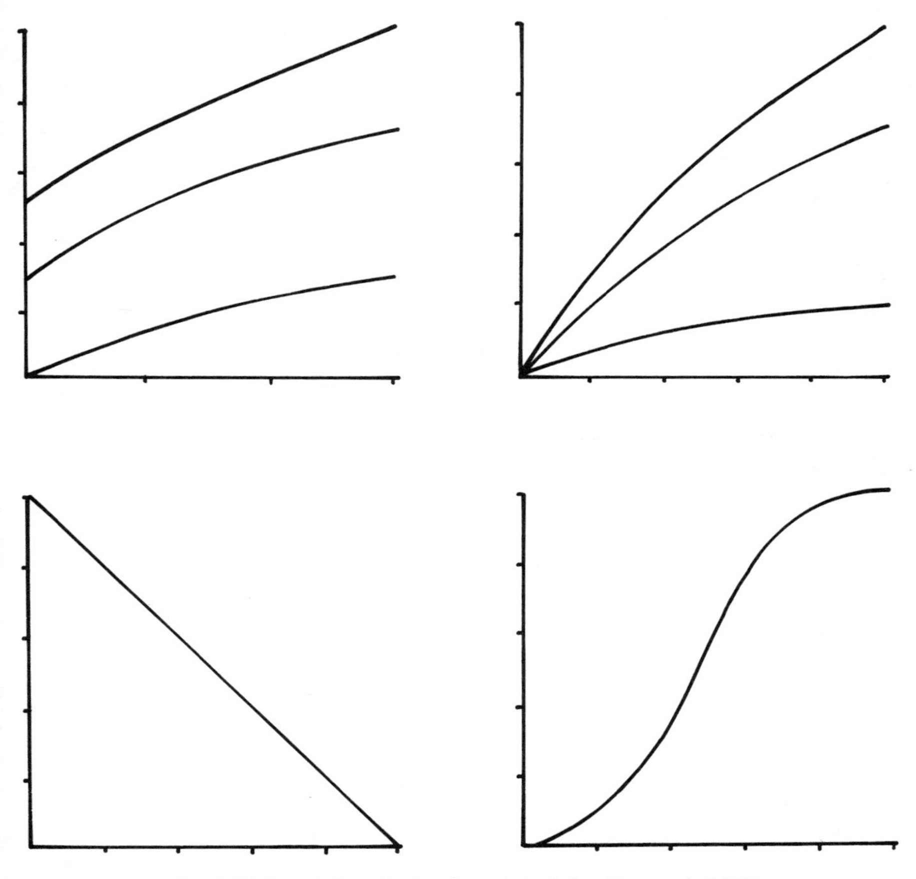

Fig. 3.22 Examples of value functions (after Dee *et al.*, 1973).

parameters considered by Dee *et al.* Some examples are shown in Fig. 3.22. These predefined functions can be applied over a wide geographical area but may not be universal. A predicted change in an environmental parameter as a result of development leads to a change in environmental quality, which can be determined directly from the graphs.

The approach of Dee *et al.* not only normalizes all environmental effects, but also provides a means of aggregating impacts. It is acknowledged that impacts are not of equal significance and a weighting scheme has been derived to accommodate these differences. Weighting factors totalling 1000, were distributed amongst 78 environmental parameters with individual weighting factors ranging from 2 to 31. Changes in environmental parameters, therefore, can be converted into changes in environmental quality and multiplied by the weighting factor to give a numerical representation of impact. If the weighted values for all impacts are aggregated a numerical representation of impact is obtained. When several alternatives are being considered it is more efficient to determine the weighted environmental quality score for each of the 78 parameters and aggregate them. This gives a numerical representation of environmental quality for each alternative, aggregate impact being the difference in environmental quality between alternatives.

Although this method appears to be objective, much subjective assessment is subsumed within it. The value functions and relative weights which are key elements in normalizing aggregating impacts are derived subjectively. Individual value functions and weights, can be criticized particularly as they represent environmental rather than, for example, social concerns. The rigid system provides one numerical value of aggregate environmental quality for each alternative. Unlike the approach of Odum *et al.* (1975), however, no attempt is made to assess the credance which can be placed in the estimates by establishing confidence limits.

3.4.5 Modelling

Litigation initiated under the *National Environmental Policy Act* of 1969 has emphasized the importance of investigating all likely impacts of a proposed development (Clark *et al.*, 1978). This has encouraged an encyclopaedic approach to EIS preparation where the major consideration is to show that all potential impacts have been assessed. The assertion that more information leads to better decision-making has been refuted by those who see the major function of impact assessment as identifying the critical impacts of development and subjecting them to rigorous appraisal (Holling, 1978).

Holling has argued that the concept of an ecosystem inexorably linked through its myriad components is not applicable in many systems. In reality, subsystems may be relatively insulated from changes in components within other subsystems. Analysis must take account of this functional isolation. The effects of development may vary considerably and many impacts do not show linear relationships between intensity of development action and environmental impact. The problem of predicting impacts is compounded by the response of ecosystems to environmental change. Uncertainty in the behaviour of ecosystems, results in unexpected changes leading to the displacement of systems from existing equilibria. Resilience within a system may lead to its eventual return to equilibrium, but in some instances thresholds are exceeded so that the original equilibrium cannot be attained. Consequently, new equilibria are established.

The use of simulation modelling, adaptive environmental assessment, has been advocated as a means of assessing a wide range of development proposals and environmental management schemes as a means of accommodating the behaviour of ecosystems. It has been used to assess forest pest management, fish stock management, tourist development, regional development (Holling, 1978), hydroelectric development (Walters, 1974; Munn, 1979) and a water-resource study (Frenkiel and Goodall, 1978). The essential difference between adaptive environmental assessment is the integration of impact prediction with the method. In addition, it is argued that use of the method should not terminate with completion of the assessment, but should continue throughout the monitoring of the scheme. Monitoring data should be used to refine the model and be used to predict further changes that are likely to occur.

Munn (1979) has defined the criteria influencing the decision to use a computer-based simulation model. The advantages of modelling lie in accommodating complex relationships between environmental variables, time-dependent relationships, explicitly stating the nature of relationships between environmental components within a model and exploring the implications of uncertainty. A model must be constructed for each proposal under consideration, although in practice components of existing models may be directly transferred from other situations.

3.4.6 Discussion

The issue which faces those preparing an impact analysis is to decide which of the available methods to use. A major indictment of many methods is that they seldom have been used in practice, indicating a failure to appreciate the practical difficulties facing those within the planning process who must operate within severe constraints on time, money, information and other resources. Predominately, those preparing impact statements have resorted to an *ad hoc* assessment or have fabricated an approach appropriate to their needs.

The impact statements which have been produced within the UK have largely ignored the theoretical methodological base that exists. Attempts have been made to use some of the methods. An independent assessment of the Drumbuie oil production platform fabrication yard was based upon the Leopold matrix (Polytechnic of Central London, 1973). An investigation of alternative sites for similar development in the Loch Carron area used a modified matrix approach which attempted a quantitative assessment of alternative sites (Sphere Environmental Consultants, 1974). An assessment of alternative sites for water storage facilities in the Lake District (Nelson *et al.*, Section 6.4) utilized the approach to impact assessment devised by Clark *et al.* (1976). Simulation modelling has been used to assess the effects of urbanization upon water availability and quality in part of the River Ouse (Frankiel and Goodall, 1978).

Other considerations, apart from the consideration of practicability and resource requirements should influence the selection of an appropriate method. Environmental analysis is only part of the decision-making process. It is intimately linked with the remaining components. It has been argued that the function of impact assessment should be to reduce the range of information available to decision-makers. Quantification and aggregation provides a system in which all data are combined into a single numerical value of composite environmental impact. Consequently, only data which can be expressed numerically can be considered. This may involve the rejection of certain data and the reduction of intangible impacts to numerical terms if they are to be incorporated within an assessment. Cost–benefit analysis has been discredited already within planning, because of the need to quantify qualitative factors. Those advocating quantification methods for impact assessment should be aware of this precedent.

Bisset (1978) has argued that quantification methods may usurp the role of the decision-maker within the political process of development planning. The selection of impacts for detailed investigation and the derivation of weights and normalizing factors significantly affect the results of an appraisal. Yet, in the case of the environmental evaluation system (Dee *et al.*, 1973), the decision-maker is excluded from these activities as they are predefined for water-related development. It has been suggested that the public should be involved in establishing weights to different impacts (Sondheim, 1978). The situation, however, may not be totally unacceptable to the decision-maker, who is reluctant to be seen to make an overtly political decision and attempt to employ technical data to justify decisions which are essentially political (Clark, 1975). The major function and capability of the decision-maker is in handling disparate data and assessing priorities. Qualification seeks to provide a 'right answer' in a situation where none can exist. The technical aspects of impact assessment can not be divorced from the broader political context.

The attraction of simulation modelling lies in its flexibility. As a new model must be constructed for each development proposal, it is possible for the decision-maker to be involved in the identification of impacts for detailed consideration. Computer-based simulation can also be used to generate alternatives in which the ramifications of different decisions can be explored both spatially and temporally. The major impediment to the widespread use of modelling, however, is its current low credibility amongst decision-makers and the public; a factor which advocates should not dismiss lightly.

Simulation modelling can be repeated if the same assumptions are built into the model, unlike other methods where the results are user dependent. As these assumptions are built into the model, rather than explicitly stated, they can not be subjected to rigorous scrutiny, except at the time of model formulation, even though they must be justified at this stage. There is a danger that the model will not be subjected to detailed scrutiny by those not involved in its construction. This could lead to a feeling of alienation, a situation which would be aggravated if the public was seen to be excluded from appraisal by an alliance of politicians and technicians.

REFERENCES

Alabaster, J.S. (1970) *J. Fish Biol.*, **2**, 1–13.

Armitage, P.D. (1976) *Freshwat. Biol.*, **6**, 229–40.

Armitage, P.D. (1977) *Freshwat. Biol.*, **7**, 167–83.

Armitage, P.D. (1978) *Hydrobiol.*, **58**, 145–56.

Armitage, P.D. & Capper, M.H. (1976) *Freshwat. Biol.* **6**, 425–32.

Arnold, G.P. (1974) *Biol. Rev.*, **49**, 515–76.

Banks, J.W. (1969) *J. Fish Biol.*, **1**, 85–136.

Birtles, A.B. (1977) in *Proceedings of the IAWPR.* Specialised Conference on River Basin Management, Essen, Paper No. 20.

Birtles, A.B. & Brown, S.R.A. (1978) in *International Symposium on Modelling the Water Quality of the Hydrological Cycle*, Baden, Austria.

Bisset, R. (1978) *J. Environ. Mgmt.*, **6**, 200–16.

Brooker, M.P. & Hemsworth, R.J. (1978) *Hydrobiol.*, **59**, 155–63.

Brooker, M.P. & Morris, D.L. (1980) *Freshwat. Biol.*, **10**.

Brooker, M.P., Morris, D.L. & Hemsworth, R.J. (1977) *J. Appl. Ecol.*, **14**, 409–17.

Brooker, M.P., Morris, D.L. & Wilson, C.J. (1978) in *Proc. EWRS 5th Symposium on Aquatic Weeds.*

Brooks, D.R. (1976) in *Landscaping and Land Use Planning as Related to Mining Operations* (ed. Aust. Inst. Min. Metall.), AIMM, Adelaide.

Brooks, D.R. (1976) Rehabilitation following mineral sand mining on north Stradbrooke Island, Queensland. In *Landscaping and Land Use Planning as Related to Mining Operations* (ed. Aust. Inst. Min. Metall.), AIMM, Adelaide, pp. 93–104.

Brown, D.J.A. (1971) *The effects of temperature and diet on the metabolism of Salmo gairdneri*, PhD Thesis, University of Nottingham.

Bussell, B.R. (1978) *Changes of River Regime Resulting From Regulation Which May Affect Ecology: a preliminary approach to the problem*, Central Water Planning Unit, UK, pp. 48.

Chelmick, D.G. (1979) *Provisional Atlas of the Insects of the British Isles Part 7: Odonata*, Institute of Terrestrial Ecology, Huntingdon.

Clapham, A.R. (1969) *Flora of Derbyshire*, County Borough of Derby, England.

Clark, B.D., Bisset, R. & Wathern, P. (1980) *Environmental Impact Assessment*, Mansell, London.

Clark, B.D., Chapman, K., Bisset, R. & Wathern, P. (1976) *Assessment of Major Industrial Applications: A Manual*, DOE Research Report, No. 13, Department of the Environment, London.

Clark, B.D., Chapman, K., Bisset, R. & Wathern, P. (1978) U.S. *Environmental Impact Assessment: A Critical Review*, DOE Research Report, No. 26, Department of the Environment, London.

Clark, B.D., Chapman, K., Bisset, R. & Wathern, P. (1979) in *Land Use and Landscape Planning* 2nd edn (ed. D. Lovejoy), Leonard Hill, Glasgow.

Clark, S.D. (1975) in *EIS Technique* (ed. Australian Conservation Foundation), Australian Conservation Foundation, Melbourne.

Consolidated Gold Fields & University of Liverpool (1978) *Ecological and Conservation Value Assessments and their Integration with Proposed Alluvial Tin Mining*, Consolidated Gold Fields, London.

Crisp, D.T. (1977) *Freshwat. Biol.*, **7**, 109–20.

Cullimore, C. & Johnson, M.S. (1978) *Reclaimed Fluorspar Mine Wastes: Dead Landscape or Wildlife Refuge?*, Int. Rept. Laporte Industries Ltd, Derbyshire.

Cummins, K.W. (1975) Macroinvertebrates, in *River Ecology* (ed. B.A. Whitton), Blackwell, Oxford, pp. 170–98.

Dee, N., Baker, J.K., Drobny, N.L., Witman, I.L. & Fahringer, D.C. (1973) *Water Resources Res.* **9**, 523–35.

Edington, J.M. (1965) *Verh. Int. Verein. Limnol.*, **13**, 40–8.

Edington, J.M. (1968) *J. Anim. Ecol.*, **37**, 675–92.

Edwards, R.W. & Crisp, D.T. (1982) in *Gravel-bed Rivers* (eds R.D. Hey, J.C. Bathurst and C.R. Thorne), University of Wales, pp. 843–65.

Edwards, R.W., Densem, J.W. & Russell, P.A. (1979) *J. Anim. Ecol.*, **48**, 501–7.

Edwards, R.W., Oborne, A.C., Brooker, M.P. & Sambrook, H.T. (1978) *Verh. Int. Verein. Limnol.* **20**, 1418–22.

EEC (1975) *Proposal for a Council Directive Relating to the Quality of Water for Human Consumption*, Commission of the European Communities.

Egglishaw, A.G. & Mackay, D.W. (1967) *Hydrobiol.*, **30**, 305–34.

Elliott, J.M. (1967) *J. Appl. Ecol.*, **36**, 243–362.

Elliott, J.M. (1975) *J. Anim. Ecol.*, **44**, 805–21.

Environment Canada (1974) *An Environmental Assessment of Nanaimo Port Alternatives*, Environment Canada, Ottawa.

Fraser, J.C. (1975) FAO Fisheries Tech. Pap., No 143, FIRS/T143.

Frenkiel, F. W. & Goodall, D. W. (1978) *Simulation Modelling of Environmental Problems*, SCOPE Report No. 9, Wiley, Chichester.

Furet, J. (1979) Unpublished PhD Thesis, University of Wales.

Gee, A.S. & Edwards, R.W. (1980) *EIFAC International Symposium on Fishery Resources Allocation*, Vichy, France.

Gee, A.S., Milner, N.J. & Hemsworth, R.J. (1978) *J. Anim. Ecol.*, **47**, 497–505.

Gilliland, M.W. & Risser, P.G. (1977) *Ecol. Modelling*, **3**, 183–209.

Goldsmith, F.B. (1975) *Biol. Cons.*, **8**, 89–96.

Hammond, C.O. (1977) *The Dragonflies of Great Britain and Ireland*, Curwen Press, London.

Harrison, J. (1974) *The Sevenoaks Gravel Pit Reserve*, WAGBI, Chester.

Hemsworth, R.J. & Brooker, M.P. (1979) *Holarctic Ecol.* **12**, 130–6.

Hjulstrom, F. (1935) *Uppsala Univ. Geol. Inst. Bull.*, **25**, 221–528.

Holdgate, M.W. (1976) in *Symposium on Environmental Evaluation*, Canterbury, 1975, Planning and Transport Research Advisory Council, Department of the Environment, London.

Holliday, R.J., Johnson, M.S. & Bradshaw, A.D. (1979) *Biol. Cons.*, **16**, 245–64.

Holling, C.S. (1978) *Adaptive Environmental Assessment and Management*, Wiley, Chichester.

Hopper, F.N. (1978) Unpublished MSc Thesis, University of Wales.

Humpesch, U.H. (1980) *J. Anim. Ecol.*, **49**, 317–33.

Johnson, M.S. (1977) *Establishment of Vegetation on Metalliferous Fluorspar Mine Tailings*, PhD Thesis, University of Liverpool.

Johnson, M.S. & Bradshaw, A.D. (1979) *Appl. Biol.*, **4**, 141–200.

Johnson, M.S., Bradshaw, A.D. & Handley, J.F. (1976) *Trans. Inst. Min. Metall.*, **85A**, 32–7.

Johnson, M.S., Putwain, P.D. & Holliday, R.J. (1978) *Biol. Cons.*, **14**, 131–48.

Jones, A.N. (1975) *J. Fish. Biol.*, **7**, 95–104.

Kalleberg, H. (1958) *Inst. Freshwat. Res. Drott.*, Sweden Report No. 39, 55–98.

Lennon, M. (1979) Unpublished MSc thesis, University College of Wales, Aberystwyth.

Leopold, L., Clarke, F.E., Hanshaw, B.B. & Balsley, J.R. (1971) *A Procedure for Evaluating Environmental Impact*, US Geological Survey Circular 645, US Geological Survey, Washington DC.

Lilley, A.J., Brooker, M.P. & Edwards, R.W. (1979) *Nature in Wales*, **16**, 195–200.

Logan, P. (1980) Unpublished MSc Thesis, University of Wales.

MacKey, A.P. (1977) *Oikos*, **28**, 270–5.

Mann, K.H. (1975) in *River Ecology* (ed. B.A. Whitton), Patterns of Energy Flow, Blackwell, Oxford, pp. 248–63.

Moore, N.W. (1962) *J. Ecol.*, **50**, 369–91.

Morris, D.L. & Brooker, M.P. (1979) *Freshwat. Biol.*, **9**, 573–83.

Munn, R.E. (1979) *Environmental Impact Assessment*, 2nd edn, SCOPE Report No. 5, Wiley, Chichester.

Nature Conservancy Council (1977) *Nature Conservation and Agriculture*, Nature Conservancy Council, London.

Nebeker, A.V. (1973) *J. Kansas. Ent. Soc.*, **46**, 160–5.

Oborne, A.C. (1981) *J. Hydrol.*, **52**, 59–70.

Oborne, A.C., Brooker, M.P. & Edwards, R.W. (1980) *J. Hydrol.*, **45**, 233–52.

Odum, E.P., Zieman, J.C., Shugart, H.H., Ike, A. & Champlin, J.R. (1975) in *Environmental Impact Assessment* (ed. M. Blisset), University of Texas at Austin, Austin.

Owens, M., Garland, J.H.N., Hart, I.C. & Wood, G. (1972) *Symp. Zool. Soc. London*, **29**, 21–40.

Perkins, D.F., *et al.* (1978) *Production Ecology of British Moors and Montane Grasslands* (eds O.W. Heal & D.F. Perkins), Springer-Verlag, Berlin, pp. 289–95.

Peterken, G.F. (1974) *Biol. Cons.*, **6**, 239–45.

Peterson, R.H. (1978) *Fisheries and Marine Service Tech. Report* No. 785, p. 28.

Pollard, E., Elias, D.O., Skelton, M.J. & Thomas, J.A. (1975) *Entom. Gaz.*, **26**, 79–88.

Polytechnic of Central London (1973) *Impact: A Study of the Effects of Proposed Oil Platform Construction in the Western Highlands*, Polytechnic of Central London, London.

Ratcliffe, D.A. (1974) *Proc. Roy. Soc. Lond.*, **339***a*, 355–72.

Ratcliffe, D.A. (1977) *A Nature Conservation Review* (ed. D.A. Ratcliffe), Vols 1 and 2, Cambridge University Press, Cambridge.

Richards, A.J. & Swann, G.A. (1976) *Watsonia*, **11**, 1–5.

Sambrook, H.T. (1976) Unpublished MSc Thesis, University of Wales.

Sharrock, J.T. (1976) *The Atlas of Breeding Birds in Britain and Ireland*, British Trust for Ornithology, Tring, Herts.

Smith, R.A.H. (1973) *The Reclamation of Old Metalliferous Mine Workings Using Tolerant Plant Populations*, PhD Thesis, University of Liverpool.

Solomon, J.J. (1978) *J. Fish. Biol.*, **12**, 571–4.

Sondheim, M.W. (1978) *J. Environ. Mgmt.*, **6**, 27–42.

Sorensen, J.C. (1971) *A Framework for the Identification and Control of Resource Degradation and Conflict in Multiple Use of the Coastal Zone*, Masters Thesis, University of California, Berkeley.

Sphere Environmental Consultants (1974) *Loch Carron Area: Comparative Analysis of Platform Construction Sites*, Sphere Environmental Consultants, London.

Stover, L.V. (1972) *Environmental Impact Assessment: A Procedure*, Sanders and Thomas, Pottstown.

Symons, P.E.K. & Heland, M. (1978) *J. Fish. Res. Bd. Can.*, **35**, 175–83.

Tubbs, C.R. & Blackwood, J.W. (1971) *Biol. Cons.*, **3**, 169–72.
Turk, F.A. & Turk, S.M. (1976) *A Handbook to the Natural History of the Lizard Peninsula*, University Press, Exeter.
Walters, C. (1974) *Technology Forecasting and Social Change*, **6**, 299–383.
Ward, S.D. & Evans, D.F. (1976) *Biol. Cons.*, **9**, 217–33.
Water Resources Board (1973) *Water Resources in England and Wales*, HMSO London.
Wathern, P. (1976) *The Ecology of Development Sites*, PhD thesis, University of Sheffield.
Westlake, D.F. (1967) *J. Exp. Bot.*, **18**, 187–205.
Wetzel, R.G. (1975) *Limnology*, W.B. Saunders Company, pp. 743.
Williams, D.D. & Hynes, H.B.N. (1974) *Freshwat. Biol.*, **4**, 507–24.
Wye River Authority (1972) *Survey of Water Resources and Demands*.

CHAPTER FOUR

Planning Procedures for Environmental Impact Analysis

4.1 OVERVIEW

This section deals with the legislative and administrative procedures for the incorporation of ecological information into the decision-making process. Whilst many planning procedures are available and have been used, the papers in this section are concerned mainly with formalized EIA procedures in view of their widespread use and applicability as a framework for impact analysis.

Whilst there is nothing new about the consideration of environmental impacts, the formalization of the process as a structured planning procedure is a relatively recent development. Environmental Impact Assessment (EIA) stems from the USA *National Environmental Policy Act* (1969) and subsequent, legal rulings and practices. This requires the proponent of an action (e.g. legislation, a policy, plan or project) significantly affecting the environment, to demonstrate that they have carried out an assessment of the environmental consequences of their proposal. The assessment is presented as an Environmental Impact Statement (EIS) which describes the environmental impacts likely to arise from the action and from alternatives to it. The EIS forms the basis for subsequent discussion, public participation and the final decision-making process which progress through a series of clearly-defined steps.

Many other countries have been attracted to formalized EIA systems and have either implemented (Australia, Canada, France, Germany, Ireland, Japan, New Zealand, Spain) or are in advanced stages of considering (Belgium, Denmark, Italy, Luxemburg, Netherlands) EIA procedures. Other countries, including the UK have avoided the introduction of a standardized EIA system, preferring instead to operate on an *ad hoc* basis with the procedural detail of the decision-making process selected to meet the individual needs of each situation.

In countries where EIA has been adopted, the tendency has been not to import the provisions of NEPA wholesale, but rather to modify existing planning and environmental legislation and management processes to take account of EIA. NEPA was introduced into the USA as an instrument of policy and planning redirection stemming from a neglect felt by some groups of integrated social, economic and environmental planning. The pre-NEPA fragmentation of the US project and policy planning institutions may be less prevalent in other countries, such that they are able to cope with the implications of a major new development without having to rely on an environmental pressure group to force a consideration of the scheme in an EIA, or on a NEPA-style approach to EIA (Wandesforde-Smith, 1978).

National approaches to EIA, therefore, reflect the politics and existing nature of environmental and planning legislation and administrative processes within each country. There is, therefore, considerable variation amongst EIA systems particularly in relation to the scope (public or private, policy or project), scale (national or local) and content (physical, biological and social parameters). There is, however, common agreement that the fundamental aim of EIA is not to determine the balance placed by the decision-maker on environmental compared to economic, social or other considerations but to ensure that the decision is made on the basis of informed knowledge of the environmental consequences of that decision. EIA is, therefore, a process of sound environmental management. Its aims, on the one hand, are to raise developer's and administrative authorities awareness of the essential environmental issues which deserve careful attention and, on the other, to ensure efficient coordination of administrative action and public opinion. Contrary to some expressed opinions, EIA is not concerned at setting up new environmental standards but rather at ensuring that existing standards and protective measures are well-adapted to the specific conditions of the proposal in question.

There also appears to be reasonable agreement about the nature of the procedure required to obtain these aims. Most EIA systems involve the preparation (through the co-operation of the proposer and regulatory authorities) of a detailed impact analysis of the proposal, alternatives to it and the presentation of the analysis as an Impact Statement. This then serves as a basis for discussion with statutory bodies, voluntary organizations and the general public prior to a final decision-making step (Fig. 4.1). In most countries the EIS would include information on:

(a) A description of the proposed project and alternatives for the site.
(b) A description of the environmental features likely to be effected by the proposal, or alternatives to it.
(c) An analyses of the likely environmental consequences of the project on the environment and alternatives to the proposed action.

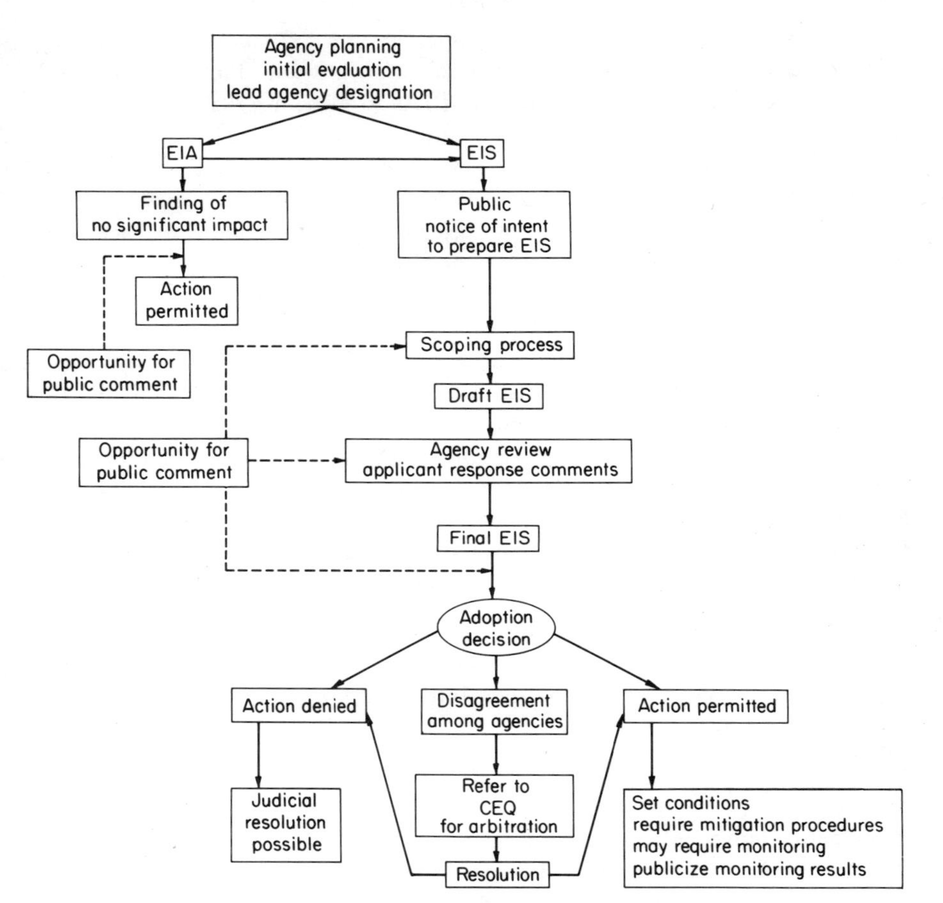

Fig. 4.1 Summary of procedural steps in the NEPA review process.

(d) A description of the measures envisaged to mitigate those consequences.
(e) A review of the relationships between the proposal and existing policy and plans.

In some procedures other information may also be required including a detailed analysis of the need for the proposal and of its consequences in relation to resource use. Legore (Section 4.2), Effer (Section 4.3) and Lee and Wood (Section 4.4) review EIA procedures in North America and Europe in relation to national settings and examine the strengths and weaknesses of EIA as a formal planning procedure.

The UK, in contrast to many other countries, has not introduced formal EIA procedures but has preferred to rely on a flexible approach to planning based on the system of developmental plans (see Booth, Section 2.1). A number of impact analyses have been carried out within the present UK process with the procedures and methodologies selected on an *ad hoc* basis to suit the requirements of individual proposals (Clark and Turnbull, Section 4.5). However, from the review of UK EIAs in Chapter 7, it appears that British attempts differ from those in countries with formalized EIA procedures in that little, or no, attention is paid to alternatives to the proposals. The real danger relating to the use of EIA in the UK is that it may become a technical exercise that is carried out after many of the important decisions have been made. Furthermore, experience indicates that the environmental consequences are rarely analysed in a quantifiable manner, most studies preferring instead to rely on simple check-lists or matrices. Whilst these techniques may be useful for summarizing and communicating impact predictions, their value in presenting impact data in a form which allows comparison with other planning considerations is unclear.

The future for EIA in the UK is uncertain. Central government whilst tacitly approving the aims of EIA seems opposed to the introduction of a formalized procedure although events may be overtaken by legislation in Brussels. Opponents to EIA in the UK point to a number of potential disadvantages (for example CBI, 1979) including:

(a) An undesirable rigidity would be introduced into the planning system and the analysis could be taken as an infallible basis for correct planning.
(b) Further delays in an already over-burdened planning system.
(c) Increased costs.
(d) Few, if any, authorities could maintain resources or expertise to analyse the relevant data.

Proponents, however, argue that North American experience indicated that these problems can be avoided by careful selection of EIA procedures and that the advantages far outweigh potential disadvantages (for example Dean and Graham, 1978; Birks, 1979). Such advantages might include:

(a) Speedier processing of applications particularly where lengthy public inquiries can be avoided.
(b) Reduction in environmental control costs where EIA is implemented at an early stage in project design.
(c) Avoidance of unnecessary conflicts on conservation or environmental grounds.
(d) Improvements in the environmental quality of the decision.
(e) Reduced dependence on the good-will of the proposer.

Grove-White (Section 4.6) discusses the relative advantages and disadvantages of EIA in the UK political context. The author draws attention to the value of EIA systems when considering policy decisions but also to the advantages of the present system of public inquiry at project level. Collaboration between the proposer and planning authority might mean that they conclude that all relevant impacts have been considered before the EIA becomes a public process resulting in an under-rating or even exclusion of impacts which have not already been identified or considered significant. Indeed many developers see this as a major advantage of EIA. In contrast Grove-White argues in favour of the adversorial component of the present public inquiry system. However, the difficulties here are that as expert witness at a public inquiry appears on behalf of one or other party, the capacity of providing information in a neutral role is severely limited. The witness must therefore either withhold the information or subsume the process by making a decision before appearing. Furthermore, the onus of post-development auditing rests, *a priori*, with the expert witness rather than with a coordinating, funded organization. The learning of lessons for future applications is therefore restricted in the present public inquiry system. Any advantages accruing must be weighed against these disadvantages.

4.2 EXPERIENCE WITH ENVIRONMENTAL IMPACT ASSESSMENT PROCEDURES IN THE USA

S. Legore

4.2.1 Introduction

I have heard it said that all government programmes have three distinct parts: a beginning, a 'muddle', and no end. This observation certainly applies to the US Environmental Impact Assessment (EIA) programme.

The origins of Impact Analysis in the United States can be traced to the *National Environmental Policy Act* of 1969 (NEPA). Although this act was largely only a declaration of national policy and intent regarding the safeguarding of environmental resources, and did not set any environmental standards, it nevertheless set forth specific provisions with far-ranging implications. NEPA created a three member Council for Environmental Quality (CEQ) to review environmental programmes and progress, and to advise the President on environmental policy issues. More significantly, however, it required federal agencies to prepare an Environmental Impact Statement (EIS), and to subject it to review prior to proceeding with any major federal action 'significantly affecting the quality of the human environment . . .'. This EIS was to serve as a tool providing guidance to governmental decision-makers so that they would be compelled to consider environmental impacts of their decisions. The term 'major federal action' has subsequently been interpreted to include the issuance of federal permits for industrial and governmental activities, under provisions of the *Clean Water Act*, the *Resource Conservation and Recovery Act* (RCRA), the *Coastal Zone Management Act* and the *Fuel Use Act* (Quarles, 1979) (Table 4.1).

When NEPA was passed in 1969, it was considered a patriotic but unimposing statement of policy. Indeed, one of the initial antecedents of NEPA, Senate Bill 1075, which introduced CEQ and 'a declaration of national environmental policy', passed in the Senate by unanimous consent on 10 July 1969 (Anderson, 1973). It is interesting to note that not even the declaration of war on Japan after the Pearl Harbor bombing attack was passed unanimously!

The requirements of NEPA were not taken seriously until a court case decided in 1972 that agencies are bound actually to consider information in

Table 4.1 Primary federal environmental permitting programmes presently covered by EIS provisions of NEPA (after Quarles, 1979)

Programme	*Purpose*	*Programmes covered by NEPA*
Clean Water Act	To maintain environmental integrity of US surface waters	(a) New point-source discharges (b) Dredge and fill operations
Resource Conservation and Recovery Act	To regulate land disposal of solid and liquid wastes, and handling of hazardous materials	All programmes are potentially subject to NEPA
Coastal Zone Management Act	To provide for State coastal zone management planning	State programmes are subject to NEPA, but once they are approved, individual industrial projects revert to State authority and are exempt from NEPA
Power Plant and Industrial Fuel Use Act	To allocate use of fuels relative to national policy	Although temporary fuel use decisions are sometimes exempt, permanent exemptions from fuel use restrictions are subject to NEPA

the EIS and to examine environmental factors. Specifically, the construction of a large electrical generation plant was delayed by this decision. Such delays are very expensive, and this court action compelled all interested parties to reconsider their attitude toward NEPA.

Time and progress have altered the public perception of the NEPA process. In the beginning, the impact analysis was regarded as a separate function from planning. It later became a source of information for planners, and presently the analysis is an integral component of planning and evaluation of alternatives (Andrews, cited by Viohl and Mason, 1974). NEPA has become established and accepted in the public mind, and more recently promulgated environmental regulations are now bearing the brunt of critical commentary.

4.2.2 EIA and EIS Procedures

Guidelines for the preparation of Environmental Impact Statements were issued by the Council for Environmental Quality in 1970 and again in 1973. The process from the beginning was marked by confusion and delay: the guidelines were too narrow in scope, and were interpreted by federal agencies to be discretionary, rather than mandatory, standards. Consequently revised and expanded EIS guidelines were issued in 1978 (43 Fed. Reg. p. 55977–56007, 29 Nov. 1978). By virtue of being termed 'Regulations' these were expected to impose uniform standards binding on all Federal agencies.

The basic EIA procedure stipulated by the CEQ Regulations is presented in Fig. 4.1. The new Regulations require that the EIA procedure be initiated 'no later than immediately after the (formal) application is received.' It calls for the immediate designation of a 'lead agency,' which is responsible for coordinating the review process, and encourages an early determination of whether an Environmental Impact Assessment (EIA), or a more complex Environmental Impact Statement (EIS) will be required for the proposed action (Fig. 4.1). If the programme will clearly entail significant environmental impact, an EIS is called for; however, if the need for an EIS is questionable an EIA may be initially required. The EIA consists of a preliminary examination of possible impacts, and serves as a primary tool for deciding if a 'Finding of No Significant Impact' is appropriate. In such cases the need for EIS is abrogated and, following public discussion, a permit may be granted. On the other hand, if the EIA discerns potentially significant impacts, an EIS is usually required.

Once the decision has been made to produce an EIS the lead agency is required to proclaim its 'Notice of Intent,' and the 'Scoping' process begins. The lead agency is to coordinate efforts to (1) determine the scope of study and identify the significant issues to be analysed in depth, (2) identify insignificant issues and eliminate them from study, (3) allocate assignments of responsibility, (4) plan schedules, set page limits, adopt special procedures, and (5) identify environmental review and consultation requirements so the lead and cooperating agencies may prepare required analyses concurrently rather than consecutively, which had been a common practice. A fundamental aim of the scoping meetings is to define the boundaries of the EIS investigations, and to avoid the imposition of altered 'final hour' requirements that have commonly occurred in the past.

After the EIS requirements are developed, a draft EIS is produced and is reviewed by all relevant agencies and interested groups, who provide their comments. A final EIS is subsequently produced, and its evaluation becomes a major step in the process to decide whether the applicant's proposed action will be denied or permitted. As indicated in Fig. 4.1, the decision may be made by the lead agency, the CEQ, or by judicial or oversight bodies, depending upon circumstances.

The EIS consists of several very specific parts, although the content of these parts may vary substantially depending upon the type of action or project under consideration. The CEQ Regulations specify that the EIS shall consist of the following:

(1) Cover sheet

(a) List of responsible agencies and their functions
(b) Title of the proposed action
(c) Agency contact individual's name, address, telephone number
(d) Designation of 'draft', 'final' or 'supplemental' EIS
(e) A one paragraph abstract
(f) The deadline date for receipt of comments

(2) Summary (usually less than 15 pages)

(a) Major conclusions
(b) Areas of controversy
(c) Issues to be resolved, with alternatives for consideration

(3) Statement of purpose and need

(a) A brief specification of the underlying purpose and need to which the agency is responding in proposing the alternatives, including the proposed action

(4) Alternatives, including the proposed action

This is the 'heart' of the EIS, because it indicates, discusses and evaluates the options available to decision-makers. It should include:

(a) Discussion of all options of the proposed action, including rejected options
(b) Discussion of the 'no action' option
(c) Identification of the agency's preferred alternative(s)
(d) Discussion of possible 'mitigation' measures

(5) Affected environment

(a) Environmental description of the area to be affected or created by each alternative action (baseline)

(6) Environment consequences

This section forms the basis for conclusions presented in number 4, above titled: 'Alternatives, including the proposed action.' It includes for *each* programme or project alternative:

(a) The environmental impact of the proposed action and each alternative
(b) Any adverse environmental effects which cannot be avoided should the alternative action be implemented
(c) The relationship between local short-term uses of man's environment vs. the maintenance and enhancement of long-term productivity
(d) Any irreversible and irretrievable commitments of resources which would be involved in the proposed action if implemented

Specific impacts to be addressed include:

(i) Direct effects and their significance
(ii) Indirect effects and their significance
(iii) Possible conflicts between the proposed action and objectives of federal, regional, state, or local land use plans, policies and controls
(iv) Energy requirements and energy conservation potential of each alternative and each mitigation procedure
(v) Depletable resource requirements and conservation potential of each alternative and mitigation procedure
(vi) Effects upon urban quality, historic and cultural resources
(vii) Mitigation mechanisms

(7) List of preparers

(8) Appendices

(a) Supporting data
(b) Literature references
(c) Data analyses
(d) Other relevant information

Fundamentally then, the EIS consists of an evaluation of each project alternative's extent and degree of impact. As described, the process seems straight-forward and sensible, as in every other human endeavour, however, politics and inefficiency sometimes intrude upon the state to confuse the script.

4.2.3 Impact analysis

The specific methods used for environmental impact analysis vary considerably, and ideally are agreed upon between the applicant and regulatory agencies prior to beginning the studies. Nevertheless, basic areas for investigation may be identified which have virtual universal applicability, although some areas may be eliminated by the study scoping process.

(a) Aquatic systems

The EIS must include detailed information on the pre-project aquatic systems, including biological, chemical, and hydrological characteristics of the area that may be impacted.

Biological considerations include the flora and fauna occurring in or which are dependent upon the local surface water systems, with special consideration being given to identifying species that are commercially important, scientifically unique or interesting, rare or endangered, or comprise 'key' constituents of the ecosystem. Extensive investigation is therefore required of all biotopes over a representative span of time, which is usually interpreted to be a minimum of one year.

Water chemistry must also be studied to determine how the chemical constituents of the water bodies relate to regulated water quality and drinking water standards, and how they may influence the biological systems of concern. Chemistry investigations generally begin with broad spectrum analyses, with subsequent narrowing of scope to emphasize specific parameters of concern. The sources of unusually high chemical concentrations should be ascertained to the extent possible, and unusually low concentrations should be documented.

Numerous hydrological characteristics should generally be considered for investigation, including water current regimes and their influence on biota and possible chemical sinks, high and low streamflow, flood history, streambed, watershed and flood plain characteristics, groundwater occurrence and movement, and land drainage as it relates to water retention, erosion or water chemistry.

Finally, all of this information should be integrated to provide as thorough an understanding of the existing aquatic systems as possible. The various developmental alternatives must then be theoretically superimposed upon the system in an attempt to predict their possible environmental impacts. Some of the alternatives that may be considered include water intake or discharge structure locations, various waste treatment procedures, alternative facility site locations, road construction alternatives, etc.

(b) Terrestrial systems

Study of terrestrial ecosystems is generally conceptually similar to that of aquatic systems. The organisms living in or utilizing the system are inventoried and their relationships are clarified, with the proposed project being subsequently superimposed upon the existing systems to attempt prediction of impact. Land-use inventories are often required as an aid to determining impacts upon present human uses, zoning policies, and land-use planning.

(c) Air quality

Before impacts upon air quality may be predicted, the pre-existing atmospheric environment must be documented. Specific parameters are usually monitored for a period of at least one year, including oxides of nitrogen, sulphur dioxide, total suspended particulate, ozone, and sometimes carbon monoxide. Meteorological towers are often installed and instrumented to monitor wind speed and direction, temperature, solar radiation, precipitation and relative humidity (or dewpoint temperature), so that resulting data may be used numerically to model dispersion patterns and possible impact areas.

Information on the expected types and quantities of air pollutants that will be emitted from the proposed project is then incorporated into a model to ascertain whether violations of air quality standards may be anticipated. If they are, then mitigation measures may be subjected to the model(s) to determine which measures are most promising. Typical mitigation measures include such actions as altering stack heights to allow effective dispersion of emissions in the immediate area, redesigning the facility to accommodate an alternative type of fuel, rescheduling operational activities to redistribute pollutants either spatially or temporally, or installing alternative pollution control equipment.

(d) Noise

The impact of noise from a proposed action must often be analysed. Existing and anticipated land-uses near the project site or route that have a sensitivity to noise should be identified, such as residences, hotels, hospitals, schools, recreation areas and wildlife habitat. Existing noise levels adjacent to the project site or route may have to be documented, and the development site may be selected in consideration of its proximity to noise sensitive surroundings. Furthermore, estimates of the maximum expected noise levels at the nearest sensitive locations should be made for each stage of construction activity and for the operational phase of the project. Violations of existing noise standards, and the expected impacts of the various types of noise should be predicted to the extent possible.

(e) Solid waste

When a project will produce solid waste, the quantities and types of waste must be identified, and the methods of disposal considered. Waste storage area, containers, potential for hazard and possible environmental impacts must be evaluated. Handling of solid waste is largely an engineering function, but to the extent that its mishandling or accidental discharge may impact environmental considerations, it must be evaluated by qualified environmental scientists.

Numerous regulations must be responded to in an EIS, particularly if it is written for a controversial project, but a complete inventory of them is far beyond the purview of this paper. Each of these regulations must be considered in the scoping process to ascertain the extent to which they affect the proposed project. If they do, then the EIS studies must discuss the project's anticipated degree of compliance with them.

4.2.4 Has NEPA been beneficial?

We must ask whether all of the effort expended on NEPA and EIS review has benefited our society, and on balance the answer is probably yes. In stating this, however, NEPA must be divorced from US standard setting regulation. NEPA provides a mechanism for analysing anticipated compliance with standards; the appropriateness of the standards themselves is a separate issue, which is much more difficult to analyse.

NEPA has injected environmental issues into governmental policy and planning dealing with almost all aspects of our relationship with the world around us. To a significant extent it has lessened our dependence on environmental altruism and good will on the part of developers, has decreased our subjection to environmental arrogance or indifference by some government agencies, and has compelled society as a whole to consider ecosystem dynamics whether it is inclined to do so or not. It is essentially an 'environmental full disclosure law' (Arbuckle *et al.*, 1974) which aspires to provide adequate information to allow effective and appropriate decision-making.

On the other hand, NEPA has experienced numerous problems which have retarded its enthusiastic acceptance and its effective application.

Unreasonable delay in the evaluation of EIS documents is often cited as NEPAs major problem. In the first few years of the law's existence, an enormous backlog of applications was caused by its lack of a 'grandfather' clause, resulting in a requirement for EISs to be developed for projects already underway. In the first two years alone, 3635 EISs were submitted to the CEQ (Anderson, 1973). These documents had to be reviewed by inadequate agency staffs and without the benefit of criteria for guiding judgemental decisions. Given the unspecific wording of the law and the lack of preparedness of most regulatory agencies, this unfavourable situation was probably inevitable, but time and experience have greatly reduced delays from this cause (CEQ, 1976).

Numerous other causes of delay are attributable to procedural problems associated with the 1970 and 1973 CEQ guidelines for EIS preparation. Since the guidelines were interpreted by federal agencies as discretionary, rather than mandatory standards, they were not treated with sufficient seriousness. From 1970 to 1978, approximately 70 different sets of EIS regulations and procedures evolved among the agencies (CEQ, 1978), creating an incomprehensible morass of regulatory redundancy, conflict, and inconsistency. Not only did applicants for construction and development permits fail to understand the requirements imposed upon them, but many of the regulators themselves had severe difficulties with the interpretation and implementation of their duties. Furthermore, the production of a

Table 4.2 Recognized causes of delays in the previous EIS review procedure, with expected corrective action imposed by CEQ regulations (1978)

Historical reason for delay	*Corrective action*
1. Confusion regarding required scope of environmental investigation	1. Required 'scoping' process, early establishment of requirements
2. Lack of guidance on required permits	2. Lead agency *must* identify required permits
3. Copious irrelevant data collection	3a. Emphasizes options between alternatives 3b. Limits size of EIS
4. Overextended review process	4a. Mandates time limits 4b. Stipulates concurrent vs. consecutive reviews by interested agencies
5. Permitting agencies intervening in late stages of review process	5. Agencies with permitting responsibility are required to cooperate with lead agency from earliest stages

'successful' EIS tended to become an end in itself, with its acceptance as an adequate document by the appropriate regulatory agency being tantamount to receiving permission to proceed with the proposed development or action. Clearly delays were inevitable and the EIS was not properly utilized as a tool by decision-makers, and was therefore not fulfilling its intended purpose.

The 1978 EIS regulations included measures designed to ameliorate these delay-causing factors (Table 4.2) but it is too soon to judge whether these measures will be effective. The new regulations were intended to promote efficiency, standardize terms, and reduce paperwork, the latter relating to an electoral campaign promise made by President Carter. While developing the new Regulations, the CEQ solicited the views of almost 12 000 private organizations, individuals, and state, local and federal agencies, and held numerous public hearings. The commentators were reportedly unanimous in stating that NEPA had exercised a net public benefit, and was therefore desirable, but that the EIS process had become a needlessly cumbersome deterrent to development that required substantial streamlining. The 1978 Regulations are CEQs response to these sympathies. Because they did not become effective until 1 July 1979, however, insufficient time has passed to evaluate their effect.

Beyond procedural problems, however, some delays are attributable to intentional obstructionism by some special interest groups. The NEPA review process includes several opportunities for public participation, which is entirely proper and desirable, for it is the public that the law exists to benefit. The safeguard is amenable, however, to effective use by individuals and organizations seeking to delay or halt projects for reasons other than those they publicly espouse during the review process. Subjective changes accusing 'lack of thorough evaluation,' 'inadequate consideration of all alternatives,' or 'excessive impact' are easily brought, but are difficult to counter because of the relative interpretations put to these words by opposing forces. It often appears that indefinite delays of a project is the most certain means of defeating it; thus some projects are faced with numerous charges that must be refuted, each in its turn.

To be sure, some projects *should* be delayed or even halted. Occasionally, however, the NEPA process has had to be overriden when the expressed interests of society at large demanded that an environmental cost is worth the benefit a particular project would impart. Thus construction of the Tennessee Valley Authority's Tellico Dam was allowed to proceed, despite the probable loss of the endangered snail darter. In this case a senior review panel was established with the power to override provisions of the *Endangered Species Act* under some conditions. Similarly, judicial review was required to limit NEPAs review authority in the Alyeska Pipeline case; despite this, however, NEPA forced the inclusion of several pipeline construction features that would not have occurred without such review.

Beyond delay problems, other difficulties can arise with respect to the specific methods used for the EIS environmental analysis. In historical practice regulatory agencies are often reluctant to judge the adequacy of a particular study plan in advance because this action may render them susceptible to public criticism during later stages of the EIS process. A result is that applicants often resort to overly elaborate, detailed, and expensive studies to protect themselves against what they consider arbitrary judgements part way through the review process. They often correctly anticipate that inordinately expensive environmental studies are in the long run less expensive than are extended delays in obtaining the permits necessary to develop their enterprises and to begin profit-making operations. Even where specific methods are agreed beforehand, the necessary environmental studies can be expensive and time-consuming unless an adequate biological data base already exists. Thus, for example, depending upon the literature base available, regulatory agencies may require multi-year baseline studies so that relationships and interdependencies between organism types may be adequately determined.

Adequate description of the environmental consequences of the proposed action and alternatives may be limited by available prediction technology. For example, the use of numerical models to attempt prediction of hydrological or chemical phenomena is becoming increasingly common, and in some cases this approach is effective. So far, however, attempts to interface models of physical or chemical phenomena with biological systems have been frustrated by the present state of this art. It is commonly conceded even by regulators, who sometimes hope to find simplistic systems to facilitate enforcement actions, that such attempts are much more effective in indicating to us the areas in which our information is deficient, than they are in producing the answers we would like to achieve.

4.2.5 Conclusions

The major lesson that our nation has learned over the ten year history of NEPA is that environmental decisions should not, cannot, and are not taken lightly by our citizenry. The EIS review process is not a parlour game; it emphatically affects people. It affects investment decisions, unemployment, health, recreation, the cost of living, housing, and virtually every other aspect of our lives. Costs and benefits must be weighed, therefore, by the people themselves, and the function of an EIS is not to render value judgements. Its function is to evaluate and present alternatives, so that the ultimate decision may be determined through sociopolitical processes by an informal public and its representatives. The system is by no means perfect, but at least it has served to raise the environmental consciousness of a recalcitrant American public.

4.3 ONTARIO HYDRO AND CANADIAN ENVIRONMENTAL IMPACT ASSESSMENT PROCEDURES

W. R. Effer

4.3.1 Introduction

The Canadian federal and provincial environmental impact assessment processes have used many of the features of the processes developed in the United States but have adapted them to specific situations and conditions existing in Canada. Generally, because legislation has not been favoured in Canada, administrative procedures and requirements have received more emphasis. In the 1960s increased concern for air, water and land pollution encouraged most provinces to enact legislation to control some of the more severe impacts of operating facilities. However, in the 1970s the move to control at an earlier stage of project development started. The developer or proponent was now required to identify impacts before the project was constructed and to make appropriate design changes to mitigate effects on the physical and social environment. More recently, and moving further back to the start of a project life cycle, we now have some legislation and procedures requiring the proponent to assess the relative value of alternative sites and even to assess the alternatives to the project itself. Moving still further back in the planning stage, proponents of large undertakings are now being required to consider presenting a comprehensive plan of development for regulatory and public review and assessment. Such early environmental involvement is now often considered to be the most cost-effective way to resolve the diverse concerns and effects (social, economic, as well as environmental) which become issues.

The federal government and all provincial governments in Canada have some procedures for environmental impact assessment. In most provinces the process is considered to be a necessary part of large-scale projects during the conceptual stage of development. Most provinces have developed guidelines, which in some cases only apply to provincial agencies or crown corporations, and in others to only large developments such as power plants and pipelines. The federal and the provincial legislation and practices related to EIA impact in Canada have been reviewed (CCREM, 1978) and are summarized in Table 4.3.

4.3.2 Federal environmental legislation and procedures

No Canadian federal legislation requires preparation of environmental impact assessments (EIA) for projects. However, regulatory powers to protect the environment are contained in the *Government Organization Act*, 1970 (which created the Department of the Environment), the *National Energy Board Act*, 1970 (regulation of interprovincial and international pipelines), and the *Atomic Energy Control Act*, 1970 (concerned with regulation of the entire life cycle and use of radioactive materials).

In December 1973, a Canadian cabinet established a federal Environmental Assessment Review Process (EARP) for projects either initiated by federal departments, or for which federal funds are solicited, or for those involving federal property. In 1973 the Department of the Environment was directed to establish a process to ensure that federal agencies and departments:

(1) Take environmental matters into account throughout the planning and implementation of new projects, programmes and activities.
(2) Carry out an EIA impact for all projects which may have an adverse effect on their environment before commitments or irrevocable decisions are made; projects which may have significant effects have to be submitted to the Federal Environmental Review Office for formal review.
(3) Use the results of these assessments in planning, decision-making and implementation.

Departments and agencies of the federal government are responsible for assessing the environmental consequences of their own projects and those of which they are the sponsor. The procedures provide for both preliminary and, if necessary, detailed assessment and review. The preliminary assessment stages involves an analysis of existing information, expert opinion and any necessary additional surveys and studies which can be carried out in the time available. The purposes of this preliminary screening stage are to describe the environmental consequences of the proposal, to evaluate the significance of the environmental impact and to clear (on environmental grounds) those proposed actions which do not have significant impacts. After screening the project the sponsor comes to one of four decisions.

(1) No adverse environmental effects, therefore, no action needed.
(2) Small environmental effects are identified but are considered insignificant, and can be mitigated by appropriate design. The initiator need not refer further to EARP.
(3) The potential adverse environmental effects are not fully known. A more detailed assessment is therefore required. The proponent prepares an Initial Environmental Evaluation (IEE) using the available guidelines (Environment Canada, 1976), and then decides to follow either (1) above, or (4) below. Detailed guidelines for preparing the IEE have been issued for several industrial sectors such as oil and gas exploration, linear transmission (e.g. oil pipelines, electric power, railways, highways), hydroelectric, fossil and nuclear electric generating stations, airports, seaports, mining and other industrial developments (Environment Canada, 1976).
(4) Where significant environmental effects are involved, the proponent prepares a detailed impact assessment and asks the Federal Environmental Review Office to establish a panel to review the project.

The purposes of the detailed assessment and review stage are to describe the environmental consequences (based on final concept design) in sufficient detail to permit the panel to make a recommendation on the proposed action and to provide sufficient information for a detailed review of the assessment.

Table 4.3 Summary of legislation and guidelines applicable to environmental impact assessment

Government	*Minister and departments responsible for environmental impact assessments*	*Major environmental assessment legislation/policies*	*Guidelines completed or in preparation*
CANADA	Environment Screening and co-ordinating Committees. Environmental Review Board. Environmental Assessment Panels	Government Organization Act, 1970. Cabinet decision 20/12/73 establishing Environmental Assessment and Review Process (EARP)	General guidelines for the preparation of initial environmental evaluation reports for: Oil and Gas Exploration and Production. Linear Transmission – Highways and Railways. Electrical Power Transmission Lines. Oil and Gas Pipelines. Hydro-electric and Other Water Development Projects. Nuclear Power Generation Projects. Airports. Ports. Mining Developments. Industrial Developments.
	National Energy Board	National Energy Board Act, RS 1970	Environmental Information Guidelines for Pipelines.
	Atomic Energy Control Board	Atomic Energy Control Act, RS 1970	Guide to the Licensing of Uranium and Thorium Mine-Mill Facilities

BRITISH COLUMBIA	Environment Environment and Land Use Committee Secretariat (major co-ordination). Land Management Branch (use of Crown lands). Water Rights Branch (use of water resources). Pollution Control Branch (pollution objectives). Mines and Petroleum Resources Reclamation Division	Environment and Land Use Act, 1971. Land Act, 1970. Pollution Control Act. Water Act, RS 1960 Coal Mines Regulation Act, 1969. Mines Regulation Act, 1967 (for land surface reclamation)	Guidelines for Environmental Impact Assessment of Power Projects. Coal Development Guidelines. Guidelines for Linear Developments. Guidelines for Environmental Impact. Control of Development on BC Crown Lands
ALBERTA	Environment Environmental Co-ordination Service	Land Surface Conservation and Reclamation Act, 1973. Clean Air Act, 1971. Clean Water Act, 1971	Alberta Environmental Impact Assessment System Guidelines
SASKATCHEWAN	Environment Environmental Impact Assessment Branch	Department of the Environment Act. EA Act, 1980	Environmental Impact Assessment Policy and Guidelines. Administrative Procedures being prepared. Content Guidelines being prepared

Table 4.3 (*cont.*)

Government	*Minister and departments responsible for environmental impact assessments*	*Major environmental assessment legislation/policies*	*Guidelines completed or in preparation*
MANITOBA	Mines, Resources and Environmental Management Clean Environment Commission. Environmental Management Division. Environmental Council. Manitoba Environmental Assessment and Review Agency. Mineral Resources Division (land surface reclamation)	Manitoba Environmental Assessment and Review Process established by Cabinet. Clean Environment Act, 1972. Mines Act (land surface reclamation). EA required under the Clean Environment Act for public sector	Guidelines for the Environmental Impact Assessment of electrical transmission lines, nuclear power plants, gas pipelines, heavy water plants, and highways
ONTARIO	Environment Environmental Assessment and Planning. Environment Assessment Board	Environmental Assessment Act, 1975. Mining Act (for land surface reclamation). Environmental Protection Act, 1972. Ontario Water Resources Act, 1972	Environmental Assessment Guidelines: Content of the Environment Assessment Document
QUEBEC	Environment Environment Protection Services. Advisory Council on the Environment. Quebec Planning Development Bureau	Environment Quality Act, 1972. Revision of Act in 1979 required EA's	Environment Quality Act (General) Regulations

NEW BRUNSWICK	Environment Environmental Services Branch	Clean Environment Act. Cabinet Agreement (1975) for public sector to participate	Environmental Impact Assessment in New Brunswick
NOVA SCOTIA	Environment	Environmental Protection Act, 1973. Water Act, 1967. EA's now being required under the Environmental Protection Act	Guidelines for pit, quarry, and mining operations being prepared. Guidelines for Asphalt Plants. Pesticide Control Act. Guidelines for Environmental Assessment
PRINCE EDWARD ISLAND	Environment	Executive Council Minute 14/2/73 requiring Environmental Impact Statements. Environmental Protection Act	Guidelines for Livestock Manure and Waste Management. Guidelines for Asphalt Plants. Guidelines for Gravel Pits and Sand Mining Operations
NEWFOUNDLAND AND LABRADOR	Consumer Affairs and Environment	Provincial Affairs and Environment Act, 1973. Environmental Assessment Act (1980) in Bill Form	Guidelines for Environmental Impact Statements

This stage may take from six months to two years and depending upon the available literature base, may include detailed field studies and monitoring programmes. The review panel issues the EIA guidelines which, depending upon the nature of the proposal, can include any consequence on the environment which falls within the Act's definition of the environment.

In common with procedures in the United States, the environmental assessment should include a statement of the 'need' for the proposals, an evaluation of the options available to the decision maker in addition to the proposed action, a baseline environmental description of the area to be affected and an assessment of the environmental consequences of the proposed action and each alternative. Specific impacts to be considered will depend on the type of project. For example noise and land use would be major issues in an EIS for airports and radiation levels in an EIS for a nuclear waste repository or nuclear generating station.

Where a project is of considerable interest to a province, the provincial government will share in the preparation of the specific guidelines to be followed. The panel issues the EIA guidelines, reviews the completed document, and acquires any necessary additional information, including public comments. Where there is a great deal of public interest, the Minister may establish an environmental review board whose members are drawn from outside the federal public service. The panel then reviews and recommends or advises acceptance by the Minister of the Environment. The Minister of the Environment and the Minister of the appropriate department then decide on the necessary action on the report. If Ministers disagree then the matter is referred to the federal Cabinet.

Under the EARP process 29 environmental assessments have been submitted to date. Twenty are under active review and nine have been reviewed and decisions handed down. Projects that have been completed include a nuclear generating station, a hydraulic generating station, the siting of a nuclear materials processing plant, a major highway improvement, two offshore drilling projects and a coal port expansion. Two of the nine projects were not approved. In one, the recommendation was that studies be carried out at other sites and in the other, the applicant was not prepared to proceed with the project, due to the required reduction in environmental impacts.

4.3.3 Provincial environmental legislation and procedures

In addition to federal EA processes environmental legislation has also been enacted at the provincial level (Table 4.3). However, Ontario is the only province to have enacted specific legislation on environmental assessment procedures. The *Environmental Assessment Act* 1975 formalizes the approach of considering environmental implications in the decision-making process of site selection, project design and construction.

The *Environmental Assessment Act* evolved in Ontario out of an awareness that pollution control is an after-the-fact measure which deals only superficially with improvement of environmental quality. Correct location and design of a project are seen as cost-effective and time-effective measures to prevent or mitigate environmental damage. The Act provided for a single EA procedure irrespective of project type. The Act requires that the EA procedure be initiated immediately following receipt of the application.

In 1973, a Green Paper on Environmental Assessment (OME, 1973) was published in which four procedures were described and analysed. After review of the comments received from industry, business, other government bodies and the public, one procedure was used as a basis for drafting the legislation. The draft bill was further amended, noticeably in strengthening the level of public involvement and the powers of the independent and lay environmental assessment board (Caverley, 1977). The procedure is shown graphically in Fig. 4.2.

When the Act came into effect it only covered activities of the provincial and public sector. However, municipalities and their activities have since been brought under the Act. Private industry was brought under the Act by regulation in 1977. The Act can now be applied to most major commercial or business enterprises or activities, plans or programmes in respect of major commercial or business enterprises of any person or persons individually designated by regulation. Whereas environmental assessments for all public undertakings must be prepared unless exempted, only those private undertakings specifically designated or defined by regulation are subject to the Act's provisions.

Under the *Environmental Assessment Act* in Ontario, 46 applications have been submitted. Of these 33 are under active review, nine have been reviewed and decisions handed down, one has been withdrawn and one has been exempted. Of the nine where decisions have been made, all have been approved. The projects include a sewage system, a resource management area, two highways, a waterfront park, and four class environmental assessments.

4.3.4 Application of the environmental assessment procedures

A task force which reviewed the various procedures across Canada was unable to identify a common national approach that would work (CCREM, 1978). Some of the comments were:

(a) Most jurisdictions had not defined the role of the environmental impact assessment process within the context of the overall decision-making activity.

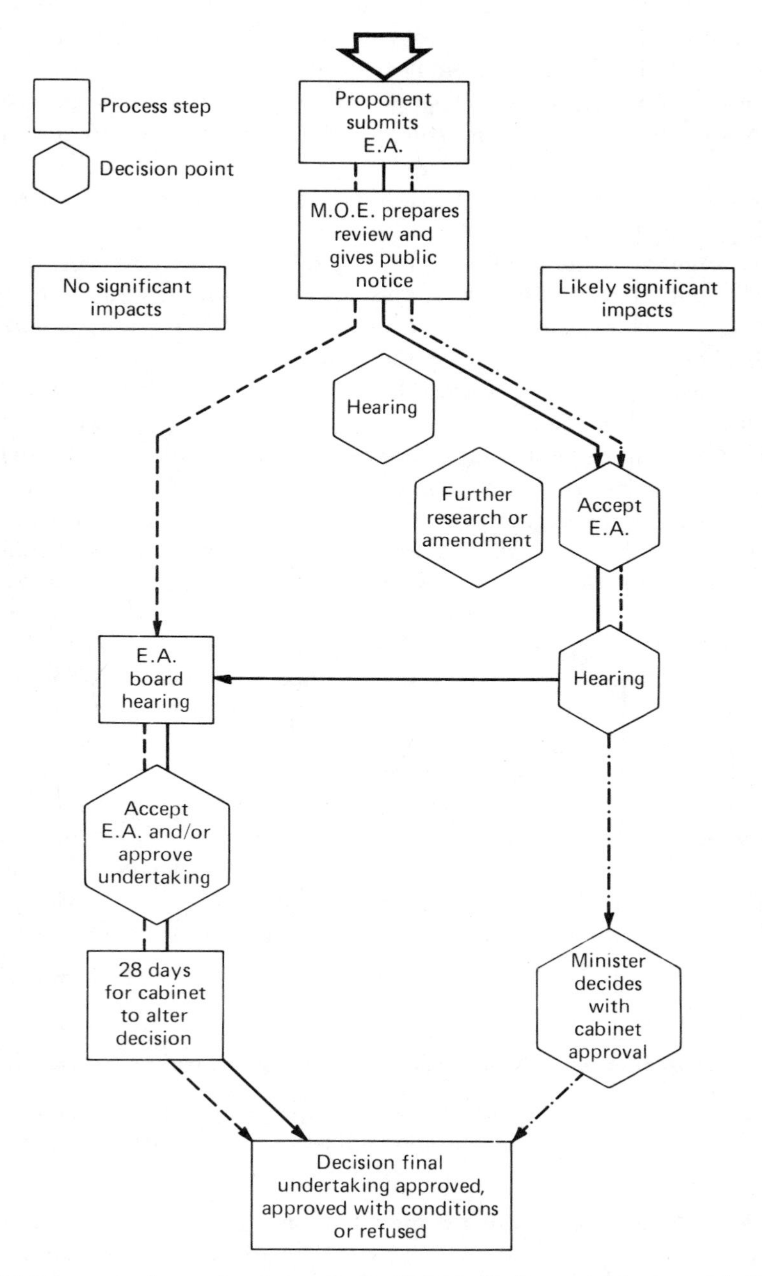

Fig. 4.2 Summary of procedural steps in the Ontario *Environmental Assessment Act* 1975.

(b) The relative roles of public, industry and government were difficult to define in a uniform way.
(c) The interface between the process and other broad policy areas such as conservation, land use and economics needs to be clarified within government, industry and the public.
(d) The need to establish different levels of application of the process. Should the approval be one of a broad programme in principle followed by a less detailed, possibly less critical assessment at the project level?
(e) The types and degree of public involvement needs to be tailored to specific projects.
(f) The difficulty of finding an appropriate design for public hearings.
(g) Lack of policy on funding of citizens and groups involved in public hearings.
(h) The difficulty in adapting the timing of the process to the schedule of the project under review.
(i) More detailed review processes reveal costs not normally attributable to the project itself. Decisions have to be reached on the responsibilities for paying such costs.
(j) The various jurisdictions are developing legislative regulatory policy procedures to ensure that the EIA process is applied without constraining economic development.
(k) Some governments are having difficulty in coordinating efforts on the assessment procedures for those projects involving more than one government.

Other publications (e.g. Environmental Canada, 1979; Munn 1979), describe the process in more detail and the application of the process in practice has been criticized (Emond, 1978). The criticism of the federal process falls into five main categories: the limited application of EARP, the principle of self (as opposed to independent) environmental assessment, the make-up and activities of Environmental Assessment Panels, the non-legislative status of EARP and the role of the public in the process.

Concern has been expressed over the possible effect of the *Ontario Environmental Assessment Act* on the process of municipal planning as embodied in *The Planning Act* which is implemented by the Ontario Ministry of Housing. After consultation with the Ontario Ministry of the Environment three principles were adopted.

(a) The natural environment should be considered during the administration of *The Planning Act* at the municipal and provincial level.
(b) The *Environmental Assessment Act* should only be applied to municipal and private undertakings when it is in the provincial interest to do so.
(c) Where overlaps are apparent an agreed simplification of the procedures should be adopted.

Under (c) the concern was mainly with private undertakings having to undergo a dual approval process. Agreement was reached to include only those private undertakings under the *Environmental Assessment Act* which are designated by Cabinet. For such undertakings approval of related municipal planning documents, previously considered under *The Planning Act*, will now become the responsibility of the decision-making body under the *Environmental Assessment Act*. No hearing before the Municipal Board will be necessary. Thus, only one comprehensive hearing will be required, and such a simplified process will also apply to any public undertakings subject to the *Environmental Assessment Act* which also require approvals under *The Planning Act*. Private undertakings under *The Planning Act* will consider the concerns for the natural environment along with the social and economic factors which it now considers (OME, 1979).

Ontario Hydro has been involved in a number of EIA studies since the inception of the *Ontario Environmental Assessment Act*. During the initial stages of putting a procedure into effect, many areas of uncertainty and vagueness need to be clarified. In this section some of the problems encountered by Ontario Hydro are briefly discussed. Although these are specific to a particular situation they serve to illustrate the range and type of problems which may very likely be encountered during development of a practical, workable process between the proponent, the regulatory agency and the public.

(a) Exemptions

At the time an environmental impact assessment process comes into effect several projects will be in various stages of planning or development. Other projects may be considered too small for inclusion under the process or may be of such an experimental or prototype nature that specific treatments are desirable. In Ontario Hydro, exemptions which have been granted included: administrative or housekeeping activities of the corporation; the activity of operating all facilities at the time the Act was proclaimed; transmission lines below a given length and voltage; upgrading and rehabilitation programme; and transmission lines and generating stations which already have been approved by a government order-in-council and were in various stages of development.

This procedure of identifying and classifying the various types of projects which should be exempted and then seeking approval of the regulatory agencies can be a very complicated and time-consuming activity. The public may question the exemption of a project which has passed through several years of design and development at the time the Act is passed, but which is not yet built and operating.

(b) Class environmental assessments

Certain small scale projects which are of similar nature and which generally have a small predictable range of effects, e.g. short lengths of transmission lines, small hydraulic developments, and roadway widening, may be treated as a class (Class EA) under the Ontario environmental assessment processes, the approval of which is designed to eliminate unnecessarily long approval times and high costs. The class EA can be applied for and may be granted after public hearings and board approval just like a full environmental impact assessment. An assessment report or memo is required for each individual project applied for within the class. The report is a minor form of environmental assessment but it does not require a public examination or board hearing. If, however, it is found that the project under the class EA does have severe environmental effects or there is a great public concern it has to be put into the full EA category. Only four class EAs have so far been approved under the Act. These are projects submitted by the Toronto Transit Authority and include commuter bus terminals, commuter bus and rail stations and storage and maintenance facilities. Ontario Hydro has submitted a class EA application for minor projects including short lengths of transmission lines and this has been under review for nearly two years. In practice, unless a large number of minor projects are included in the class EA application the proponent may not find it worth his time to go through the procedure with its requirement for public hearing at locations where all the expected class projects may be located. He still faces the possibility of some of the projects being reassigned the full EA status. In this situation more time would have been spent than if the proponent had gone directly through the normal full EA route.

(c) The need

There has been a great detail of confusion over the various definitions of 'need' and who approves which defined need. It is reasonable to assume that the environmental impact assessment processes under development in other countries will perpetuate this confusion unless definitions are agreed upon at an early stage.

In a large utility a number of matters should be addressed in support of the need (need 'A') for system expansion of the provincial electricity supply. These range from very broad aspects to the very specific and include:

(a) Further requirements for electricity
(b) New generation and transmission capacity
(c) Timing
(d) Financing
(e) Types and availability of fuel

(f) Regional location of proposed facilities
(g) Selection of appropriate sites and routes
(h) Design of facilities

In an interim approval procedure for generating stations (OME, 1978), Ontario Hydro will incorporate items (a) to (e) into a system expansion 'need report' to the government about six months before submissions of a project environmental assessment. Various government bodies will review the need report and if agreement is reached this will establish government policy. These matters can be discussed under the subsequent environmental assessment process but are not subject to reversal. Items (f) to (h) constitute the need (need 'B') to place a given type of facility at a proposed location and are both addressed and assessed as the undertakings, with alternatives, under the environmental assessment process. Although the *Environmental Assessment Act* does not contain the word 'need' the act requires that the rationale for the undertaking be addressed for both the proposed undertaking and the alternatives.

Another need (need 'C') has been discussed at great length by conservationist groups and the regulatory agencies and essentially tries to go beyond that defined by the system expansion need (A). The argument is that the goals of society in terms of quality of life, and how these goals influence long-term provincial development policy, need to be re-examined. It is postulated that such considerations will result in a modification of the demand for electrical supply which will in turn modify what Ontario Hydro will say is 'needed' in its system expansion requirements.

In summary, therefore, we have need A – to respond to an identified demand, need B – to justify an undertaking of a specific type, and need C – to question society's goals. This attempt to discuss and define the types of need may appear rather irrelevant to the studies which ecologists are required to carry out under the environmental assessment process. In fact, because most ecologists are also conservationists it is extremely important for them to sort out the role they are assuming in assessing a project. If they have a doubt that the project is needed, which 'need' is this? Working in their professional capacity as ecologists, either preparing or reviewing an environmental assessment, they should be only considering need 'B'.

(d) Lead-time and the number of steps in the process

The full development of a large-scale project like an energy centre (a site capable of supporting more than one generating station) may take 20 years. In Ontario Hydro we believe that this long time-scale makes it very difficult, if not impossible, to present just a one-step EIA in support of the site acquisition and all the generation to be eventually installed on the site.

Accordingly, we prepare a site selection EA which supports the application for a particular site, followed in subsequent years by project EIAs as required covering each of the proposed projects to be built on the site. This latter approach has been criticized as being too fragmentary for a rational decision to be made. The arguments are, among others, that the choice of the best site may depend on each of the future projects to be built on it and that no correct assessment can be made unless the whole development, including the system planning programme is considered and approved first. The counter argument is that no development programmes are certain, especially when projected 20 years ahead, and that it would therefore not only be very difficult to describe in any detail the technical aspects of the future generation but the assessment and prediction of environmental effect would be almost impossible. Ontario Hydro is presently following a two-step process in which a site selection EIA is carried out to select the most preferred site based on a range of possible future developments (Effer and Malvern, 1976). The second step is the project EIA. As mentioned previously the long-term system plan may be considered as a preliminary to these two steps.

This aspect of the EIA process should not apply in the great majority of cases where a site is being acquired for a single, specifically designed project which is to be built very soon after the site is acquired. The EIA should then include both the rationale for the selection of the site for the undertaking to be committed to that site.

4.3.5 Conclusions

Adoption of an EIA process should recognize the existence of previously established and tested planning and policy procedures. The process should therefore compliment rather than replace existing procedures.

An EIA procedure should be readily adaptable to the type, size, schedule, cost and importance of an undertaking. Where several small, comparable projects with similar predictable environmental impacts are planned, a form of Class Assessment should be adopted. Through such a process, the proponent may obtain a quicker, less costly decision than if he were to go through the normal procedure.

For large organizations having extensive long-term development plans, future problems with an EIA process would be reduced if a policy on development is first prepared which is examined by regulatory bodies, special interest groups and the public. Subsequent, more detailed steps in implementing the plan such as site selection and specific project approval can then be reviewed within the previously established policy framework. Such a step-wise procedure will prevent or at least reduce the time that is spent in discussing broader issues than are relevant to a particular stage of development. The ecologist and the many others involved in the approval process should therefore be involved initially in applying only broad principles and opinions on those aspects and concerns of the undertaking relative to his specific discipline. More detailed aspects such as field studies are not less important but are generally only appreciated at a later time in the project development.

4.4. ENVIRONMENTAL IMPACT ASSESSMENT PROCEDURES WITHIN THE EUROPEAN ECONOMIC COMMUNITY

N. Lee and C. M. Wood

4.4.1 Introduction

The implementation of the *National Environmental Policy Act* (1969) in the United States has stimulated interest in the application of environmental impact assessment (EIA) procedures and methods at both Member State and Community levels within the European Economic Community (EEC) during the 1970s. By the early 1980s the majority of the Member States had introduced some elements of an EIA system or indicated their intention to do so in the near future. Further, in June 1980 the European Commission submitted a proposal for an EIA directive, of Community-wide application, to the Council of Ministers. Accordingly, this paper outlines these developments and explores the direction of change in EIA practice which might be encouraged during the 1980s.

4.4.2 Member state developments

There has been a general tendency, during the last decade, for Member States to make increased provision for environmental impact assessments. However, they have done so in different ways: the range of activities covered has varied, the provisions relating to different elements in an EIA system (such as publication of impact studies and public participation) have differed, and their legal status has also varied.

In 1975 the Cabinet of the Federal Republic of Germany adopted a resolution describing the principles on which public measures proposed by the Federation should be assessed for environmental compatibility. By this decision, federal authorities and agencies are now required to examine and take into account at the earliest possible stage any harmful effects upon the environment caused by activities within their jurisdiction. The resolution is designed to cover only those actions for which there is no existing legal provisions for environmental protection and permits federal agencies to decide whether they need to invoke the principles and, if so, how and when to apply them.

In France, the *Protection of Nature Act* (1976) provides that studies undertaken prior to commencing significant public works or private projects requiring public authorization must include an impact assessment study. A number of application decrees were approved during 1977 which specified the coverage, content, public consultation provisions, etc. for this assessment. The provisions of the Act focus on project appraisal rather than the broader issues of government policy and apply to all projects likely to harm the environment unless they have been exempted. Projects that are exempted in the decree include maintenance and repair projects and projects costing less than six million francs, although some types must be assessed even if they fall below the monetary limit. Overseeing of the implementation of the Act has been entrusted to a central unit in the Ministry for the Environment and the Quality of Life.

In Ireland, the *Local Government (Planning and Development) Act* 1976 empowers the Minister for Local Government to require the preparation of an EIS as part of the development control procedure for large, private sector projects. Regulations to implement this procedure were issued during 1977. The Act applies to projects which have a value in excess of five million pounds and are likely to pollute but specifically excludes many developments, including roads and other public infrastructure initiated by local authorities in their own areas.

In Luxembourg, the Protection of the Natural Environment Law (1978) provides for impact analysis studies to be made for development and infrastructure projects which may interfere with the natural environment. Physical planning of individual projects outside conurbations must be subject to an impact study, if their size or the scale of effects on the natural environment could have an adverse effect on the latter. In addition, separate legislation on buildings which are dangerous, unhealthy or used for noisy or noxious trades stipulates that impact assessment may be required in respect of a range of public and private operations.

In the Netherlands, the Government announced, in 1979, its official standpoint on the basic principles for legislation on EIA and presented a bill to Parliament by early 1981. In Belgium, draft EIA legislation is also under preparation.

In the United Kingdom, there is currently no statutory requirement for any specific form of environmental impact analysis. The debate over impact analysis in Britain has focused on the need to refine the forward planning and development control system created by the *Town and Country Planning Acts*. The Department of the Environment has recommended that developers, local planning authorities and central government agencies make use of a manual prepared by the University of Aberdeen (Clark *et al.*, 1981) for the impact analysis of large-scale industrial project applications. Although various forms of EIA have been used by local authorities and developers on a voluntary *ad hoc* basis, the British government is opposed to mandatory EIA procedures though in 1978, the Secretary of State for the Environment indicated that the government endorsed 'the desirability . . . of ensuring careful evaluation of the possible effects of large developments on the environment' as advocated by Catlow and Thirlwall (1976).

Subsequently a House of Lords select committee heard a great deal of evidence on the impact of the European Commission's proposal and came down firmly in favour of its acceptance in the UK (House of Lords, 1981). The government, however, while accepting the virtues of assessment, remained opposed to its mandatory use:

> This Government therefore supports the use of environmental assessment studies for major classes of development in appropriate circumstances. Major projects can have complex and important consequences for the environment. These consequences rightly concern people, and it is essential that they should be examined thoroughly and efficiently. A systematic approach is the best way of achieving this, and that is why we see a positive role for environmental assessment studies within our existing planning system. Properly conducted, and in the appropriate circumstances, they strengthen the planning system in the eyes of everyone – the developer, who wishes to see the benefits of his project realized quickly; the local community, who may be anxious to safeguard their local environment; the wider public, who expect to see national interests given due weight. And such studies can help to highlight environmental opportunities as well. . . We must avoid legislation which could be difficult to enact, hard to implement and, by virtue of its uncertainty, a source of litigation and dispute. (House of Commons, 1981, Wood, in press.)

4.4.3 European Community developments

In the first Programme of Action of the European Communities on the Environment, approved in 1973, the general principle was adopted that 'the best environment policy consists in preventing the creation of pollution or nuisances at source rather than subsequently trying to counteract their effects' (*OJEC*, 1973). To this end it was agreed that 'effects on the environment should be taken into account at the earliest possible stage in all technical planning and decision-making processes' (*OJEC*, 1973). Thus, the 'preventive philosophy' of environmental protection was established at the outset.

The possible contribution of some form of EIA system in the implementation of this preventive philosophy was acknowledged in the second

Programme of Action in 1977 (*OJEC*, 1977). It indicated that 'the application at the appropriate administrative levels of procedures for assessing environmental impact meets the need to implement the objectives and principles of Community environmental policy'. Accordingly, the Council of Ministers instructed the European Commission to conduct studies on the subject and submit proposals to it. A number of such studies were carried out during the late 1970s (EC, 1976, 1978 *a*, *b*, *c*) and these formed the basis for the Commission's current proposals.

During the latter part of 1978 the European Commission prepared its preliminary draft proposals and these were submitted for examination to a meeting of a Group of Experts, comprising officials for each Member State. Subsequently these have been examined in more detail within individual Member States (see, for example, House of Lords, 1981 and above) in bilateral discussions with the European Commission and at a second meeting of the Group of Experts held in June 1979. The original proposals have been amended and refined, as a result of these successive stages of consultation, before they were submitted to the Council of Ministers (*OJEC*, 1980).

In the preparation and subsequent refinement of these proposals it has been necessary to take into account different and potentially conflicting requirements. For example,

(a) To learn and benefit from N. American experience with EIA but, at the same time, to develop proposals which are appropriate to the European situations.
(b) To provide the opportunity to apply EIA to all types of action likely to have a significant impact on the environment but, at the same time, to ensure that proposals are not over-ambitious from a practical and operational viewpoint.
(c) To allow a degree of flexibility in the interpretation of EIA requirements, to accommodate different institutional arrangements and traditions within the Member States but, at the same time, to guard against problems arising from widely differing interpretations of EIA principles or from non-compliance with them.

These, and similar considerations, have influenced the form of the Commission's proposals. A lengthy process of consultation over the provisions of the draft Directive is now under way; a favourable opinion has been received from the Economic and Social Committee and the European Parliament. A revised version of the draft Directive has been published (*OJEC*, 1982) and the detailed text is being examined by a working group established by the Council of Ministers.

The current proposals accept the principle that EIA should be applied in such different areas of decision-making and planning as the approval of individual projects, land-use plans, regional development programmes, sectoral and other economic plans (Lee and Wood, 1978). The draft third action programme spells out the European Commission's thinking:

> EIA is the prime instrument for ensuring that environmental data (are) taken into account in the decision-making process. It should be gradually introduced into the planning and preparation of all forms of human activity likely to have a significant effect on the environment such as public and private development projects, physical planning schemes, economic and regional development programmes, new products, new technologies and legislation (*OJEC*, 1981).

However, largely on grounds of practicality, the Commission's proposals are restricted, in the first instance, to the use of EIA in the authorization of individual projects. It therefore proposes that a general obligation be placed on each Member State to ensure, before planning permission (or similar authorization) is given, that projects likely to have a significant environmental impact are subject to an EIA. No distinction is drawn between public and private projects.

The determination of whether an EIA is required in any particular case is likely to depend upon the type of development which is proposed. The Commission proposes that:

(a) EIA is to be obligatory for a specified list of projects which are likely significantly to affect the environment under any circumstances, regardless of their scale or size, but with provision for exemptions below a threshold size set by Member States in agreement with the Commission.
(b) EIA is also to be obligatory for a further list of projects subject to certain criteria (e.g. threshold size etc.) determined by the appropriate authorities in the Member States.
(c) Other projects are to be subject to EIA at the discretion of the Member State in accordance with its general obligation as indicated above. Such projects may be described *a priori* as unlikely to produce significant effects on the environment except in exceptional circumstances: for example, if the project were located at a site with a particularly sensitive environment.

The Member States are likely to have a real measure of discretion, varying according to the type of project under consideration, in determining when an impact analysis is required. At the same time the Commission would expect to review periodically with Member States their lists of projects and their criteria for selecting projects in order to promote consistency in the application of EIA within the Community.

Where projects are subject to EIA, the developer (with the assistance of the appropriate public authorities) would be required to prepare an EIS. This would normally contain information relating to:

(a) A description of the proposed project and where applicable, of the reasonable alternatives for the site and/or project design.
(b) A description of the environmental features likely to be significantly affected.
(c) An assessment of the likely significant effects of the project on the environment.
(d) A description of the measures envisaged to eliminate, reduce or compensate these impacts.
(e) A review of the relationships between the proposed project and existing environmental and land-use plans and standards.
(f) In the case of significant effects on the environment, an explanation of the reasons for the choice of the site and/or project design, compared with reasonable alternative solutions having less effect on the environment, if any.
(g) A non-technical summary of the items above (*OJEC*, 1980, 1982).

The study would subsequently be made available for comment to relevant authorities with responsibilities for environmental matters (including, where applicable, those in other Member States) and the competent body (which authorizes the project) would fix, as appropriate, a suitable time limit for replies. Similarly, the competent authority would make the impact study available to the public and obtain their views within a suitable time limit. Finally, the various comments would be brought together and summarized. This summary would be attached to the document upon which the final decision by the competent authority would be based and this would also be made available publicly.

The form of these provisions relating to the preparation of the impact study and for consultations upon it is designed, as far as possible, to enable them to be assimilated with minimal disturbance into existing procedural arrangements within individual Member States. One of the purposes of the discussions that have taken place has been to try to facilitate this assimilation.

4.4.4 EIA practice

The various developments described above have mainly been concerned, at least initially, with the strengthening of *procedural* requirements for EIA. As such they provide the legal, organizational and administrative framework within which environmental impact statements are prepared and subsequently used. However, these do not directly determine the *quality* of the analyses which are made: this depends upon such matters as the availability of environmental and other relevant data, the assessment methodologies selected for use, and the skill and experience of the personnel who perform impact analysis (Lee and Wood, 1980, Lee, 1982). In discussions, at both Member State and Community levels, matters relating to the improvement of assessment practice are now being given greater attention and it is in this particular area that professional and scientific expertise should have an important role to play.

The Commission's draft proposals provide for a transitional period of two years, following their approval by the Council of Ministers, before they are required to take full effect. This allows some time for preparations to be made but the period is relatively short and it would seem desirable to prepare priority programmes for the improvement of assessment practice as soon as possible.

In developing such programmes we should suggest that it may be helpful to take the following considerations into account:

(a) Methods of impact analysis must be capable of practical application and must operate within and contribute to an efficient decision-making system for project authorization. Therefore, they must be realistic in terms of the man-power and other resources likely to be available for the studies. The benefits of more detailed and accurate information depend upon the type and quality of information needed to reach a sound decision and, in any case, have to be compared with the costs associated with any delays incurred in obtaining information. Particular attention should be paid to the use of screening methods to avoid the mistakes of some environmental impact studies which have gone to unnecessary extremes by exploring, in too great a degree of detail, too many alternatives and too many categories of possible impact.

(b) Although it is obvious that scientific understanding of the likely environmental impacts of many types of project is incomplete, it is unrealistic to programme for the correcting of many of these deficiencies in the short-term. Therefore it is desirable to distinguish short and longer-term strategies for the improvement of EIA practice, although the two should complement each other. In the short-term, the main priorities should probably be to identify what is 'best available practice' in impact analysis, to ensure that it is adequately disseminated and that appropriate training is given in its use, and that existing environmental and related data are readily accessible to those who need to use them. In the longer term, it is desirable to obtain a clear indication of the improvements to the environmental data base which are desirable and practicable, and of the most efficient means of achieving these. Additionally, an equally clear indication is needed of research priorities so that resources are deployed within the EIA area where they are most likely to yield the greatest benefit (Lee and Wood 1980; Lee, 1982).

Since the ecological component of EIA practice is one where improvements should be sought within both short- and longer-term strategies, the involvement of those professionally active in this area is desirable both in helping to define these strategies and through actively participating in their implementation.

4.5. PROPOSALS FOR ENVIRONMENTAL IMPACT ASSESSMENT PROCEDURES IN THE UK

B. D. Clark and R. G. H. Turnbull

4.5.1 Introduction

Within the past two decades there has been a growing concern in the United Kingdom about the effects of new development on the quality of the environment and the ability of the planning system to take these effects into account at the project-submission stage and in the context of forward planning. The concern is by no means new, but until recently the projects which have aroused opposition have been intermittent and geographically dispersed. Since the discovery of dry gas in the southern part of the North Sea in 1965, and the discovery of oil and gas beneath the North Sea off the Scottish coast in 1970 this has no longer been the case.

It is against this background that the concept of environmental impact assessment (EIA) has evolved in the UK. It is a concept which has become politically sensitive both here and abroad. As a political instrument, the role of EIA in the UK has also become controversial because of the European Economic Community (EEC) proposals to introduce a mandatory system of EIA for certain specified types of projects.

At this stage, it is important to stress that there is no general and universally accepted definition of EIA. The following examples selected at random from a number of authorities illustrate the great diversity of definitions which exist:

(a) '.... an activity designed to identify and predict the impact on man's health and well being, of legislative proposals, policies, programmes and operation procedures, and to interpret and communicate information about the impacts' (Munn, 1979).
(b) '.... an assessment of all relevant environmental and resulting social effects which would result from a project' (Battelle Institute, 1978).
(c) '.... assessment consists in establishing quantitative values for selected parameters which indicate the quality of the environment before, during and after the action' (Heer and Hagerty, 1977).

In this paper EIA is taken to mean the systematic examination of the environmental consequences of projects, policies, plans and programmes. Its main objective is to provide decision-makers with an account of the implications of alternative courses of action before a decision is made. The results of the assessment are assembled into a document referred to as an environmental impact statement (EIS). When an EIS has been prepared it is used by decision-makers as a contribution to the information base upon which a decision is made. It is also suggested that EIA should not be thought of as some kind of scientific 'black-box' which produces definitive answers. Instead it should be thought of as a concept which allows decision-makers to be made aware of likely impacts should a proposal proceed. Whether or not it does proceed may be influenced by other factors such as energy needs, employment levels or technological necessity, which legitimately may have greater priority at any given place or time. In these circumstances EIA may be able to suggest mitigating actions or the form of monitoring which should take place. It may also allow various 'trade offs' between competing objectives to be more fully understood.

This paper will consider how EIA has been developing in the United Kingdom in respect of project appraisal. Attention will be focused on the growth of procedural methods for the implementation of EIA and future prospects for the development of EIA in Britain will be evaluated, particularly with regard to the EEC draft directive.

4.5.2 The development of EIA in the United Kingdom

Since the *Town and Country Planning Acts* became law in England and Wales and Scotland in 1947, UK planners have gained considerable experience in the technical assessment of the likely consequences of major development proposals. From the late 1960s, the UK also witnessed an upsurge in public concern for the protection and enhancement of the environment which, in turn, intensified public interest in the impact of major developments.

In this context, the proposal to build a third London airport in south-eastern England can be considered a watershed. Considerable time and resources spent in adapting cost/benefit analysis in project appraisal led to a realization that other assessment methods were required if a balanced appraisal of likely environmental and social impacts were to be achieved.

Although the third London airport proposal was an important event in the history of project appraisal in the UK, the main impetus resulted from the discovery of North Sea oil and gas. The extraction of oil and gas required large production platforms whose construction necessitates access to deep water. As exploration and production moved to the north of the North Sea, oil province planning authorities in North Scotland began to receive applications for platform construction sites. Most of these sites were situated in areas noted for their outstanding visual quality and containing small rural communities characterized by a distinctive way of life. In many cases planning authorities were unfamiliar with the characteristics of platform construction yards and often did not possess the resources to undertake thorough appraisal of the development.

To help overcome some of these problems, the Scottish Development Department (SDD) issued an advice note to planning authorities entitled 'Appraisal of the Impact of Oil-Related Development'. The advice note pointed out that oil-related developments need '. . . rigorous appraisal' and also stated that other non-oil related '. . . large-scale and unusual projects . . .' require similar attention. This was the first recognition by a UK central government department that major developments necessitated special appraisal if the UK planning system were to meet the demands of developers and the public. A suggested approach to the assessment of major proposals was set out including the commissioning of an impact study, by a planning authority, if it was considered necessary.

From 1973 onwards, a number of environmental impact studies of major oil developments have been undertaken (Manners, 1978). Many were financed jointly by the Scottish Development Department and planning authorities and carried out by independent consultants whilst others were financed by developers. Reports presenting the results of these impact studies are similar, in many respects, to the EIAs prepared in the US under the *National Environmental Policy Act* of 1969. These reports were initiated on an *ad hoc* basis and not as a mandatory planning requirement.

As well as oil-related impact studies a number of other impact assessments were initiated in the 1970s. These include a study for the National Coal Board (NCB) in connection with its proposal to mine coal in the Vale of Belvoir, an EIA for the Craigroyston pumped storage electricity scheme at Loch Lomond and a number of EISs of motorway and truck road proposals. Again it should be noted that these studies were non-mandatory.

Additional support for the concept of EIA was given in the Dobry Report on the Development Control System (Dobry, 1975). The report advocated that, in the case of specially significant development proposals, the planning authority (PA) should be able to insist that an applicant submit an impact study. The main topics to be covered in such a study were outlined together with information that would be required from the developer. The Royal Commission on Environmental Pollution (1976) endorsed Dobry's proposal that 'developers provide an assessment of the effects of air, water, wastes and noise pollution of certain major developments', and stated that the developer would need the assistance of planning authorities and other bodies in providing information and that rules should be made to ensure this. The Stevens Committee on Mineral Workings adopted a different attitude. They recommended that a standard form for mineral applications be introduced which would require far more information than is currently requested. It would be comprehensive enough to allow planning authorities to carry out an environmental appraisal of any mineral proposals and the Committee therefore 'see no need to apply to mineral applications the special impact statement procedure proposed by the Dobry Report'.

Experience of assessing platform sites, the likelihood of major oil and gas developments and other types of large-scale developments showed that various aspects of project appraisal required investigation. Central government took the initiative and set up two separate, but complementary studies to investigate certain aspects of project appraisal. In 1973, the Project Appraisal for Development Control (PADC) research team at the University of Aberdeen, was commissioned to prepare a manual to assist planners in assessing the environmental impacts of major industrial developments. The appraisal procedure developed was published as *The Assessment of Major Industrial Applications – A Manual* and issued as an interim report to all UK planning authorities (PADC, 1976). A revised, extended and updated version of the manual has now been published (Clark *et al.*, 1981).

In 1974, the Secretaries of State for the Environment, Scotland and Wales, commissioned an investigation into the '. . . desirability of introducing a system of impact analysis in Great Britain, the circumstances in which a system should apply, the projects it should cover and the way in which it might be incorporated into the development control system under the *Town and Country Planning Acts* as it applies to both private and public sector development'. Both reports deal with assessment of projects and pay little attention to policy appraisal and review.

One further area where EIA is increasingly being used is in the field of forward planning. In Scotland central government issued national planning guidelines (Scottish Development Department, 1977). The 1977 guidelines circular described their function in terms of defining the level at which there was a 'national interest' in development control matters, but it also made clear that development control could not be considered independently of development planning. The guidelines also contained suggestions for positive action. In the guidelines on sites for large-scale industry, site selection criteria were set out with an indication of the areas most likely to meet the criteria. Planning authorities were encouraged to examine the potential for large-scale industry in these areas and to frame their plans and development control policies accordingly. Similarly the guidelines for petrochemical developments stated that it would be prudent for planning authorities to establish the potential for petrochemical development in their area and to frame their structure plans, local plans and development control policies so that sites can be readily identified if and when required. In both cases environmental impact assessment has become an important part of the evaluation procedure undertaken and this will be illustrated by looking at the example of forward contingency planning for petrochemicals (Turnbull, 1979).

The development of the North Sea offshore petroleum reserves has already led to demands for land, labour and infrastructure, with consequent environmental, social and economic impacts. As development moves into the production stage the opportunities created by these resources are of increasing significance in areas of Scotland which are not at present major centres of oil- and gas-related activity. Experience suggests that the ability of central and local government to influence development in these areas is dependent on many factors but not least upon a continuous information flow and forward assessments of changing capacity and capability, and desirability of development in particular areas.

Since the publication of the 'Coastal Planning Guidelines' in 1974, which defined preferred development and conservation zones, the Scottish Office has maintained a close working relationship with planning authorities, the purpose of which has been to exchange information and to identify the potential of the areas involved to accommodate oil- and gas-related developments. The formal background to this work was reinforced by the national planning guideline for petrochemical developments issued in 1977, and the seven mainland regions most concerned with contingency planning for petrochemicals now have a fairly well developed appreciation of the potential and environmental capacity in their respective areas.

The terms of reference, approach and methods adopted by the regional working groups which were established to produce contingency plans have differed according to local conditions, attitudes and the availability of information. In some cases, the initial approach has been made purely from a technical standpoint whereas other approaches have been tempered by political considerations. The working groups have carried out, or have been associated with, studies aimed at establishing the physical and environmental capacity for petrochemical development within their area and to date forty-six search areas have been examined. Some of the regional authorities concerned have commissioned consultants to undertake special studies ranging from the economics of feedstock and product transport to site selection criteria, environmental capacity and safety aspects.

Through the application of a blend of site suitability, and constraints to development and local authority policy, the original list of 46 sites has been refined and reduced to 13 which are thought to have the requisite potential to support petrochemical development. On present evidence, most of the 13 sites will be presented in structure plan submissions. The establishment of 13 preferred petrochemical sites from the original 46 does not mean that the residual 33 sites have no industrial potential, for several of the sites were identified as being suitable for large-scale non-petrochemical development.

One regional authority commissioned consultants to carry out an examination of the engineering and other physical factors which would be involved in developing industrial plant on the sites identified. A study of the environmental problems and environmental costs of possible development was undertaken within the Physical Planning Department. The general conclusion reached was that none of the sites examined was considered suitable for petrochemical development in terms of the environmental consequences. Development on almost all of the sites would cause serious damage to landscape, ecology, the tourist industry, agricultural and recreational use. This combined with the potential hazard to existing development and existing population was considered to be an unacceptably high price for the region to pay.

It is important to note that conclusions drawn from this initial environmental assessment led on to a more comprehensive review of alternative sites appropriate in environmental terms for petrochemical development. In this and approaches by other regions, the preliminary application of environmental assessments was clearly of value, in that at a very early stage it was possible to identify the main environmental impediments to the use of the identified sites for petrochemical development. Of no less importance is the need to carry this work forward to promote an atmosphere in which industrialists can invest freely in new capacity or add to existing facilities, yet at the same time ensure that the interests of the public and environmentalists have been taken fully into account.

4.5.3 Procedural methods of EIA in the UK

In the UK, there is no single method of impact assessment which is used generally for identifying impacts, information collection and presentation of results. There has been no central government requirement which specifies a particular procedure for assessing impacts. Consequently, local authorities, public sector developers and privately owned concerns have been free to adopt methods which they consider will help achieve their particular objectives.

An examination of impact studies implemented in the UK since 1973, shows that most procedures incorporate impact identification, measurement, interpretation and communication but there has been considerable variety in methods used. One trend has been the use of matrix methods accompanied by the quantification, weighting and aggregation of impacts (Sphere Environmental Consultants Ltd, 1974; Leonard and Partners, 1977). Other impact studies have been carried out using methods which avoid quantification and aggregation of impacts. Interaction matrices have been used to structure impact identification, the collection of information and the presentation of results. Matrices such as the Leopold matrix (Leopold *et al.*,

1971) have formed the basis for particular adaptations formulated for specific assessment exercises. Most EIAs have presented information in a quantitative, descriptive manner. Ordinal rankings of alternatives, supplementing written descriptions of impacts, have been produced, for example, the study of alternative oil production platform sites in the Clyde estuary carried out by Jack Holmes Planning Group (1974).

The PADC method contained in the manual was developed to fit into the existing planning framework (PADC, 1976). It was formulated in an attempt to resolve two current problems encountered by planners assessing proposals, namely, the difficulty of obtaining sufficient detailed information from prospective developers and the lack of a framework for the systematic appraisal of proposals.

Components of the assessment procedure are grouped into three sets of related activities (Fig. 4.3). While the approach is seen as a sequence of events, many of the activities may be performed concurrently. The approach incorporates certain activities which are already features of the assessment of planning applications in the UK. However, certain innovations are introduced to help identify potential impacts, predict impacts and present information to decision-makers.

Two elements of the method are useful in identifying potential impacts. These are the Project Specification Report (PSR) and the Impact Matrix. With a planning application, prospective developers are asked to submit a PSR giving detailed information on a proposed plant and its processes to ensure that planners receive a wide range of detailed information at the outset. Potential impacts are identified by relating data on a proposal to the characteristics of the site and its environs. Using a matrix, cells representing likely interactions are identified. Each of these interactions is given consideration during appraisal.

The method provides planners with a means of predicting impacts. It advocates the collection of detailed information on the proposed development and data on the existing area. An analysis can be made of the scale and significance of potential changes for each likely impact identification in the Impact Matrix. These predictions can be structured using Technical Advice Notes included in the manual. The manual also contains a list of questions identifying the range of factors to be resolved during assessment.

Detailed information on potential impacts can be drawn together into an impact statement which can be made available to both decision-makers and the public. In an impact statement each impact is described briefly and assessed in relation to the five criteria identified above. The 'no go' alternative can also be considered.

An examination of procedural methods used in EIA in the UK shows that quantified approaches to measuring relative impact importance and magnitude have not been prominent. Most impact studies have considered impact

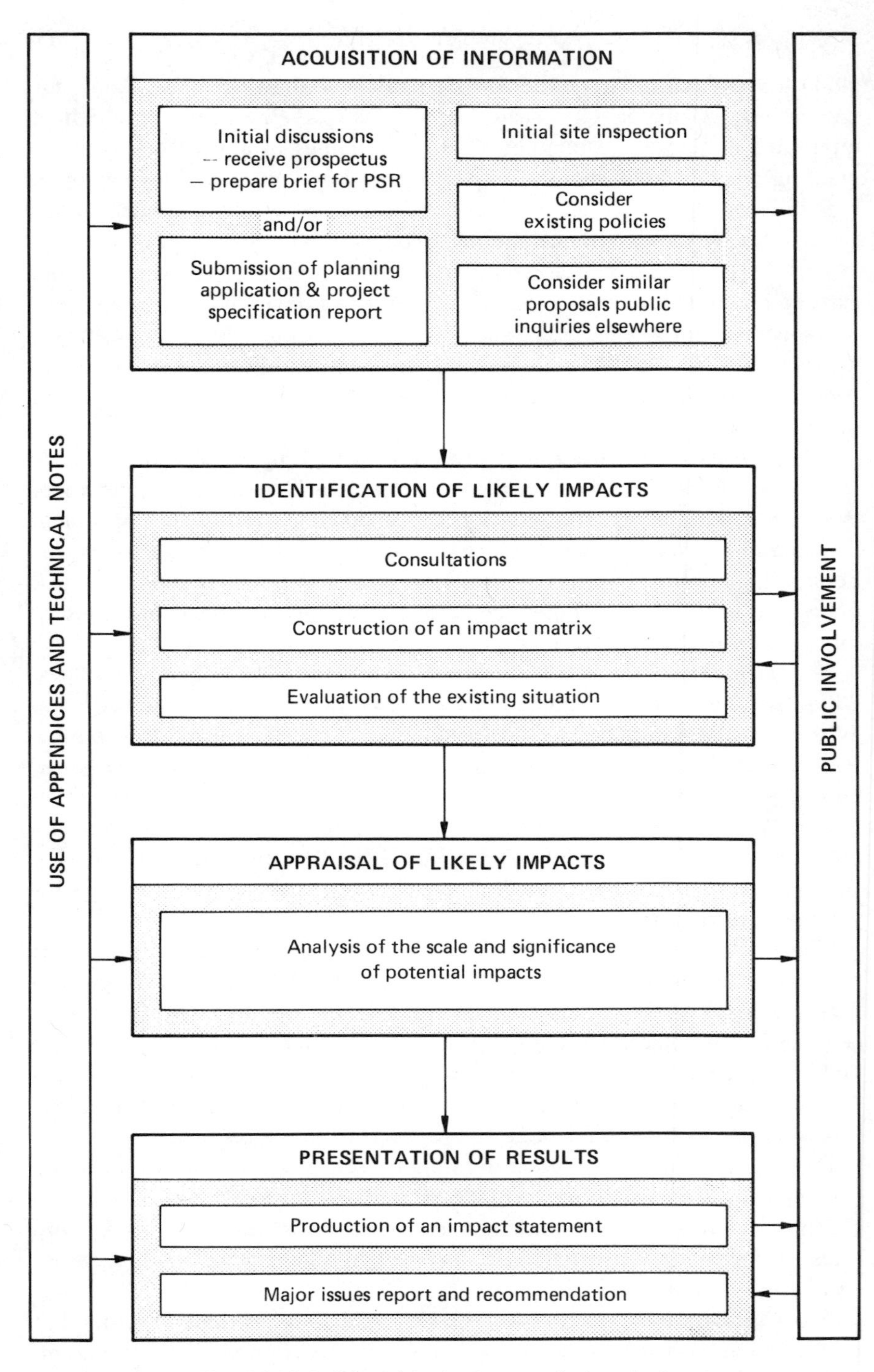

Fig. 4.3 Linked activities in the appraisal method.

data in a qualitative manner and have provided written accounts of the characteristics of all likely impacts.

4.5.4 Future prospects for EIA in Britain

During the 1970s, UK planning authorities have acquired considerable experience in the appraisal of major projects. The appraisal of these developments has not taken place, however, within an environmental impact assessment system requiring implementation of systematic study of all likely impacts. Although many projects have been subject to thorough environmental appraisal there is no administrative or planning mechanism which requires it. The extent and nature of appraisal carried out is discretionary and dependent on the attitudes and perceived needs of the decision-making authorities.

The UK planning system can incorporate both detailed and partial environmental assessment procedures. The type of appraisal, if indeed it is undertaken at all in a structured manner, depends to a large extent on the circumstances of each case. For example, the thermal oxide fuel reprocessing plant (THORP) at Windscale was not subject to a detailed environmental appraisal prior to the public inquiry.

The present status of environmental assessment seems to indicate that central government takes a cautious view of its role in UK planning. It recognizes that detailed appraisal will be required for major developments, but is content to leave decisions on whether it will be carried out to planning authorities and/or developers.

Confirmation of the Labour government's view that project appraisal should continue to be implemented on a discretionary basis was contained in a speech on public inquiries given in 1978 by Peter Shore MP, then Secretary of State for the Environment. He gave partial encouragement to the use of environmental impact assessment in the appraisal of major contentious projects such as the proposed Vale of Belvoir coalfield and the first commercial fast-breeder reactor (Shore, 1978). It is suggested that reports presenting the findings of these assessments may be submitted to public inquiries. There is no mention of the need for a mechanism for screening major developments to ascertain whether they will require impact assessment or for formal procedure and guidelines for implementing such appraisal. As yet the present government has made no major public announcement as to its attitude towards EIA. However, in a recent speech Malcolm Rifkind (Minister of State, Scottish Office) has said that two of the major objectives of planning must be 'the stimulation of economic activity' and 'safe-guarding the rights of the individual' (Rifkind, 1980). A key question, therefore, must be the role that EIA might play in helping to resolve these often conflicting objectives.

There is, however, one event which may alter present thinking on EIA in the UK. Its systematic introduction may occur if the Commission of the European Communities introduces a directive requiring preparation of EISs for a wide variety of private and public sector developments. A draft directive has been produced and has been the subject of discussion between representatives of the nine Member States of the European Communities in Brussels. It proposes that initially only projects would be subject to impact assessment. Since a directive is mandatory on Member States, a formal requirement for EISs could be introduced into UK planning by the Commission. There has been considerable debate on the form that the directive might take to make it acceptable to all the nine Member States.

Many points of interest, indeed some would say of concern, have arisen in the ongoing debate which is taking place during the preparation of the drafts of the directive. The following questions, arising from the proposals in the latest draft, appear to be key considerations. Should the onus be on a developer to prepare an EIA, albeit in co-operation with 'competent authorities', or should prime responsibility rest with decision-making authorities whether they be local or central government? Should a developer be expected to incorporate his interpretation of planning policies for a proposed development site in his EIA? Will the public have full confidence in an EIA prepared by a developer and accept it as a balanced assessment of both beneficial and adverse impacts? Will a mandatory list of projects which in all circumstances must be subject to EIA lead to an EIA system which may be legally sound and enforceable in all Member Countries but which nevertheless may generate unnecessary EIAs, raise irrelevant issues, collect unnecessary data and in the long-term do little to enhance or maintain the quality of the environment?

It is very difficult, therefore, to talk meaningfully about future procedures for EIA in the UK without knowledge as to whether they will have to take account of a mandatory EEC directive, or whether, as has happened up to 1982, procedures will continue to evolve in an *ad hoc* manner to meet particular needs and circumstances.

The consideration of certain recent proposals has also led to a recognition that EIA could play a role in the formation and evaluation of policies. The public inquiry into the THORP proposals at Windscale has shown the problems involved when both projects and policies are considered in an examination of the likely effects of a specific development required for implementation of a particular policy. The inquiry ranged over local, regional and national issues. Also, it considered project and policy implications of proceeding with the reprocessing plant. This resulted in a lengthy, costly public inquiry which satisfied few of the participants. It would appear that lessons learnt from the Windscale inquiry have had an important effect on the attitude of central government to policy review. It has encouraged public discussion of the need for policy review, and administrative mechanisms to implement it, and in this EIA may well begin to play a key role.

4.5.5 Conclusion

In view of the above comments it is difficult to draw anything but the most tentative conclusions about the current state of environmental impact assessment in the UK. Indeed, in a period when there is considerable debate, and uncertainty, about the future role of EIA, it is perhaps best not to draw any firm conclusions as these may be overtaken by events.

There can be no doubting, however, that with increasing public interest in environmental protection, the impact of a proposed development on the environment has become a significant factor in determining whether or not it will proceed. Projects which are expected to have a detrimental effect may be delayed by objections and in the end may have to be abandoned. The need for more information about the effects of major installations on the environment has, therefore, been instrumental in the development of procedures of impact appraisal. EIA may be used by authorizing or funding agencies to determine whether developments should proceed. In these circumstances the appraisal would be undertaken when the planning of the project is well advanced. EIA also has a role earlier in the project planning process when used by developers to ensure that environmental issues are considered at an early stage. The image of EIA as a slow, bureaucratic procedure is largely based on the years immediately following the implementation of NEPA in the US. As experience has been gained in the US and UK, the value of EIA as a planning tool has been recognized. In response to demands for more effective assessment of projects, plans and policies EIA is likely to play an increasingly important role in the UK planning system.

4.6 THE ROLE OF ENVIRONMENTAL IMPACT ASSESSMENT IN DEVELOPMENT CONTROL AND POLICY DECISION-MAKING

R. Grove-White

4.6.1 Introduction

The underlying principle behind EIA, which I fully endorse, was in NEPA's 1969 words 'to create and maintain conditions under which man and nature can exist in productive harmony for present and future generations'. But what grounds for confidence are there, in the UK context, on available evidence, that the introduction of EIA will assist such 'productive harmony' here? My view of the process in a UK context is one of suspicion: because of our established institutional framework, we may be losing sight of the original conception of EIA, as found in NEPA. I should qualify this at once by saying that I, and CPRE, favour any measures to improve the environmental effectiveness of development control – so in no sense do I reject EIA *per se*.

So far this book has concentrated on ecological and socio-economic aspects of EIA; I should like to broaden the issues and focus on some political dimensions.

4.6.2 NEPA

EIA in the US must be seen as *one part* of the NEPA package. The overall intention of this legislation was both broad and far-sighted. More than one authority has characterized its intention as having been not only to get Federal agencies to reflect environmental priorities in their decisions, but also to harness the creative energies of the US political system to begin pushing towards new ends and new ways of doing things. This latter challenge in particular was a response to the wholly new dimension of environmental problems seen to be arising in industrial society.

Because of its implications for the very manner of making executive decisions, NEPA has also been characterized as a measure of administrative reform. A central part of this process, which accorded with NEPA's very broad aspirations, was the feeding-in and developing of new, alternative ideas about policy. In the American context, this inevitably, and intentionally, involved bodies in the official nexus, with the help of the courts.

Wandesford-Smith (1978) and Legore (Section 4.2) have noted that the volume of litigation during the first five years of NEPA should not be seen simply as obstructive. Rather it should be recognized as the working-out of a creative dynamic, to realize the overall ambitions of NEPA. These, as I have said, were to transform public policy. It must of course be accepted that the full consequences of NEPA's EIA provisions were not widely foreseen. Nevertheless, these flowed from the words of the statute and, given the mechanics of US constitutional relationships, could have been anticipated to a large degree.

But it cannot be said too often that the political design and rhetoric of NEPA – and hence the context from which the original conception of US EIA sprang and developed – was bold and imaginative. It is far too early to reach final conclusions about the Act's success. There are reports that the second five years have been far less litigious than the first – and that the scope, and limits, of the EIA process are now clearly defined. In this context the 1978 amendments have been beneficial.

But it was always going to be difficult to get NEPA to bite. Its purpose after all was to get to grips in quite a new way with what can be seen as the new structural problems of a sophisticated and dynamic industrial economy.

4.6.3 EIA

(a) The political context

From this consideration of NEPA it is clear that the political context of EIA is important. And my disquiet about EIA proposals on this side of the Atlantic also arises from this context. I should like to consider first some relevant features of the political background against which our discussions of EIA are taking place in the UK.

Increased numbers of major developments are stemming from government policy initiatives – energy, transport/roads, agriculture, airports, even minerals. But Government departments, through the public inquiry system are becoming increasingly, judge and jury in their own cause. The recent Bushell judgements in the House of Lords have added weight to this inclination.

UK government has a total distaste for statutes which enable outsiders to sue it – regardless of whether or not the public interest might benefit from litigation.

There is substantial resistance by Government, to which DOE is attentive, to the introduction into the planning system of impediments to industrial development – apparently because of the recession.

There is resentment in many quarters at the 'success' of environmental bodies in the 1970s and a corresponding wish to reduce this influence. Recent Government statements on public inquiries provide evidence of this.

As a result of fears about the litigation spawned in the early years of NEPA, discussion of EIA in the UK has become increasingly specialized and 'methodological'. In my view, this has led to a limited and restrictive view of the impacts which EIA ought to consider and of EIA's potential role.

As a corollary of this latter tendency – unintended, no doubt – non-specialist outsiders have been in practice excluded from discussion of EIA, as this has concentrated ever more on 'scientific' methodology, and less and less on the political dimension – the underlying policy dimensions which lay behind EIA and NEPA in the United States.

Collectively, these factors do not encourage one to conclude that thinking of the kind that underlay NEPA will find a warm welcome in Britain in 1980.

(b) The planning context

There are two basic approaches to EIA, in relation to present planning processes, by its advocates in the UK. First, there are those who see EIA as a *supplement* to existing planning procedures – to fill gaps or deficiencies in those procedures without major alterations. These range from Brian Clark's team at Aberdeen University to initiatives of companies such as the British Gas Corporation, BP and BNFL. There is a spectrum of approaches and intentions to be found here – but all see EIA as a supplement, whether because it's good for planning or good for business. These moves are admirable. More description of impacts is plainly desirable; I shall return to the shortcomings as I see them in a moment. Second, there are those advocates who see EIA as potentially *substituting* for aspects of existing planning procedures, for example Catlow/Thirlwall and others. They would like a much higher degree of local authority/developer collaboration on EIA in advance of a proposal.

I have worries about both approaches. They arise from my reading of the political factors I described earlier.

Take the first approach. If the EIA process is seen as being descriptive only (a checklist or something more prosy), it stands to make only a relatively small contribution to the kinds of purposes (and problems) identified in NEPA. Does not the need for EIA arise, in this country as in the US, from a recognition of a quite new order of problems for our political systems – from conflicts between man and nature, between the escalating demands of industrial society and the limits of the biosphere and its resources? If so, won't the conception of EIA as a checklist of impacts at project level make only a tiny contribution to addressing these problems?

I stress, CPRE wholely favours such reforms for their own sake – and we have consistently pressed for them. But their very modesty (some would say 'realism') must call into question their effectiveness for addressing the broader problem.

Let us move then to the second approach to EIA. This is more ambitious – a way of substituting for the capriciousness of our existing planning system, and in particular for its lack of exhaustive and systematic elaboration of the full impacts a proposal can have and of alternatives to it.

This causes quite new problems in a UK context. For the greater the stress placed on early collaboration between local planning authorities and developers in advance of public inquiries, the greater the potential erosion of public influence and indeed of established political processes. EIA will be dangerous if it is seen as a means of taking the heat out of adversarial inquiries. And this could well be the case with this approach.

Let me illustrate this by pointing to one or two of the problems that could arise.

Collaboration might tend to mean that what the developer and planning authority believe to be the 'relevant' impacts will already have been identified before the EIA becomes a public process. There could be a consequent tendency to underrate, or even exclude from serious examination, impacts which have not been so identified – but which individuals and organized groups outside the EIA process might wish to pursue.

EIA on this basis would tend to encourage the promotion of pseudo-scientific quantification of impacts, which would in turn be presented to the inquiry as matters of fact. We need to look no further than Government proposals for motorways to see how damaging and phoney such quantification can be.

Or again, collaborative EIA will appear to involve rigorous examination of alternative sites or methods. The sheer volume and weight of paper generated by EIA will tend to pre-empt discussion of alternatives at the public inquiry. But there will be no guarantee that such canvass of alternatives has in fact been rigorous or comprehensive.

Finally, where a particular local planning authority favours a development, it will have no incentive to press for a truly exhaustive EIA.

The greater the collaborative element, the greater the trust required (on the part of environmental groups and the public at large) in the competence and *far-sightedness* of the local authority.

But remember, the likelihood is that EIA on this basis would not be introduced in the context of binding and enlightened statutory duties like those of NEPA, recognizing 'the profound impact of man's activity in the interactions of all components of the natural environment, particularly the profound influences of population growth, high-density urbanization, industrial exploration, resource exploitation and new and expanding technological awareness'. There would thus be no way of ensuring, through the courts for example, that such intentions were being addressed fully by the LPA and/or the developer. And indeed, realism suggests that they would almost certainly *not* be addressed comprehensively.

Furthermore, for the reasons I have given, the risk would be created of undermining the effectiveness of the sole adversarial process (the public inquiry) for which UK procedures *do* now provide. The adversarial element in public inquiries is valuable. The developer and his witnesses are subjected to cross-examination and alternatives can be put forward and parties compelled to look at them. These are crucial roles of the public inquiry, which ought not to be put at risk for hypothetical advantages of EIA of the collaborative kind. Time and again the adversarial elements in our planning processes have proved their value.

4.6.4 Alternative solutions

It is possible therefore for a believer in the profound importance of EIA to be sceptical about both approaches to it which have so far been advanced authoritatively in the UK.

What more promising initiatives might be proposed? I think we should restore to the centre of our thinking NEPA's recognition that the need for new procedures for the assessment of the environmental implications of industrial developments arises from the quite new situation into which industrial societies have led themselves. What have been called 'the politics of development' are putting increasingly severe strains on our national – and global – physical and social resources. Yet it is still not public policy in this country to address systematically the implications of this fact for future development policies. The truly central importance of environment is still given minimal recognition by our political system, where future investment is at issue.

Increasingly, it is the policies of national governments and the EEC which are the prime movers of industrial development in the UK. Policies for energy supply, for roads, for regional development, for airports, for tourism, for agriculture – the state is now behind them all.

EIA should be directed to these *policies*. We need a process whereby systematic consideration of the full environmental dimension of these policies can be explored and unravelled, at a stage when options are still open. It would need to be an open and public process, possibly attached to Parliamentary processes, but requiring candour and open-mindedness by the executive and its agencies. By addressing environmental impact at the formative stages of policy-making, Government could begin to create a new consensus about the direction of industrial society.

It is no coincidence that EIA at the *policy* level has been advocated repeatedly by bodies like my own, to Catlow and Thirlwall, to the EEC, to central government and to Parliamentary Select Committees. It is a sad reflection that the implications and potentialities of policy-level EIA have barely been addressed by the professional and administrative communities in Britain.

4.6.5 Conclusions

I conclude therefore by expressing the hope that such discussion on EIA at the policy level will gain ground now. By all means let us have more information about environmental impact at a project level. But let us not pretend that this addresses significantly the fundamental problems the authors of NEPA had in mind in 1969. A response by our political system to those problems lies ahead of us. Its importance has never been greater.

REFERENCES

Anderson, F.R. (1973) *NEPA in the Courts: A Legal Analysis of the National Environmental Policy Act*, Resources for the Future, Inc., Johns Hopkins University Press, Baltimore and London.

Arbuckle, J.G., Schroeder, S.W. & Sullivan, T.F.P. (1974) *Environmental Law for Non-lawyers*, 2nd edn, Government Institutes Inc.

Battelle Institute (1978) *The Selection of Projects for EIA*, Commission of the European Communities Environment and Consumer Protection Service, Brussels.

Birks, J. (1979) *J. Petrol. Technolog.*, **31**, 287–290.

Canadian Council of Resource and Environment Ministers (1978) *Canadian EIA Process*, EIA Task Force, April 1978.

Catlow, J. & Thirlwall, C.G. (1976) *Environmental Impact Analysis*, DOE Research Report 11, Department of the Environment, HMSO, London.

Caverley, D.S. (1977) in *Proc. Conf. EIA Instit. Environ. Studies*, University of Toronto, Toronto.

CEQ (1976) *Environmental Impact Statements. An Analysis of Six Years' Experiences by Seventy Federal Agencies*, Report of the CEQ to the President of the United States, March 1976.

CEQ (1978) *National Environmental Policy Act. Implementation of procedural provisions; Final Regulations*, Federal Register, Wednesday, November 29, Part VI, 43(230): 55977–56007.

Clark, B., Chapman, K., Bisset, R. Wathern, P. & M. Barrett (1981) *A Manual for the Assessment of Major Industrial Proposals*, Department of the Environment, HMSO, London.

Dean, F.E. and Graham, G. (1978) in *Proc. UNECE Symposium on the Gas Industry and the Environment*, Minsk, USSR.

Dobry, G. (1975) *Review of the Development Control System: Final Report*, HMSO, London.

European Commission (1976) The introduction of environmental impact statements in the European Communities (ENV/197/76) Brussels.

European Commission (1978*a*) The selection of projects for environmental impact assessment (ENV/579/78) Brussels.

European Commission (1978*b*) Environmental impact assessment of physical plans in the European Communities (ENV/37/78) Brussels.

European Commission (1978*c*) Methods of environmental impact assessment for major projects and physical plans (ENV/36/78) Brussels.

Effer, W.R. & Malvern, R.J. (1976) *Environmental Aspects of Nuclear Power Plant Siting*, Engin. Inst. Canada Regional Tech. Conf., Winnipeg.

Emond, D.P. (1978) *Environmental Assessment Law in Canada*, Amond-Montgomery, Toronto.

Environment Canada (1976) *Guidelines for Preparing IEE*, Issued by Environmental Assessment Panel, Environment Canada.

Environment Canada (1979) *Revised Guide to the Federal EARP*, Issued by Environmental Assessment Panel, Environment Canada.

Heer, J.E. & Hagerty, D.J. (1977) *Environmental Assessments and Statements*, Van Nostrand Reinhold, New York.

House of Commons (1981) Parliamentary Debates official report, 9 June 1981, HMSO London.

House of Lords Select Committee on the European Communities (1980) Second report, *Environmental assessment of projects*, HMSO, London.

Jack Holmes Planning Group (1974) *An Examination of Sites for Gravity Platform Construction on the Clyde Estuary*, Glasgow.

Lee, N. (1982) *J. Environ. Mgmt.*, **14**, 71–90.

Lee, N. & Wood, C.M. (1978) *J. Environ. Mgmt.*, **6**, 57–71.

Lee, N. & Wood, C. M. (1980) *Methods of Environmental Impact Assessment for Use in Project Appraisal and Physical Planning*, Occasional Paper No. 7, University of Manchester.

Leonard and Partners (1977) *The Belvoir Prospect*, London.

Leopold, L., Clark, F.E., Hanshaw, B.B. & Balsley, J.R. (1971) *US Geological Survey*, Circular 645, Washington DC.

Manners, I. R. (1978) *Planning for North Sea Oil: The UK Experience*, Policy Study No. 6, Center for Energy Studies, University of Texas.

Munn, R.E. (ed.) (1979) *Environmental Impact Assessment: Principles and Procedures*, John Wiley & Sons, London.

Official Journal of the European Communities (1973) *C122*, 20 December 1973.

Official Journal of the European Communities (1977) *C130*, 13 June 1977.

Official Journal of the European Communities (1980) *C169*, 9 July 1980.

Official Journal of the European Communities (1981) *C305*, 25 November 1981.

Official Journal of the European Communities (1982) *C110*, 1 May 1982.

Ontario Ministry of the Environment (1973) *Green Paper on Environmental Assessment*, September 1973.

PADC (1976) *The Assessment of Major Industrial Applications – A Manual*, DOE Research Report, No. 13, Department of the Environment, HMSO, London.

PADC (1978) *Environmental Impact Assessment in the USA: A Critical Review*, DOE Research Report, No. 26, Department of the Environment, HMSO, London.

Quarles, J. (1979) *Federal Regulation of New Industrial Plants*, Morgan Lewis and Bochius, Washington DC.

Rifkind, M. (1980) Speech at Hamilton, Scottish Office Press Notice 46/80.

Royal Commission on Environmental Pollution (1976) *Fifth report*, Cmnd. 6371, HMSO, London.

Scottish Development Department (1977) *National Planning Guidelines for Large Industrial Sites and Rural Conservation* and *National Planning Guidelines for Petrochemical Developments*, Edinburgh.

Shore, P. (1978) Speech at Manchester, September 1978.

Sphere Environmental Consultants (1974) *Loch Carron area: Comparative Analysis of Platform Construction Sites*, SEC, London.

Stevens Committee (1976) *Planning Control over Mineral Working*, HMSO, London.

Turnbull, R.G.H. (1979) *Environmental Appraisal for Future Petrochemical Developments at National and Regional Levels in Scotland*, UNECE, Villach, Austria.

Viohl, R.C. & Mason, K.G.M. (1974) *Environmental Impact Assessment Methodologies: An Annotated Bibliography*, Council of Planning Librarians, Exchange Bibliography No. 691.

Wandesford-Smith, G. (1978) *Environmental Impact Assessment in the European Community*, International Institute of Environment and Society, Berlin.

Wood, C. (1983) *Planning Outlook* (in press).

CHAPTER FIVE

Ecological Considerations in Rural Planning

5.1 OVERVIEW

Both planners and ecologists make a broad range of contributions to the management of the rural environment. However, it is only in a relatively few areas that these activities overlap. This is largely because the responsibility of planners under the *Town and Country Planning Act* (1947) are fairly limited in rural areas. Nevertheless, this interaction may increase considerably if the increasing need for arbitration of sectional interests in the countryside (such as agriculture, forestry, conservation, recreation etc.) is met by an extension of the planning legislation as many have suggested (Shoard, 1981).

Rural planning in the UK has largely been preoccupied with the problems of containing physical growth. In fact the *Town and Country Planning Act* of 1947 was largely a response to the uncontrolled urban sprawl of the 1918–1939 period. This Act gave planners only limited powers in rural areas as it was thought that the traditional practices of agriculture and forestry would not conflict with other interests such as conservation, recreation etc. Planning legislation covers rural infrastructure (roads, village developments etc.) and control over farm buildings where the ground area is $>465\ m^2$ or the height is above 12 m. The main planning consideration, however, was that high grade agricultural land should not be taken for development purposes. To assist planners, the Ministry of Agriculture, Fisheries and Food (MAFF) developed the Agricultural Land Classification which graded land on the principle that, in times of changing farming priorities, the high quality land was related to the flexibility it allowed in supporting a broad range of crops. Local authorities are bound to consult with MAFF where proposed developments on green field sites are over 10 acres. Unfortunately, this constraint on development has often been misunderstood (e.g. development allowed on all but Grade I land despite Grade II or III land being very productive for specific uses). Consequently, urban encroachment has continued and agriculturalists argue that the current rate of land loss ($\sim$50 000 ha per annum) is unacceptable. About 30% of this has been attributed to urban development (of the remaining, 50% has been lost to upland forestry and 20% due to readjustment of the statistics). However, the loss to urban or industrial development has largely been on the better quality land. To compensate for this loss, farming has intensified and also extended into marginal areas. Both these processes have heightened conflicts with other rural interests and both of these land-use changes are largely outside the framework of present planning legislation.

Conflicts with conservation interests have arisen as intensification of better quality land has led to (a) removal of hedges, (b) machine maintenance of remaining hedges, (c) uncontrolled use of agrochemicals leading to river eutrophication, pesticide and herbicide toxicity on non-target organisms etc. (see Green, Section 5.2). Extensification of agriculture has resulted in reclamation of marginal land of high conservation value, particularly heathland, moorland and ancient woodland (see Goode, Section 5.4). Forestry also competes for land in marginal areas and the recent proposals to double the forestry area has focused attention on the consequences of large-scale afforestation. These consequences include (a) loss of wildlife habitats, (b) effects on fisheries through increased siltation and acidity of streams and freshwater lakes and (c) loss of landscape or recreation value.

Changes in both agriculture and forestry practices have produced conflicts with recreational interests through reduced accessibility, visual amenity or landscape value of rural areas.

Ecological principles play a central role in the modern scientific approach to farming and forestry. The concepts of sustained yield, the instability of monocultures, the inter-relationships between climatic, edaphic and biotic factors are examples of the application of ecology to maximizing productive output. In addition, the effects of fertilizer runoff, pesticide and herbicide drift, food-chain accumulation of pesticides, effects of afforestation on the hydrological cycle etc., are all predictable given a sound understanding of ecological processes.

Ecological considerations in the planning process have not, however, received so much attention by professional ecologists in the past. This may have been because planners have not articulated their need, or ecologists have perceived the interaction to be subjective assessment and therefore, not subject to rigorous scientific understanding. Nevertheless, there are planning matters in the rural context (outside of strategic considerations and specific developmental control dealt with under Chapter 6 and 7) which should utilize ecological information. The more obvious of these include (a) management of local nature reserves and countryside recreational areas, (b) planting and afforestation schemes in relation to landscape design, (c) woodland management, (d) road-verge management, (e) effects of waste disposal, (f) reclamation and restoration schemes and last but not least (g) environmental education.

The four main levels at which ecological information can be utilized are described by Bunce and Heal (Section 5.9). Environmental description (including climatic and edaphic constraints, current land-use and potential for a range of uses) is an integral part of sound environmental management and planning optional land-uses. Monitoring the distribution and abundance of plant and animal communities in areas of different land-uses and the responses to changing land-use are an essential basis for positive management. Prediction of changing priorities, constraints and developments in rural land-use are in the realm of forward planning but prediction of the environmental consequences must be based on sound ecological information. Optimization of land use by zoning for priority uses, or planning for multiple-uses must be based on an understanding of the interactions of the main land-uses. About half the county councils employ ecologists whereas the remainder incorporate ecological information by consultation with the relevant body (e.g. NCC, Countryside Commission, Local Trusts etc.).

At present, ecological information on rural areas has been incorporated in the planning process at the Structure Plan level by a variety of environmental assessment methods. A recent survey (Selman, 1981) showed that half the Counties in England and Wales had habitat maps, 80 % had some form of ecological evaluation and 75 % had the Nature Conservation Review (Ratcliffe, 1977). However, most planning applications are dealt with by the District Councils which only have a statutory requirement to consult the NCC where NNR's or SSSI's may be affected. According to Selman (1981), less than 30 % of District Councils have standard criteria for the incorporation of further ecological information. Ecological information is often utilized solely in the protection of conservation sites in the context of development control; with the consequence that planners often see environmental considerations as a restraint on land-use planning. In addition, conservation groups are criticized for not being able to reduce the complex value judgements involved in designating areas for conservation to simple indices (see Goode, Section 5.4).

The main conflicts in rural land-use stem from changes in management practices outside statutory planning control rather than from problems of positive uses of ecological information in the planning process. The balance between the main rural land-uses (agricultural, forestry, water resources, conservation and recreation) are largely decided by sectional interests of single-purpose agencies at Central Government level. This has resulted in government grants being available to destroy habitats which other agencies are trying to protect. This situation has arisen primarily as there is no assessment of the consequences of policy decisions effecting the rural environment. Various approaches have been proposed to minimize these conflicts (see Green, Section 5.5). Some see that the conflicts operate at the local level and should be dealt with by an extension of responsibility of the planning authorities to cover major land-use changes in rural areas. These authorities would then operate by zoning areas for priority uses or the development of multiple-use policies. Others propose a zoning of agriculture/forestry priority areas and a strengthening of the NCC/Countryside Commission to manage zones for conservation and amenity use. Meanwhile, the Wildlife and Countryside Bill (see Goode, Section 5.4) has improved the consultative procedures between government agencies and given support to management agreements for the protection of important sites. It will be interesting to see how this works in practice.

5.2 THE IMPACTS OF AGRICULTURE AND FORESTRY ON WILDLIFE, LANDSCAPE AND ACCESS IN THE COUNTRYSIDE

B. H. Green

5.2.1 Introduction

There can be little doubt that the practice of farming and traditional woodsmanship moulded the British countryside into an exceptionally pleasant environment; one arguably richer in wildlife, landscape and recreational opportunity than the almost continuous tracts of woodland, marsh, fen and peat bog which was, and still would be, its natural state. It is equally certain that modern agriculture and forestry are destroying this environment and its amenities. It is true that people have always looked back to a rural arcadia and lamented its passing (Williams, 1973). But the current changes represent the culmination of a long process of erosion of our natural and semi-natural ecosystems which is taking place just at the time when there is a widespread appreciation of wildlife and countryside and a growing demand for rural recreation and conservation (Table 5.1).

5.2.2 The impacts of modern agriculture

The 1947 *Agriculture Act* guaranteed prices for the major farm products. Grants and subsidies were also later made available for materials such as fertilizer and improvements including drainage, ploughing, scrub and hedge clearance and new buildings. Scientific and technological developments in crops, machines, husbandry and crop protection provided the tools for the expansion catalysed by the state support for the industry. As a result agriculture has been successful in not only keeping up with increasing population and nutritional standards but in increasing our self-sufficiency in temperate foodstuffs, now over 70% compared with about 30% immediately pre-war (Beresford, 1975). Increased self-sufficiency remains a prime objective of agricultural policy (MAFF, 1979*a*) despite our entry in 1972 into the European Economic Community with its surpluses of many foodstuffs.

Table 5.1 Membership of Countryside Recreation Organizations (taken from Digest of Countryside Recreation Statistics (1978) Countryside Commission, CCP86 Cheltenham)

	1950	1960	1965	1970	1972	1975	1976	1977
County Nature Conservation Trusts	800	3 006	20 960	57 000	74 815	106 759		115 328
Royal Society for the Protection of Birds	6 827	10 579		65 577	128 528	104 997		244 841
National Trust	23 403	97 109	157 581	226 200		539 285		613 128
Ramblers' Association	8 778	11 300	13 771	22 818		31 953		29 541
National Federation of Anglers			394 653	354 401				446 136
Royal Yachting Association	1 387	10 543	21 598	31 089		36 368		52 140
British Field Sports Society	27 269	20 250	18 401	20 965		43 000		55 000
Wildfowlers' Association of GB and Ireland				21 255		30 815		34 412
British Horse Society	4 000	6 000	10 000	17 000			22 500	25 500
Pony Club	20 000	30 000	29 000	33 300		45 500		49 500

Increased production has been brought about in two main ways. Production on the better soils has been intensified, and wild or marginal agricultural land, like wood, marsh, heath, moor and down have been enclosed and reclaimed to arable or improved pasture. Increased efficiency is most readily achieved by taking land out of livestock production into tillage. Energy losses along food chains mean that as much as ten times as much food per unit area can be gained from plants as from animals. Abundant subsidized inorganic fertilizer has both freed the arable farmer from the necessity of keeping stock for their manure and enabled light, infertile soils to be cultivated. Greater tractive power and new machines have overcome both problems of ploughing heavy soils or poor terrain and drainage difficulties which could not be tackled in the past. Hedgerows have lost their main function on farms without livestock to be contained, or where they are kept mostly indoors, or grazed under new systems controlled by electric fences. On arable farms they are just costly hindrances to the big machines. New pesticides and herbicides have controlled diseases and enabled long periods of continuous cultivation which have replaced traditional rotations. There have also been big changes in farm size, ownership and the working of the land. Much more land is now farmed in larger holdings, more holdings are owner-occupied and over 300 000 people were shed from the agricultural workforce between 1945 and 1977.

5.2.3 Loss of hedges

In becoming more productive there have thus been great changes in the pattern of farming and this has meant equally substantial changes in the pattern of the countryside. In England and Wales cropland has expanded from less than 25 % of the land area in 1939 to 38 % in 1971, and permanent grass has declined from 42 % to 26 % in the same period (Best, 1976). It has been estimated that in 1946 there were 800 000 km or 160 000 ha of hedge in England and Wales which was grubbed out at a rate of 8000 km per annum to 1963 (Pollard *et al.*, 1974). There has been a great deal of debate over the loss of hedgerows and its impact on wildlife and landscape. Although a substantial part of our lowland flora and fauna occur in hedges, particularly woodland species, none seem to be exclusively dependent on them. Nor is there much evidence that they act as routes through the countryside for wildlife. Where they still survive they are commonly burnt in stubble fires and sprayed with agricultural chemicals, either deliberately against pests or weeds like couch grass, or accidentally by spray and fertilizer drift, so that vigorous nitrophilous species like nettles (*Urtica dioica*) and herbicide resistant species like cleavers (*Galium aparine*) commonly predominate in them to the exclusion of most other herbaceous species. 'We must conclude that while hedges are essential for the survival of many species of birds on many individual farms in larger areas of countryside without woods, they are not essential for the survival of species in the countryside as a whole, either as providing breeding habitat or connecting corridors between suitable habitats' (Pollard *et al.*, 1974). The impact of hedgerow loss on landscape value may be more significant, but it cannot be readily quantified in the same way as wildlife losses.

The value of hedges to the farmer as sources of insects to pollinate crops and control crop pests is ambivalent. They do give some benefit but they also harbour pests and diseases like the broad bean aphid (*Aphis fabae*) which infests crop from spindle (*Euonymus europaeus*) and crown rust (*Puccinia coronata*) from buckthorn (*Rhamnus catharticus*). Their value as windbreaks is also limited and mainly confined to tall screens of exotic conifers around orchard crops (Pollard, 1969; Huband, 1969).

5.2.4 The effects of herbicides, pesticides and fertilizer

The use of agrochemicals has had far more serious impacts on wildlife than hedgerow removal. The widespread use of herbicides has not only been responsible, together with improved seed cleaning for the decline in the cornflower (*Centaurea cyanus*), corn marigold (*Chrysanthemum segetum*), the corn cockle (*Agrostemma githago*) and numerous less showy plants, but also of the insects and birds dependent on them as well (Fryer and Chancellor, 1974). The diverse swards of untreated pastures may contain as many as 30 or 40 species of vascular plants and up to 20 species of butterfly; the near monocultures of ryegrass (*Lolium perenne*) support no butterflies at all (Nature Conservancy Council, 1977). The decline of the common partridge (*Perdix perdix*) has also been attributed to the affects of herbicides on the food plants of the sawfly larvae on which its chicks depend heavily (Potts, 1971).

Agricultural use of pesticides, particularly the persistent chlorinated hydrocarbons like DDT have been responsible for some major reductions in wildlife populations. When these compounds came into general use in the late 1950s and early 1960s there were mass deaths of seed-eating farmland birds and their predators. Dieldrin, used as a seed dressing to protect against the wheat bulb fly, was largely responsible for the decline in sparrow hawks (*Accipiter nisus*) which became rare in much of lowland England. Aldrin, used in sheep dips led to the collapse of the golden eagle (*Aquila chrysaetos*) population in Scotland. By 1963 the UK population of the peregrine falcon (*Falco peregrinus*) was only 44 % of the 700 pairs breeding in 1939. (Ratcliffe, 1972; Newton, 1974). Industrial pollution from polychlorinated biphenyls has since been shown to have been a contributing factor in some of these population crashes. Voluntary agreements restricting, and eventually phasing out, the use of these chemicals have been implemented through the aegis of a government Advisory Committee on Herbicides and other Toxic Chemicals representing manufacturers, users and conservationists. An Agricultural Chemicals Approval Scheme and Pesticides Safety Precautions Scheme now give clearance to products which they have screened. As a result some species like the golden eagle have recovered well. Others like the peregrine falcon and otters have been much slower to recover and have yet to return to much of their former range. In the USA and many other countries the decline of wildlife populations due to pesticides and fungicides has been much greater than in Britain.

The third category of agricultural chemical to have had widespread environmental impact is the synthetic inorganic fertilizer. The use of nitrogenous fertilizers in England and Wales increased threefold between 1955 and 1970 and application rates on grassland are still less than optimum for maximum yield response (Tomlinson, 1971). The effect of the increased fertility is to encourage a few vigorous, competitive plants to the detriment of most others. The fall in pasture diversity which results is beautifully demonstrated in the classical Park Grass plots at Rothamsted (Fig. 5.12). Though grassland retains most of the nitrogen applied, crops may take up no more than 50% and the remainder commonly leaches into watercourses and aquifers. Here it has the same effect on aquatic plant diversity and that of associated animals. The decay of the greatly increased biomass of aquatic plants may cause deoxygenation and further reduce animal and fish populations.

Nitrate can be reduced to nitrite in the intestines of young babies and cause a sometimes fatal condition known as methaemoglobinaemia. The World Health Organization has recommended a limit of 50 ppm NO_3-N for water to be fed to young babies. This has been exceeded in British rivers, where sewage and nitrogen losses from converting pasture into arable greatly add to that from fertilizers. Even a small, seemingly unpolluted river like the Frome in Dorset discharged more than 500 tonnes of NO_3-N in 1972, an increase of 42% over the previous seven years (Natural Environmental Research Council, 1976). More than twice the WHO recommended limit was found in boreholes of the Eastbourne Waterworks Company on the South Downs in 1967. Although farmers were not exceeding Ministry of Agriculture recommended fertilizer-application rates, 25% of the nitrogen applied to the catchment was abstracted in the water supply (Greene and Walker, 1970). Recent studies have suggested that percolation through aquifers may be very slow indeed (Royal Commission on Environmental Pollution, 1979). The nitrate concentrations now emerging might only represent the contamination which took place some 25–30 years ago when fertilizer-application rates were only just beginning to increase.

5.2.5 Reclamation of marginal land

Until recently discussion on agricultural impacts on the environment has centred mostly on the intensification described above. Surveys have, however, begun to show the great extent of enclosure and reclamation of down, heath and moor to improved pasture or arable land. Nearly 50 % of Wiltshire downland (26 000 ha) was lost between 1937 and 1971 and 25 % of Dorset downland (4500 ha) between 1957 and 1972. Even in National Parks there has been substantial loss of the heath and moorland which is their main wildlife and landscape resource. Exmoor lost 20 % (5000 ha) between its designation in 1954 and 1975 and the North York Moors lost 20 % (15 000 ha) between 1952 and 1975 (Shoard, 1976). More recent studies by the NCC show the losses mounting. Downland in Hampshire and Sussex has declined by 20 % since 1966 and heathland in the south even more. This loss of habitat leads directly to loss of species and the fragmentation and isolation of habitats which remain threaten others which require specific habitat features or large areas. In 1930 the snakeshead fritillary (*Fritillaria meleagris*) occurred in 116 10 km squares; with loss of meadows it now occurs in only 17. In the last ten years sand lizard (*Lacerta agilis*) localities on southern heathlands have fallen from 159 to 42 (Goode, 1981).

Agricultural drainage and infilling of wetlands ranging from farm ponds to the huge schemes which now threaten the Norfolk Broads and Somerset Levels has also led to loss of wildlife. It is probably for this reason that the common frog (*Rana ridibunda*) is no longer at all common over much of lowland England and why many waders, wildfowl and other species of wetlands are also less common. Habitat loss, particularly that attributable to agriculture and forestry, is responsible for much of the substantial decline in the British flora and fauna which has been widely documented (Perring, 1974; Hawksworth, 1974; Sharrock, 1976). Not all species have declined. Some common, adaptable species like the woodpigeon (*Columba palumbus*) have been favoured by agricultural change and increased; some, like this species, to pest status.

5.2.6 The impacts of modern forestry

It is to forestry that we owe the sustained yield concept of conservation, developed by Pinchot, the first director of the United States Forest Service. Forestry there and in European countries like France and Germany has been primarily concerned with the planned exploitation and regeneration of the natural woodland with which they are well endowed. Traditional woodmanship in Britain was similar but modern forestry here has been much more concerned with the planting and subsequent clear-felling of rapidly growing trees, usually exotic species introduced from other parts of the world. Its practice is much more like farming than traditional woodmanship and there is the same dependence on monocultures and use of herbicides, pesticides and fertilizers to minimize competition and maximize production. Forestry is, however, a much longer-term enterprise than agriculture. Calculation of whether the effort and resources expended will have proved worthwhile when the crop matures in the distant and unpredictable future is difficult. It is often said that when afforestation began in Britain in the sixteenth and seventeenth centuries it was to provide timber for the navy, but by the time it was mature the navy had gone over to iron ships.

Some have argued that it will not prove economic to grow our own timber and they see the functions of state-forestry and state-supported private forestry as the creation of employment in depressed hill areas or the provision of recreational and tourist facilities in the forests (Richardson, 1970). Considerable provision has recently been made for these interests but the growing of wood and development of the home timber industry remain the overriding objectives of forestry policy. Only 8% of the UK demand for wood products is supplied from British forests and the import bill of £2370 million for the remainder is the third most expensive item in our overseas trading account. Recent studies suggest that up to 2·1 million ha of poor hill land could be afforested with little loss to agricultural production (Forestry Commission, 1977; Centre for Agricultural Strategy, 1980). It would seem unlikely therefore that there will be any major change in the policy to maximize home timber production.

There are three ways in which production can be increased. Firstly, output from existing forests can be boosted by improved silvicultural techniques, new crop species, tree breeding and the use of fertilizers and other more intensive methods of production. Secondly, areas of 'unproductive scrub' can be converted into productive forest. Thirdly moorland, heathland and other open areas can be afforested. All these procedures involve the use of conifers if production is to be maximized. A crop of native hardwoods like oak (*Quercus* spp) or beech (*Fagus sylvatica*) may be worth as much, or more than one of conifers such as Corsican pine (*Pinus nigra*) or Sitka spruce (*Picea sitchensis*), but they grow at only half their rate (4–6 cf. 10–20 m^3 ha^{-1} an^{-1}). The hardwood crop is harvested in about 140 years and the conifers at 60 years. Discounted revenues can thus make the conifer up to 10 times as profitable (Aldhous, 1972).

Norway spruce (*Picea abies*), silver fir (*Abies alba*) and European larch (*Larix decidua*) formed much of the Forestry Commission's early plantings; now Sitka spruce is replacing Norway spruce as the most widely planted tree on the peaty moorlands and poor grazings in the north and west where much afforestation has taken place. Other species from western North America, including Douglas fir (*Pseudotsuga menziesii*), western hemlock (*Tsuga heterophylla*), and lodgepole pine (*Pinus contorta*) have also been used. Lodgepole pine does well even under the most extreme conditions of north Scotland.

There has been much criticism of this afforestation in the hills. Early objections were on landscape grounds reflecting a widespread dislike of the 'serried' and 'regimented' ranks of dark, even-aged monocultures marching over the hills in straight-edged blocks. There have consequently been calls for planning control of forestry, at least in National Parks (Sandford, 1974; Price, 1976). However, recent plantings have been much more sympathetically contoured and disguised with 'cosmetic strips' of hardwoods or fire-resistant conifers like Japanese larch (*Larix leptolepis*), and there has been much provision of nature trails, picnic sites, holiday cabins and other recreational facilities in the forests. It is worth remembering too that Forest Parks are the oldest designated amenity lands in Britain. New plantations on open ground are rich in small mammals and are thought to have greatly helped the recovery from game-keeping persecution of species like the hen harrier (*Circus cyaneus*) and short-eared owl (*Asio flammeus*). Mature plantations are dense-canopied with a sparse, limited and similar ground flora under a variety of crop tree species (Hill and Jones, 1978; Anderson, 1979). The spread of conifer woodland cover has, however, led to increases in the population of species such as pine marten (*Martes martes*), polecat (*Mustela putorius*), wild cat (*Felix sylvestris*), sparrow hawk (*Accipiter nisus*), capercaillie (*Tetrao urogallus*) and crossbill (*Loxia curvirostra*).

The more recent concern of conservationists over afforestation of the uplands has been with the great expanses which it is now proposed should be planted. Although much of the land is of low agricultural and wildlife value fears have been expressed that the open moorland habitats of birds like golden eagles and merlins (*Falco columbarius*) will be lost, and that habitats never previously forested like blanket bogs will be drained and planted. Forestry has also been attacked on other grounds. It has been argued that conifers build up mor humus leading to podsolization and loss of soil fertility exacerbated by the clear felling of the crop and its nutrient load. The evidence on this is conflicting (Ovington, 1965; Ford *et al.*, 1979). It is well-established that forest catchments yield much less of the precipitation falling on them as runoff than does open vegetation, conifers less than hardwoods (Ovington, 1965; Centre for Agricultural Strategy, 1980). Although the flow from afforested catchments is less flashy and may help prevent floods, the loss of water may mean expensive additions to hydroelectricity and water supply projects.

Existing woodlands which are converted into conifer plantations lose most of their associated wildlife and amenity value. Rich assemblages of woodland plants and animals are mainly associated with deciduous trees and with the ways by which deciduous woods were traditionally managed. A large variety of plants, and animals dependent on them, thrive under the open conditions of coppiced woodland, probably because the open seral and edge conditions of the coppiced compartments in various stages of regrowth resemble the glades in natural woodland created by windthrow, fire, flood and avalanche and maintained by grazing animals. Plant species like bluebells (*Endymion non-scriptus*) and wood anemones (*Anemone nemorosa*) which complete much of their photosynthesis in the spring before the canopy closes are particularly favoured for the dense coppice canopy later in the year penalizes vigorous competitors like bracken (*Pteridium aquilinum*) and brambles (*Rubus fruticosus*). The diversity of plants and the layering and structure of coppiced woods makes them rich in invertebrates and birds. Nightingales, for example, reach their highest breeding densities in the earlier stages of regrowth in coppiced woods (Stuttard and Williamson, 1971). Wood pastures with pollards are rich in lichens, for epiphytic communities take a long time to develop and the pollard boles, having been spared felling, are often very old. They therefore also often bear holes and deadwood, both vital resources to many birds and invertebrates. If fallen and decayed trees are removed from a wood the ecosystem is likely to be deprived of 20 % of its fauna (Elton, 1966).

Conversion into conifer production removes nearly all these features of traditionally managed woodlands. It is not quite such a devastating loss as the grubbing of entire woods for agriculture, but a much more serious impact than the deceptive retention of a tree cover may lead one to believe. It has been estimated that one third to a half of the ancient woodland extant in 1945 has now gone; more woodland destruction than in the previous 400 years (Rackham, 1976). This loss has to be balanced against the gains in natural secondary woodland which have resulted, in this era of modern agriculture, through the abandonment of uneconomic heathland, moorland and downland grazings. The seral scrub communities which develop through natural succession can be important habitats, particularly for small birds like warblers, and are amongst the most popular habitats for picnicking and other kinds of informal recreation. But secondary woodlands of this kind are not a scarce resource and take a very long time to become colonized by many of the more typical species of ancient woodland, though their mobility varies (Peterken, 1974*a*). Conservation effort has, and should, concentrate on maintaining the much rarer heathlands and downlands from which they originate.

5.3 LANDSCAPE EVALUATION AND THE IMPACT OF CHANGING LAND-USE ON THE RURAL ENVIRONMENT: THE PROBLEM AND AN APPROACH

R. G. H. Bunce and O. W. Heal

5.3.1 Introduction

Rural land-use covers 85 % of the land surface of Britain and is continuously variable in its composition, reflecting the heterogeneity of the underlying environment and the varying interaction between management and that environment. Herein lies the problem, in that changes are subtle and can occur imperceptibly in both time and space. Moreover the small-scale mosaics in Britain are in marked contrast with some of the wide patterns present in the European plains.

In the present paper the traditional ways of analysing change are described, and a recent quantitative approach developed in the Institute of Terrestrial Ecology (ITE) is discussed because it is particularly suited for examining changes on a strategic scale.

The rural land-use patterns in Britain have been determined by the interaction of man with the vegetation cover which developed since the Ice Age, in response to the changing climate. The initial pattern was set by the limitations of the natural environment. Since the Industrial Revolution, however, other pressures have modified the basic pattern and although many of the underlying correlations between land-use and the physical environment still exist they have been distorted by industrial, social and economic factors. The detail of the present pattern results from the major land-uses of agriculture and forestry, the developments of which are determined by the interests of the land owners and managers and the policies of major agencies e.g. the Forestry Commission (FC), the Ministry of Agriculture (MAFF), the Water Authorities, the Nature Conservancy Council and the Countryside Commission.

The fundamental problem is that the rural land is not capable of satisfying the national demands and the conflicts between forestry, agriculture and amenity are therefore likely to increase, particularly in relation to predicted forest requirements (Centre for Agricultural Strategy (CAS), 1980). Rural land-use, to a large extent, is outside the statutory planning procedures. Thus, while at government level it is necessary to develop national plans for agriculture, forestry, water, amenity and conservation, the compatibility of such plans and the possibility of alternative combinations of development is not formally assessed. Nor is there an overall assessment of the impact of changes in rural land-use. There are many independent detailed studies of the effects of particular processes, e.g. the use of pesticides or herbicides or the removal of hedgerows and small woodlands, as well as studies of the effects of changing land-use in particular localities. Such studies represent an uncoordinated assessment in response to the many subtle, gradual and often long-term changes (e.g. in grazing pressures) and the effects of continuous arable cropping or forestry. However worthwhile these uncoordinated assessments, they have little influence on major trends in land-use. Concern is reflected in various papers and reports considering interaction between land-uses (Strutt, 1978; Westmacott and Worthington, 1974; Porchester, 1977; ITE, 1978; CAS, 1980; Blacksell and Gilg, 1981). Whilst it is necessary to consider the consequence of site specific changes in land-use, as in the Exmoor National Park (Porchester, 1977), the focus of attention on individual cases ignores the general conflict between the supply and demand for land. Each national agency predicts what its requirements will be (MAFF, 1979*b*) but they rarely specify *where* changes will or should occur.

Thus it seems necessary to assess the combined land-use requirements, then to compare those with the land-use potential and determine the consequences of selecting particular options. Ecology can help in the initial definition of the land-use potential, as well as in the analysis of impacts because the production potential for agriculture, forestry and game, and the

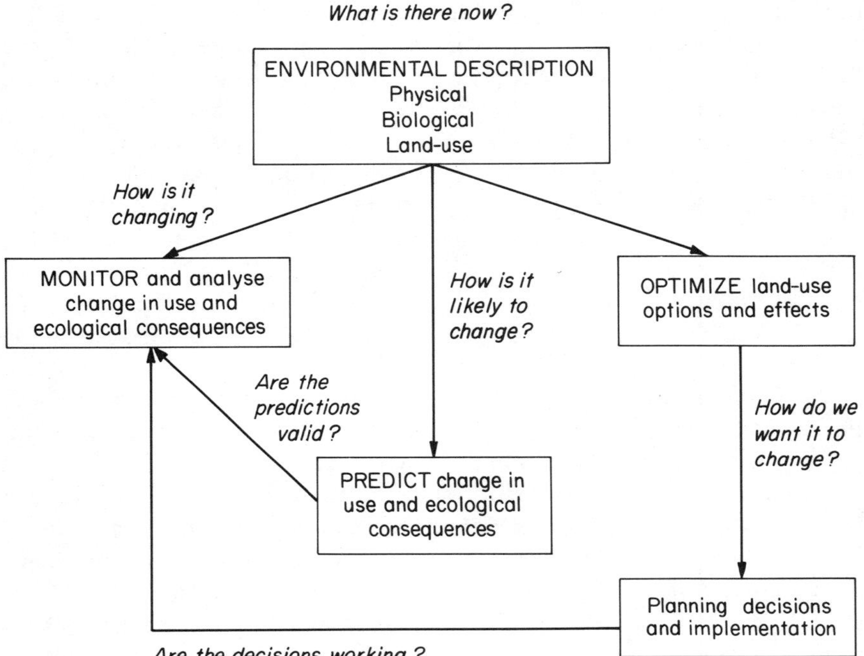

Fig. 5.1 Interrelationships between stages in procedures of environmental assessment.

carrying capacity for recreation, is determined by the same physical and biological characteristics of land which determine the response of the flora, fauna and soil to management practices. Four distinct processes, with associated questions, are recurrent in land planning (Fig. 5.1), and are used as a framework for the present paper: environmental description, monitoring, prediction and optimization.

5.3.2 Environmental description

The first process in environmental assessment is to define the current state of the system as a baseline on which to assess the likely changes and their impacts. In rural environments this involves defining:

(1) The physical environment which is relatively unchanging and which determines the response of the system to an impact.
(2) The land-use, mainly agriculture and forestry, which represents the impact factor.
(3) The biological environment which is relatively flexible and which responds to the impact factor.

In Britain, current national coverage of these attributes is so diffuse that a comprehensive ecological land-classification system, based on existing data, has not been produced. What is required is a procedure using explicit environmental criteria as a baseline for field survey to facilitate definition of the composition of rural landscapes on a national and regional scale.

On a national scale the individual features of the physical environment are reasonably well described. Physiographic information is given in the Ordnance Survey (OS) maps; climatic factors are covered by meteorological maps (e.g. Chandler and Gregory, 1976), geology by maps at a scale of 1:625 000 and soils in England and Wales by maps at a scale of 1:1 000 000 (Avery *et al.*, 1975). In Scotland, with a slightly different soil classification, strategic and regional mapping at 1:250 000 will be published in 1982 (Bibby, 1980).

Thus there is a reasonable description, at a low level of resolution, of the basic physical features of the land. However, the pattern of national and regional variation resulting from the association of these features has received little study although it is this combination which determines the biological and land-use characteristics. Selected features of slope, climate and soil have been combined into an assessment of agricultural potential over England and Wales (MAFF, 1974). Such capability classifications are now widely used in many countries but, as emphasized by Bibby (1980), the field of capability interpretation is still in its infancy, and the Agricultural Land Classification of England and Wales is being revised.

The Forestry Commission has its own system of assessment of forestry potential based on climatic and soil characteristics (Toleman and Pyatt, 1974) and although maps are not generally available, the environmental criteria associated with yield classes of different crops are reasonably well established and therefore can be derived from maps of basic characteristics. The agricultural and forestry potential classifications are interpretations of the original physical environment data. Whilst they help in assessment of the potential for specific land-uses they are of little help for assessment of the potential for alternative land-uses or different management regimes because they do not retain the independent basic data.

The assessment of potential, based on environmental description, has been widely used in the analysis of impacts as in the case of the North West Water Authority study of alternative reservoir sites (Nelson *et al.*, Chapter 6) and in analysis of motorway routes and urban development in rural areas. In contrast to the physical features, data on the biological environment at the regional or national level are very sparse.

Traditionally, a great deal of effort has been put into descriptions of ecosystems and much of the ecological work at the beginning of the century (e.g. Smith, 1900), was concerned with vegetation surveys. However, despite the long experience, no national picture has yet been produced and in many ways much current ecological descriptive work is on a narrower basis than originally. At an even wider scale the potential vegetation map of Europe (Ozenda, 1979) shows what can be done once it is recognized that a low level of detail is acceptable in order to obtain consistent coverage over large areas for strategic planning. In terms of natural vegetation the coverage of Britain is poorer than in many European countries, for example in France, where there is a series of monographs describing the vegetation in sub-regions for the whole of the country. Some regions of Britain are described in detail (e.g. the Highlands of Scotland, McVean and Ratcliffe, 1962). The current National Vegetation Classification project at Lancaster University aims to provide a dictionary of vegetation types but it is not designed to establish the relative proportions of the types in the country. There are many studies of smaller areas but these are at different scales and use diverse analytical techniques, from the descriptive to the numerical, thus severely restricting comparison between areas.

Turning to land-use, there are the major Land Utilisation Surveys of Stamp (1937–47) and Coleman (1961), and although the maps of the latter are unpublished, they are available through Dr Coleman (Geography Dept, King's College, London University). These two surveys provide the most detailed extensive mapping of land-use in Britain. Quantitative data for agricultural crops are published by MAFF/DAFS (1981) derived from census data. The FC census gives details of forest crops in major regions (Locke, 1970).

Landscape represents the combination of land-use with the physical and biological environment. Description of landscape is controversial, particularly because it is concerned with value judgement. The principle in the separation of a scientific statement of affinities of components in a landscape from the subjective assessment of its value is analogous to the distinction between environmental description and land-use potential discussed above. That a correlation exists between the components and the value judgements has been shown in a number of studies (e.g. Robinson *et al.*, 1976). Thus basic environmental descriptions such as the land form classification of Scotland by Linton (1968) can be interpreted in terms of scenic interest (Duffield and Owen, 1970) even so in the selection of national scenic areas, the Countryside Commission for Scotland specifically selected an approach based on the subjective judgement of assessors in preference to a more objective analysis (Turner, 1980).

The above discussion emphasizes the disparate nature of much of the information available for the basic description of the environment and land-use, and the wide range of approaches adopted for analysis of the information. The subject is extensive and we have selected only a few examples, which indicate the range of methods of environmental description, methods which have parallels in many other countries.

5.3.3 A systematic approach to environmental description

In order to improve the existing data base for environmental description at the national level it is necessary to adopt a sampling strategy since complete coverage of any system is not efficient in terms of the time and effort involved. What is needed in effect are environmental strata analogous in many respects to the social strata used in opinion polls.

The requirement for such a system was recognized in 1975 by the Institute of Terrestrial Ecology and a project was set up to achieve the main objective of a summary of the overall environment of Britain, which could provide a series of environmental strata from which representative samples could be drawn for a variety of ecological purposes. The techniques were developed in a survey of the county of Cumbria (Bunce and Smith, 1978) which in turn developed from an earlier survey of the Lake District (Bunce *et al.*, 1975).

The project involved the use of multivariate methods to analyse the environment as expressed by a large number of attributes, rather than a few variables measured in detail. In the Cumbria survey the National Grid of 1 km squares was used as a sampling basis in the way quadrats are used to sample vegetation. It was not considered desirable to use units defined by environmental variables in the first instance because of the difficulty of maintaining a consistent standard over the whole country and of prejudging the importance of particular environmental variables. The 1 km squares are a convenient base for recording the initial data at a scale suitable for subsequent field survey and are also a convenient reference system for mapping. Such a system was adopted, for example, by Buse (1974) in an ecological survey in Caernarvonshire and, on a national scale, the 1 km squares are sufficiently small for natural boundaries between regions to emerge while providing an objective basis for dividing the land surface.

The principle behind the national project is that the major ecological parameters are associated with the underlying patterns of land. The system seeks to formalize these relationships by recording quantifiable criteria from maps, subjecting these to analysis and subsequently relating field data to them. The project is therefore in three main phases:

(1) Land classification: the recording of environmental data from maps and subsequent analysis to produce a land classification which is used for stratification.

(2) Field sampling: the strata (land classes) are then used as a basis for field survey of various land-use and ecological parameters.
(3) Characterization and prediction: the land classes are then redefined, the amount and distribution of particular features measured, and finally, land-use and ecological features are predicted for areas in which only the land class is known.

A further principle adopted at the outset of the project was that all the data used should be readily available and that although computers were to be used in analysis of data, all the major stages of the data extraction would need only simple arithmetic. Hence, although mathematical procedures are involved in the data analysis, the application of the results does not require specialist knowledge. Whilst it is necessary to use judgement in the initial selection of variables the project has been designed to reduce personal bias as far as possible, and statistical tests using independent data enable the stratification system to be validated. Further details are provided by Bunce *et al.* (1981 *a*).

In order to obtain a standard grid of squares, the basic data were extracted from the centre of 15 × 15 km intervals, resulting in a total of 1228 squares over Great Britain. An ideal balance of types of data is not possible since adequate maps are not available. The main guidelines used were to include relatively unchanging variables that provided an expression of the underlying environment and not to include many measures of the same factor (e.g. temperature) merely because they were available. Data on soil were not included as the maps on a British basis involve interpretation from climatic and geology maps.

5.3.4 Land classification

The data base for the first phase of the British survey includes the following types of information:

(a) Climatic: data from climatic maps on a scale of 1:1 000 000 representative of the range of climatic data available.
(b) Topographic: data from the 1:50 000 OS maps, incorporating features such as altitude and slope.
(c) Human artefacts: data from features available from the 1:50 000 OS maps.
(d) Geological series: data concerning the presence of the main geological series and surface drift.

The data were originally a mixture of attributes and variables, but the latter were converted into attributes by separation into four equal classes, there being no theoretical basis why any particular transformation should be used. A total of 282 attributes was eventually used in the multivariate analysis

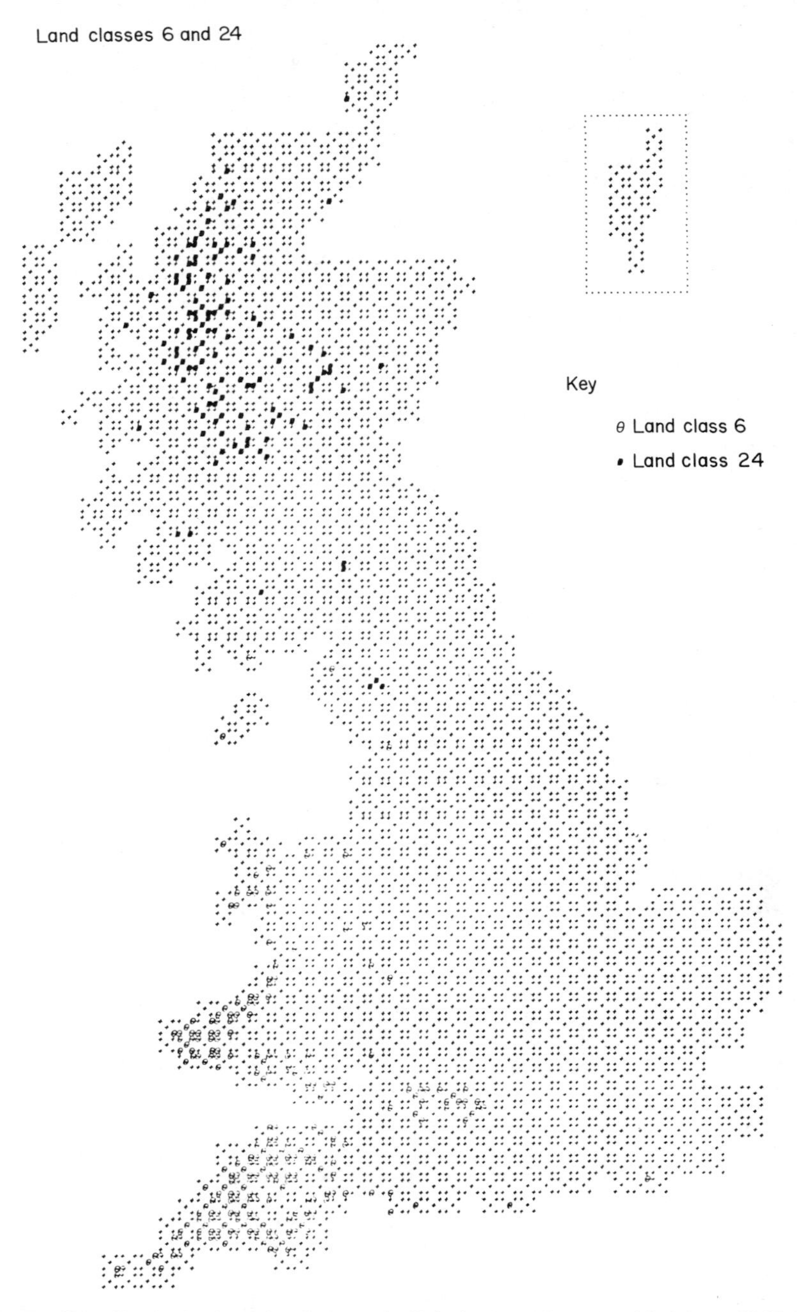

Fig. 5.2 Distributions of two land classes in Britain from data classified from 6040 1 km squares. Brief descriptions of the characteristics of the land classes are given in the text.

described by Hill *et al.* (1975) and 32 land classes were produced by successive division into 2, 4, 8 and 16 divisions, thus representing a progressive dissection of the land surface of Britain. The analysis identifies indicator attributes which may then be used to assign any kilometre square in Britain to its appropriate land class. Subsequently a further 4826 squares were assigned to their appropriate land classes, giving in total 6040 squares in Britain. Rapid estimates of the land class composition of any region in Britain can be obtained, without recourse to further computer analysis, and comparison of the variation in land type made between areas or the land characteristics of an individual site, e.g. a potential reservoir or valley, to be placed in a regional or national context.

The land classes show well-defined geographical distribution patterns (Fig. 5.2) but, as with Cumbria, distinctive unexpected patterns emerged which are interpretable in terms of their combinations of environmental features. The first division is largely associated with altitudinal and climatic features, but in later divisions features such as geology and slope become important in distinguishing land classes.

5.3.5 Field sampling

The second phase of the project has been to take eight squares at random from each of the 32 classes (i.e. 256 squares) and to use these as a sample for field survey. The following data have been recorded from each square:

(a) Species data from higher plants and a restricted list of bryophytes, from 5 random 200 m^2 quadrats.
(b) A standard soil profile description from a pit 75 cm deep in the centre of each of the 5 quadrats.
(c) The higher plants and a restricted list of bryophytes from linear quadrats, 10 m × 1 m, randomly placed along the sides of streams, roads and hedgerows, where present.
(d) The principal features present in the kilometre square, including a map of the land-use using 65 defined categories, and ecological information such as domestic animal breeds, and woodland composition.

5.3.6 Characterization and prediction

In the third phase the field data are added to the initial characteristics of the land classes to improve their definition. As examples, summaries of two classes taken at points through the series, are given below and their distribution in Britain is shown in Fig. 5.2. More detailed descriptions are provided by Bunce *et al.* (1981 *b*).

Land Class 6. Gently rolling enclosed country, varied agriculture and vegetation
Low emphasis relief with undulating countryside in which small streams form detailed local patterns. The climate is mild and moist, with pronounced oceanic affinities. Visually the land surface is broken by many hedgerows often growing on banks. Small copses are frequent, and the scenery has a markedly rural aspect. There are also many roads and the landscape therefore seems rather closed. Pastoral farming predominates (51 %) with arable farming mainly occupying only a small area (17 %). Some non-farmed land occurs, but the land is mainly intensively farmed. The soils are mainly brown earths, but with some gleys also present. Native vegetation is limited but mainly semi-natural grassland where present.

Land Class 24. Upper, steep, mountain slopes, usually bog covered
Very mountainous, with steep rocky slopes and the majority of land at high altitudes, with much surface water and bad drainage. The climate is markedly

Table 5.2 Comparison between the corresponding estimates of various land-uses derived from Callaghan (personal communication) and those derived from the land classes (Bunce *et al.*, 1981*a*). The former were adjusted to Great Britain by subtracting the figures for Northern Ireland. Figures in 10^6 ha (figures in brackets are percentages of total)

	Callaghan (personal communication)		*Derived from land classes*	
Total GB	23	(100)	23	(100)
Rural	21·3	(92·4)	20·8	(90·6)
Urban	1·7	(7·6)	2·2	(9·4)
Cultivated	12·6	(54·8)	12·0	(52·8)
Natural and semi-natural	8·7	(37·8)	8·8	(38·2)
Amenity	0·5	(2·2)	0·3	(1·1)
Other	1·2	(5·2)	1·9	(8·3)
Grassland	6·6	(28·7)	6·3	(27·7)
Arable	4·3	(18·7)	4·3	(18·8)
Forest	1·4	(6·1)	1·4	(6·1)
Orchards	0·05	(0·2)	0·05	(0·2)
Rough grazing	6·4	(27·8)	5·3	(24·9)
Woodland	0·6	(2·6)	0·6	(2·6)
Inland water	0·3	(1·3)	0·7	(3·1)
Other semi-natural	1·4	(6·2)	*	*
Leys	2·0	(8·7)	3·5	(15·2)
Permanent pasture	4·6	(20·0)	2·9	(12·6)
Cereals	3·6	(15·7)	3·6	(15·7)

* No comparable category.

montane, but with heavy rainfall and short growing seasons. Visually the landscape is very varied with many mountain features and distant prospects. In some lower areas modern plantation forestry is present, but the majority of the land (78 %) is open range with low grazing intensities by deer and sheep. The soils are mainly peats but there are some peaty podzols and peaty gleys. Native vegetation is widespread consisting of varied moorland and bogs.

It is essential to define the extent to which the data base is a representative sample of Britain. The total number of kilometre squares containing 10 % cover of land in Britain is estimated at 236 430; the 6040 squares which have

Table 5.3 Comparison between estimates of various crops (ha) derived from official statistics as produced in the Annual Review of Agriculture for the years 1977 and 1978 (the years of the field survey) modified from UK to Great Britain figures

	Official statistics	*Derived from the land classes*
Wheat	1 166 000	1 061 270
Barley	2 309 729	2 117 240
Oats	182 000	182 750
Mixed grain	18 864	32 830
Rye	10 000	5 080
Potatoes	210 161	200 923
Sugar beet	205 000	146 959
Oil seed	60 000	44 535
Vegetables	216 000	233 562
Total tillage	4 716 470	4 321 080
Total grass	6 280 898	6 385 070
Orchards	53 000	59 684
Maize	17 000	26 976
Lucerne	17 025	19 600
Fodder crops	207 667	190 444

been classified are a sample from which the total area of each land class in Britain can be estimated. The occurrence of the measured features in each land class can therefore be used to estimate the amount and distribution of the features and, for some land-uses, estimates can be compared with independently derived national figures. There are problems in comparing such data, in particular in the non-coincidence of published categories.

Estimates of land-use from the present survey correspond closely with those drawn from a combination of published sources for Great Britain (Table 5.2). Comparison with figures for agricultural crops drawn from Ministry statistics shows good agreement, particularly with the more extensive crops (Table 5.3). The permanent and temporary grass categories agree well when combined together but have a different balance since in the ITE survey the

Table 5.4 Allocation of the land surface of Britain to various major land-uses, based on estimates derived from field survey in eight sample squares in each of the 32 land classes. Conversion into a Great Britain basis is made by estimating the proportion of the kilometre squares belonging to each land class. Areas are in hectares (figures in brackets % of total)

Major land-use	Category	Area	Subcategory	Area
All crops 4 428 207 (19·6)	Cereals	3 426 148 (14·9)	Ploughed/fallow	107 398 (0·7)
	Other crops	709 013 (3·4)	Sugar beet	146 959 (0·6)
	Horticulture	293 046 (1·3)	Animal fodder	190 955 (0·9)
			Mixed crops	263 701 (1·2)
All grass 6 385 070 (27·8)	Leys	3 483 270 (15·2)	Short-term leys	2 357 204 (10·4)
			Other leys	1 098 820 (4·8)
	Permanent grass	2 901 800 (12·7)	Generally reseeded	1 447 997 (6·4)
			Older grassland	1 453 806 (6·3)
All wood 2 207 300 (9·6)	Broad-leaved wood	561 010 (2·4)	Copses	781 433 (0·3)
	Conifer	1 404 737 (6·1)	Shelter belts	34 552 (0·2)
	Scrub	241 556 (1·1)	Scrub	241 556 (1·1)
			Woodland	1 853 050 (8·1)
Semi-natural 5 514 553 (24·0)	Rough grassland	2 177 424 (9·5)	Rough grass	505 270 (2·2)
			Mixed rough grass	513 960 (1·4)
			Pteridium dom.	360 530 (1·6)
			Juncus dom.	374 358 (1·6)
			Mountain grass	621 306 (2·7)
	Moorland	3 337 129 (14·5)	*Molinia* dom.	761 907 (3·3)
			Eriophorum	669 994 (2·9)
			Calluna dom.	1 260 201 (5·5)
			General moorland	645 027 (2·8)
Unavailable 4 334 471 (18·9)	Aquatic	726 394 (3·2)		
	Human	2 992 440 (13·0)	Buildings, etc.	2 277 890 (9·9)
			Communications	719 550 (3·1)
	Unavailable	615 637 (2·6)	Inland rock	169 403 (0·7)
			Maritime	446 234 (1·9)

separation was on species composition whereas the Ministry figures are related to the time since the last reseeding. Full details of the areas and percentages of land-uses and vegetation are given in Table 5.4. Thus a carefully stratified field sampling of only 256 km² has provided realistic estimates of the current allocation of the land surface of Britain into major uses. The field survey also provides detailed information on the composition of the 5·5 million ha of semi-natural vegetation.

The matrix of land-use from which the above estimates were derived and its distribution within land classes (e.g. Table 5.5) can be used to provide estimates for any geographical area for which the land class composition is known. The more restricted the area under consideration the greater the

Table 5.5 Matrix of land uses as related to the two examples of land classes quoted in the text, the distribution of which is given in Fig. 5.2. Eight combined categories of land-uses are given (the table is derived from the same data base as Table 5.4) and separated into the appropriate land classes. The data are derived from the area measured in the field survey of the eight squares from each of the 32 land classes. Results are expressed as % of each land class

Land uses	*Land classes*	
	6	24
All crops	16·9	—
Temporary grass	27·1	—
Permanent grass	29·7	—
Woodland	11·3	8·6
Roads and buildings	11·6	0·6
Water and rock	0·9	11·0
Moorland vegetation	—	56·0
Mountain grassland	2·4	23·8

likely error in estimates of the proportions of land-uses, although this will differ since some land classes are more uniform than others. Tests in the Cumbria and national surveys show the level of accuracy that can be achieved (Bunce and Smith, 1978; Bunce *et al.*, in press).

In addition, from the matrices and the known distribution of land classes, the geographical distributions of land-use, vegetation, individual species or other sampled features in Britain can be defined. Examples are shown in Figs. 5.3, 5.4 and 5.5.

The procedure described can therefore enable a basic quantitative description of a large area to be produced within a short time which places it in its regional and national context and indicates its contribution to the resource of native species, soils or land-use in Britain. Thus the approach has an important potential in the basic descriptive phase of environmental planning in that a rapid statement of areas of interest can be obtained,

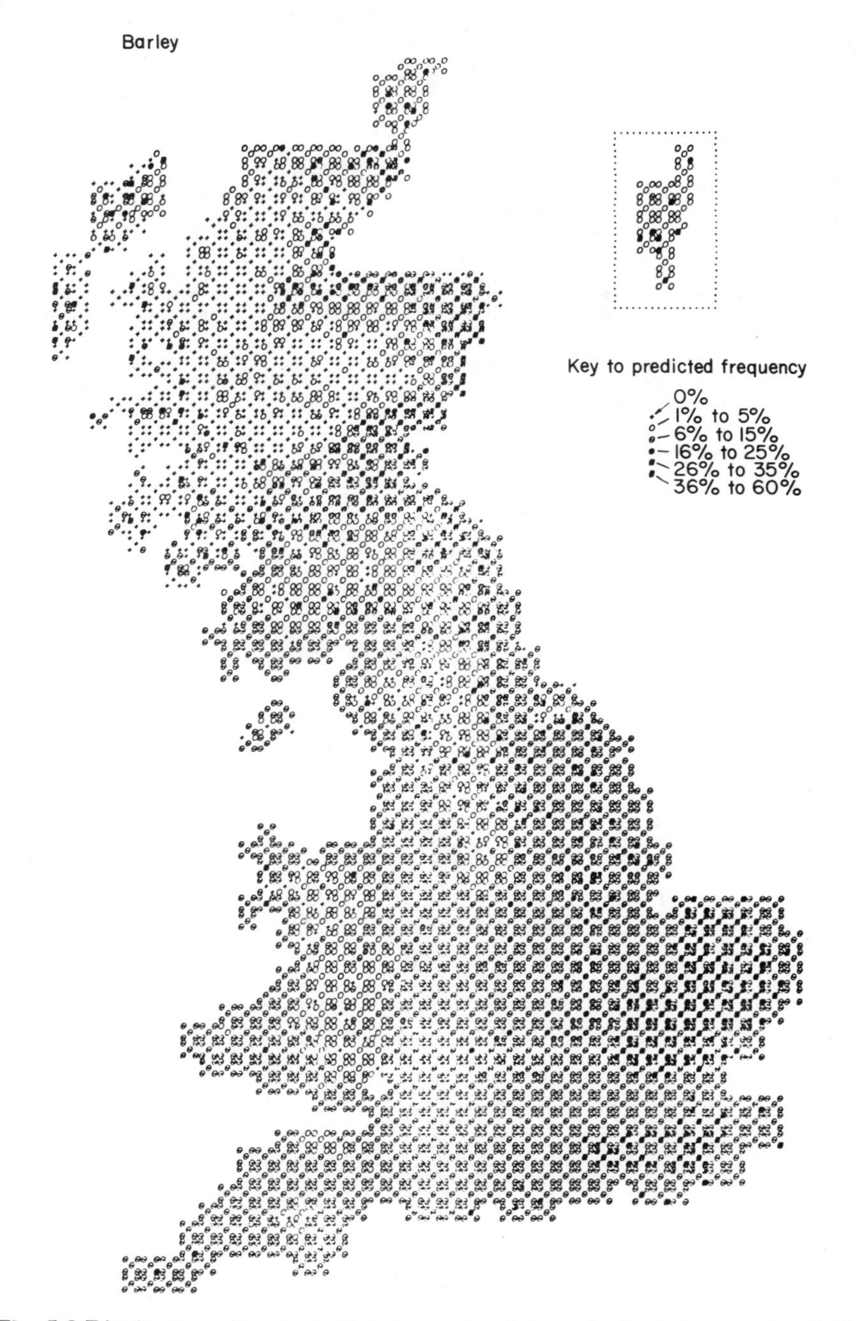

Fig. 5.3 Distribution of barley in Britain predicted from the land classes using 6040 squares. The land classes were characterized by field survey of 8 sample squares in each of the 32 land classes and the extent of barley measured by the area of the fields in which it was present.

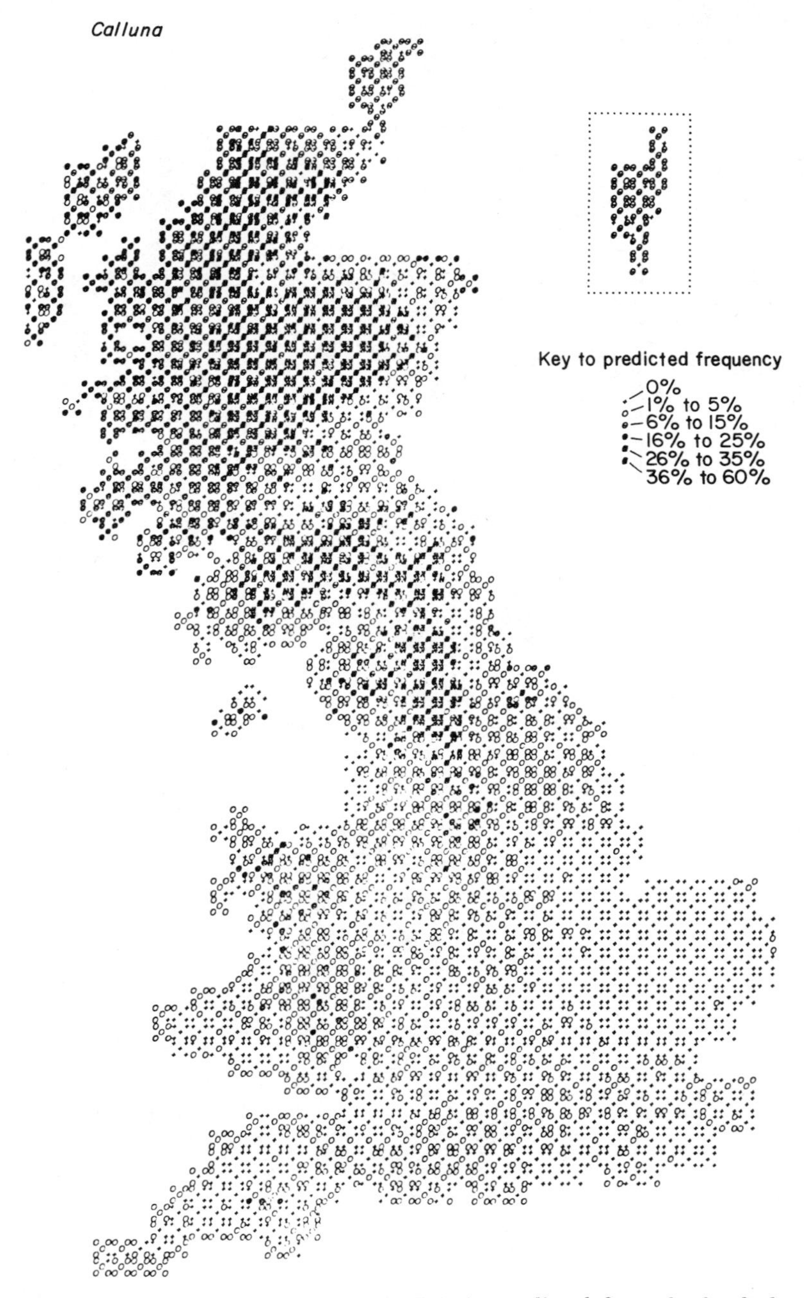

Fig. 5.4 Distribution of *Calluna vulgaris* in Britain predicted from the land classes using 6040 squares. The land classes were characterized by field survey of 5 random 200 m² plots in each of the 8 sample squares in each of the 32 land classes. The percentage cover was estimated from the cover recorded in the 40 sample plots.

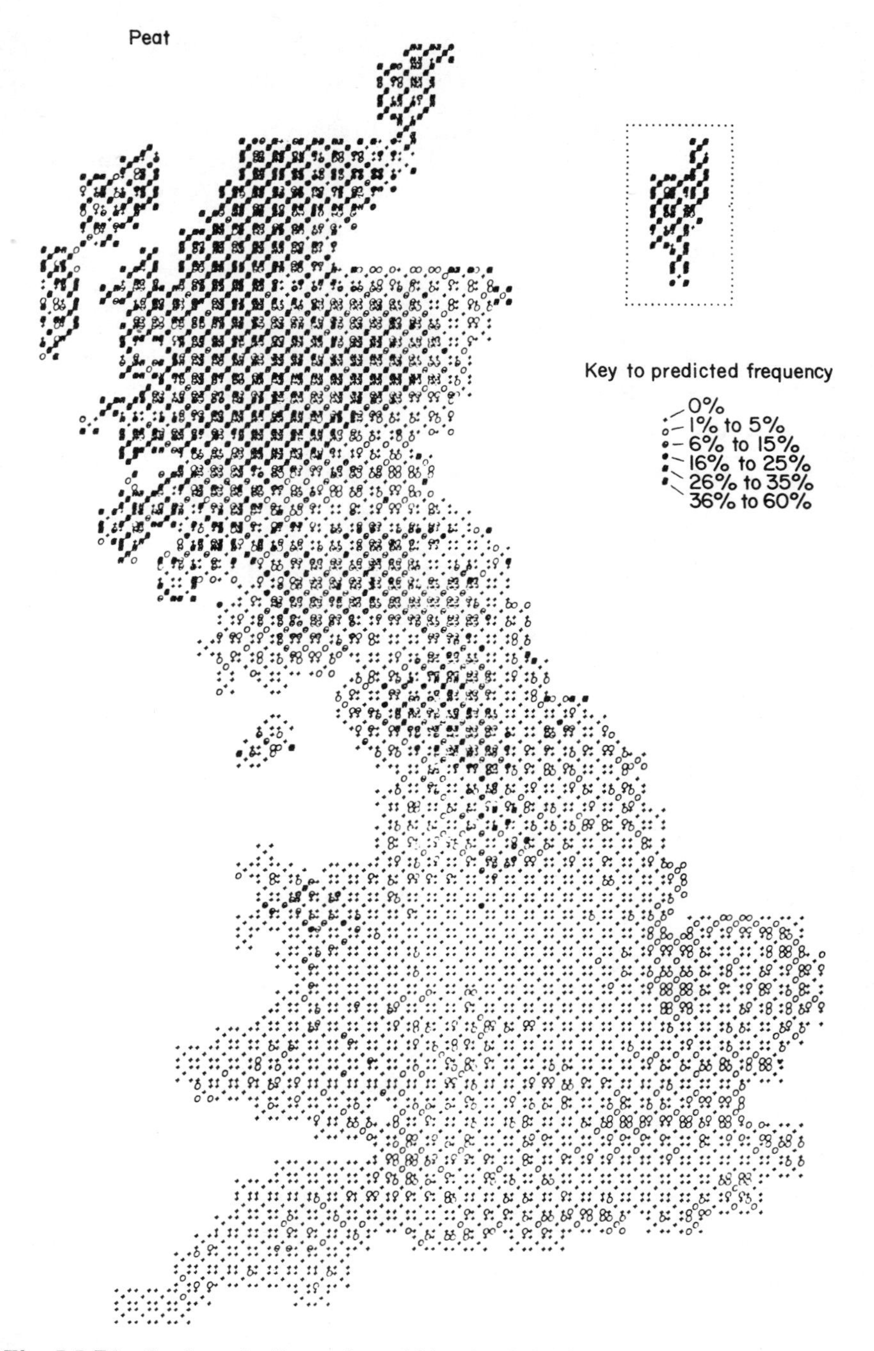

Fig. 5.5 Distribution of soils coming within the definition of peat, with an organic layer of more than 35 cm in Britain using 6040 squares. The land classes were characterized by field survey of soil pits at 5 random points in each of 8 sample squares in each of the 32 land classes. The percentage of soils coming within the definition of peat was derived from these 40 sample pits.

without recourse to new survey, and information on the extent and distribution of a feature can be obtained with minimal field visits using land classes as sampling strata.

5.3.7 Monitoring

Monitoring change in rural land-use and its effect is vital since it provides not only information on the current change in the resources, but also provides a basis for the prediction of future change and a test of our predictions. There are two main types of change:

(1) Major changes in land-use that are abrupt and are reasonably well documented, although there has been no consistent national approach covering all uses.
(2) Subtle and gradual changes, that require more intensive study to detect and which are particularly important for environmental assessment. However, data are limited because of the long-term scale of such projects.

(a) Abrupt changes

The broad changes in land-use are documented in the old OS maps and historical archives. The Land Utilisation Survey of Stamp (1937–47) formed a baseline from which Coleman (1961) has derived some estimates of change in England and Wales. Many studies of particular areas are present in the literature, e.g. Parry (1976), Blacksell and Gilg (1981), while Sheail (1980) provides a general background to the use of documentary evidence in historical ecology. Quantitative information on national and regional changes in forests are provided by the Forestry Commission censuses of 1947 (FC, 1952) and 1965 (Locke, 1970). Changes occurring within agriculture may be derived from the MAFF annual data e.g. MAFF/DAFS (1968); and there is an extensive literature on the changing national balance of land-uses, e.g. Best (1976) and Coppock and Gebbett (1978). The general trend is of progressive loss of farmland to urban growth in the lowlands and to forestry in the uplands.

Superimposed upon this national pattern are local details, e.g. conversion of woodland, agricultural improvement of moorland or the invasion of chalk downland scrub. The ecological consequences of such abrupt changes are usually readily understood. However, information on the vegetation and other aspects of ecology, prior to a change in land-use, is rarely adequate to allow accurate monitoring. Good comprehensive baseline data are necessary for monitoring, and a stratification system as described in the previous section could allow selection of a limited number of samples for recurrent survey.

(b) Gradual changes

Gradual ecological changes may result either from abrupt alterations in land-use, or from subtle modifications of factors such as grazing patterns. This is because while management can produce rapid changes in crops the underlying ecological systems, particularly soil, may be strongly buffered and show only gradual changes. However, the majority of subtle changes are brought about by equally minor trends in land-use, as in the trend from cattle to sheep grazing or a change in the distribution of grazing in the hills. Here it may be as difficult to document the land-use change, as to follow its ecological consequence. Such patterns are common in the uplands because of the low intensity of use.

Thus quantitative, spatial information is generally lacking on changes in the distribution of plant and animal assemblages, the distribution and abundance of species, and on ecological processes such as nutrient cycling directly associated with land-use change. The few exceptions are individual long-term experiments, e.g. Rawes (1981), or are opportunistic, e.g. the work of Thomas (1960) on the vegetation changes following myxomatosis, rather than the result of coordinated research.

Some comprehensive monitoring schemes are concerned with major groups of species based on distribution maps. Regular recording allows detection of population changes, but the details of associated changes in land-use are often lacking (Perring and Walters, 1962; Sharrock, 1976; Pollard, 1979). Intensive monitoring of changes in fauna is rare and the effect of modifications in land-use is usually inferred from comparison of areas under differing land use (e.g. Moore, 1962). Soil changes are particularly long-term and, although an experimental approach has been adopted (Dimbleby, 1952; Ovington, 1958; Jenkinson, 1971), analysis of chronosequences can allow detection of main trends (Maltby, 1975; Miles and Young, 1980).

A difficulty in monitoring is in maintaining the required effort over many years. A further challenge is that the recording must be sufficiently explicit, and objective, to allow repetition by later workers and be sufficiently sensitive to detect small changes. The consequence of lack of detailed monitoring is that informed opinion is normally used in environmental assessment in order to compensate for factual deficiencies. The increased pressure on the rural area requires a structured programme of monitoring of representative ecosystems. Without such a programme, ecologists are likely to remain unconvincing in their assessments of ecological changes.

5.3.8 Prediction

Prediction is the essential part of impact assessment but, inevitably, it is the area of greatest uncertainty. It is relatively straightforward to discuss, in general terms, the sort of changes in vegetation and fauna which might occur through the development of a management practice. However, the goal must be to specify the probability of occurrence, the extent and the rate and direction of change which will result from modification of management. The value judgement of whether a change is desirable or undesirable, e.g. the nature conservation value, must be separated from the estimation of the response of species, communities or processes (Goode, this Chapter).

Changes in rural land-use are rarely explicitly proposed and tend to be piecemeal. As a result there are no definitive statements of what ecological impacts are expected, where they will occur and what are the likely effects. Further, negative changes in management, such as the abandonment or reduction of the intensity of agricultural use, may cause ecological changes as great as those resulting from intensification or extension, but these are also outside the formal framework of planning.

In the absence of statements of change and assessment of consequences it is necessary to extract data informally. An initial problem is in defining what changes in use are likely to occur, and where. Taking agriculture and forestry as an example, MAFF (1979*b*), in forecasting the agricultural production in the UK, expect a land loss of about 50 000 ha yr^{-1} between 1978 and 1983. This is in line with recent figures and the major losses are of grassland and rough grazing, with little change in arable. As indicated in Strutt (1978) the losses are to forestry (about 13 %), urban (about 51 %) and 'miscellaneous, including adjustments in statistics' (36 %). These losses represent major changes in use and do not indicate the changes in intensity of management within agricultural land. In the absence of explicit statements it might be assumed that about 7000 ha yr^{-1} will be transferred from agriculture to forestry. This is well below the minimum new planting target for UK forests of c. 20 000 ha yr^{-1} up to the year 2000 (CAS, 1980). Although some of the planting will be on land already owned by the forestry industry there is a major discrepancy in supply (from agriculture) and demand (by forestry). The FC predict that about 60 % of planting will be in Scotland and the majority of the remainder in Wales and northern England. These figures match with the distribution of the main type of agricultural land being lost, i.e. marginal upland. Such general statements are important in putting into context likely future changes and help to provide national trends that are a useful base for predictions. But where are the changes in land-use and vegetation likely to occur? In the absence of specific plans for alterations in the distribution and type of rural land-use, it is necessary to resort to prediction. For example in Cumbria, the areas in which there is likely to be a decrease in intensity of agriculture were predicted on the basis of criteria of population trends, land quality and farm type and structure (Fanstone and Himsworth, 1976). A similar approach based on recent trends in land-use, was used to predict areas of agricultural and forestry change in a sample of upland areas in England and Wales (Ball *et al.*, 1982).

The type of ecological change resulting from afforestation of marginal land has received considerable attention and the use of checklists, matrices, networks and flow diagrams can help in identifying the impact and the response. Many of the responses can be quantified, e.g. changes in numbers of plant species (Hill and Jones, 1978), although changes in soil conditions

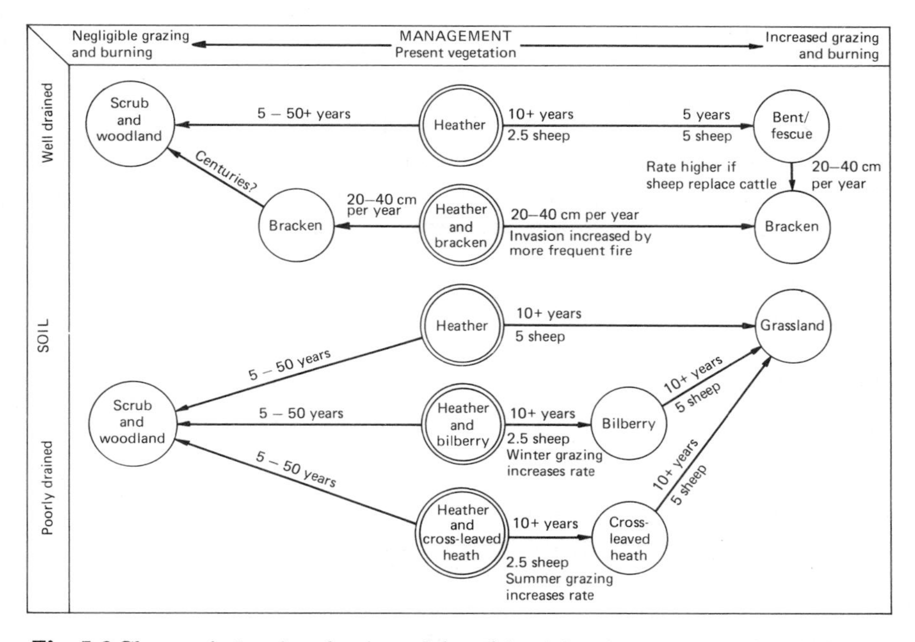

Fig. 5.6 Changes in heather-dominated dwarf shurb heaths as a result of variations in intensity of grazing and burning. Present vegetation shown in double circles: probable and possible stages in the vegetation succession are shown by single and broken circles respectively. Rate of change where known is shown above the arrows: sheep numbers are per hectare. Differing soil conditions are noted in the left hand margin (after ITE, 1978).

are less well understood (Stone, 1975; Miller, 1979). The response of land-use and ecology adjacent to a forest may be more difficult to predict. On marginal land it is likely that adjacent agriculture will be sheep and/or cattle grazing. Stock densities may increase because of reduced area of land available or decrease because of isolation of land units (Cunningham *et al.*, 1978). Reduced stock densities can affect many components of ecosystems, e.g. carrion feeders (Marquiss *et al.*, 1978), but vegetation change is the most obvious. The vegetation changes resulting from changes in grazing pressure have been described in a number of studies (Watt, 1960), which provide the basis for prediction of change in other areas. The changes observed in specific areas must be extrapolated using explicit criteria. An attempt at defining the rate and direction of vegetation change (Fig. 5.6) provides a level of generalization which is applicable to much of the uplands of Great Britain (ITE, 1978; Ball *et al.*, 1982). The inadequacies of the generalization or hypothesis are recognized, but the challenge is to improve it.

It is possible to transform the information on vegetation change into more explicit expressions such as Markov models in which the probability of transition from one state to another is estimated for each vegetation type (e.g. Horn, 1975). From the transition matrix and the known area of each vegetation type, the amount of each type can be predicted over a series of time steps. Although they make certain simplistic assumptions, such as linear change, models of this type provide a first approximation of future states. The Markov and other matrix models do not incorporate understanding of the dynamics of the processes. Dynamic models are intuitively more satisfying to the ecologist, but they are more complex and demand more data.

The prediction of vegetation change in areas where land-use is modified can be developed from the land classification and survey approach already described. Given the relationship between the cover of various plant species and the national land classes (e.g. Table 5.5) ecological knowledge of the interrelationships between these species and management (e.g. Fig. 5.6) can be used to change their proportions according to a specified impact. Thus the impact can be predicted in terms of the area and distribution of species or communities, based on explicit criteria. Such matrices allow the use of modelling techniques, as described by Wathern (Chapter 3), but they depend for their success on an adequate factual data base and an understanding of the quantitative response of the system to impacts.

5.3.9 Optimization

Because of the competition between agencies and owners for the land, and the absence of overall planning control, it is rare for a specific goal, for a combination of uses, to be set for an area. Rather, a mosaic of land-use develops, determined by history and market forces, with few attempts to produce an optimal solution of multiple use for mutual benefit. In many discussions on land-use and ecology, e.g. the Hill Land Discussion Group in Scotland, a principal objective in debate is integration between various land-uses but practical examples of planned integration are few and research into options is very limited. Many of the County Structure Plans and National Park Plans are now proposing patterns of land allocation which reflect compromise between competing demands. These plans do not contain production targets which are determined by national policy and there is little use for techniques to identify options for land-use distributions which can achieve specified objectives.

In rural land-use the developments tend to be long-term and continuous; therefore there is opportunity for exploring alternative options to meet changing demands. However, in both agriculture and forestry, the need to determine possible future uses, is seen in the production of the Agricultural Land Classification of MAFF and by the site classification of the FC. These classifications define the production potential based on environmental factors and current management methods, but it is not their purpose to analyse the options for combinations of uses for a particular area (Cunningham *et al.*, 1978). However, the Hill Farming Research Organization has shown how options for the allocation of land to single or integrated use can be explored, indicating the effect of different land-use distributions on stock densities and economics on a local scale (Maxwell *et al.*, 1979). A similar study, using a linear programming model, was based on the land classification system described earlier and applied to Cumbria (Bishop, 1978). The stages are summarized as follows:

(a) The land classes provide a definition of the variation in physical environment. The level of production which can be achieved in each land class for each land-use was defined from examples of existing use to provide a matrix (Table 5.6). Additional values for labour and energy input were defined for each land-use in each land class.
(b) The area of land available was defined as the total number of km of squares in each land class, but as forestry is not permitted on common land and in certain parts of the Lake District National Park, the area of each land class available for forestry was reduced accordingly.
(c) The minimum level of production to be achieved for each use was defined as the current agricultural and forestry output for the county.
(d) Using linear programming, Bishop then calculated, for each use, the maximum production which could be achieved from the county, within the constraints of land designation and the necessary minimum levels of production for other uses. For each calculation the optimum allocation of land-uses between land classes was specified.
(e) Constraints on forestry in specified areas were successively removed and the forest production and land-use allocation re-calculated.

The model indicated, for example, that forest production of 332 000 m^3 yr^{-1} on the current land allocation could be increased by 42% within existing constraints, and by 133% if all constraints on land-use were removed. Maximizing for forestry also had marginal bonus increases in some agricultural products.

The details of the model are not important here since it is used to demonstrate the potential of the approach where quantitative analysis can indicate the combination of uses which can meet the varying demands on the land by reallocation of types of land to produce an optimal solution. This type of analysis is a logical development from environmental description and assessment of production potential of the land and it provides a technique for exploring the compatibility of land-use policies. The potential application of the approach to local planning has been examined by Smith and Budd (1982).

Table 5.6 Forestry ($m^3\ km^{-2}\ yr^{-1}$) and meat production (tonnes $km^{-2}\ yr^{-1}$) in relation to land classes in Cumbria derived from analysis of data from the county. Land class 7 was omitted as it is largely estuarine (from Bishop, 1978)

	Land use					
	1	2	3	4	5	6
Land class	*Softwood*	*Hardwood*	*Beef*	*Dairy/beef on improved pasture*	*Sheep on improved pasture*	*Sheep*
1	1230	480	25·0	21·4	—	16·7
2	1310	510	26·5	22·7	—	17·6
3	1090	430	22·8	19·5	—	15·2
4	1110	430	19·8	17·0	—	13·2
5	1270	490	27·9	23·9	—	18·6
6	1280	500	27·9	23·9	—	18·6
8	1230	480	25·0	21·4	—	16·7
9	800	260	5·3	8·2	5·7	3·7
10	840	270	9·9	8·5	8·9	6·9
11	990	380	9·9	8·5	8·9	6·9
12	960	370	9·9	8·5	8·9	6·9
13	490	85	3·0	5·1	2·9	1·7
14	85	—	1·2	—	—	0·7
15	560	127	3·6	5·7	3·3	2·1
16	200	—	1·4	—	—	0·8

Bishop (1978) also incorporated into the calculations an assessment of the nature conservation and recreation value of each land-use in each land class. These assessments were in the form of scores derived, as far as possible, from measurable attributes, e.g. number of species or presence of metalled road for access. Although such scores are difficult to interpret, especially when summed, they provided an assessment of the extent to which different land-use options enhanced or detracted from the conservation or landscape value of the area. The emphasis was on the development of explicit, detectable criteria as a basis for applying value judgements (Usher, 1980).

Land-use options have been explored using linear programming and other forms of model, in various situations, e.g. forest planning (Dane *et al.*,

1977 and more generally by Miron, 1976) to clarify options and consequences. In the current situation in the UK where demand for land is increasing and the resource is limited, it seems logical to use these tools to help the planning even when there is no overall control over development.

5.3.10 Conclusions

Ecologists have concentrated on input to environmental evaluation in the context of site specific developments. However, the major problems in rural land-use, now and in the future, are concerned with the gradual changes in agriculture, the more obvious changes through afforestation, and the increasing demands on a limited land resource. These changes in land-use are occurring on a large scale, but are outside the statutory planning procedures. Because the proposed developments in agriculture and forestry are not stated explicitly by the relevant agencies, it is necessary to predict what changes are likely to occur as well as to predict their ecological consequences.

A limited view of the ecological aspects of environmental assessment is taken, ignoring the many specific studies on animal and plant populations that have been carried out. Four major stages in environmental assessment (description, monitoring, prediction and optimization) are explored in the development of an assessment of the general impact of rural land-use changes. The application of an ecological survey method based on the definition of environmental strata to these various stages is described.

An environmental description is the first stage of environmental assessment and should provide a definition of the current state of the physical, biological and land-use characteristics of the system as a baseline for the assessment of likely impacts. Current national coverage is of a disparate nature but a land classification system, using explicit environmental criteria as a baseline for field survey, enables estimates to be made of rural land-use and ecological parameters on a national and regional scale. The contribution of a particular area to the overall resources in Britain can therefore be estimated rapidly, with assessable accuracy.

Monitoring of land-use and ecological changes is essential since it not only provides information on ways in which the resource is changing but also enables future changes to be predicted. Major changes in land-use are quite well documented, even although the approaches adopted are fragmented. In contrast, information on the subtle and gradual changes, for example in species composition or in soil structure, is inadequate because of the long lead times involved. Long-term studies are urgently required and the land classification could provide a system for identifying representative sites.

Prediction is the essential part of environmental assessment but is the subject where the risks are greatest. Rural land-use changes are rarely explicitly proposed and are largely determined by the advocacy of the national agencies and by socio-economic factors. The prediction of future patterns tends to rely on subjective judgement, while the effects are assessed mainly by informed opinion. As with monitoring there is a more detailed level where matrix, Markov and dynamic models may be used to predict the rate and implications of change. Much of such work is at a preliminary stage but has great potential for the future.

Optimization is a topic rarely considered formally although it is a major topic in discussions on land-use, and is an important element in environmental assessment. Despite the competition between forestry and agriculture, there is little effort to produce an optimal solution for a given area of land – rather there is a trend towards separation of land-uses. In the current situation in Britain where the competition for land is increasing such research is needed to help planning. Even where there is no overall control of development the explanation of options can indicate solutions of mutual benefit which can then be implemented or encouraged by agencies.

In both optimization and prediction adequate data bases are required. Coordination of local studies can provide these in some instances but frequently quantitative data are absent, although the land classification now provides a strategic framework. As with many other ecologically related disciplines, intuition will inevitably play an important role in environmental assessment as applied to the rural environment. Technical advances have in many cases outstripped their practical application but innovation and research must continue otherwise ecological input to environmental assessment will remain qualitative and subjective.

5.4 CONSERVATION AND VALUE JUDGEMENTS

D. A. Goode

5.4.1 Introduction

In its broadest sense nature conservation is concerned with ensuring the future survival of as much of the natural world as possible in the face of the demands made upon it by man. It is concerned with the mitigation of human impacts and as such is based on both scientific analysis and value judgements. Scientific analysis is required to monitor changes in the environment and to predict as accurately as possible the likely ecological consequences of major new developments or changes in land use. Such analyses are essentially neutral, being factual statements based on the available scientific knowledge. The quite separate, though equally important, process of deciding whether such changes are acceptable or not is based on *value judgements*. In any assessment of environmental impact the question 'Does it matter?' can only be answered by reference to some scale or code of environmental values which provide the basis for decisions on environmental quality. Similarly nature conservation involves a range of value judgements which form an implicit part of the decision-making process. Indeed these values provide the underlying philosophical basis of nature conservation.

This presentation considers the role of value judgements in conservation and ways in which changes in such values affect both the theory and practice of conservation. Trends affecting the overall nature conservation resource, especially habitats, are summarized initially before discussing the various mechanisms available for protection of habitats and the role of value judgements in the selection of conservation priorities.

5.4.2 Pressures on wildlife habitats

Much has been written in recent years about the increasing rate of destruction of natural habitats throughout the world as a direct result of man's activities. The wholesale destruction of tropical forests has been particularly well documented by Myers (1979) who estimates that Africa has lost 52 % of all its Tropical moist forest, south-east Asia 38 % and Latin America 37 %, most of which has occurred within the past 10–20 years. At current rates of exploitation it has been predicted that virtually all these forests will be destroyed within the next 20 to 30 years. Destruction of such forests has already resulted in the total extinction of whole assemblages of species in certain areas especially where a high proportion were endemic species (i.e. with a limited geographical range). The further destruction of this habitat puts a very large number of species at risk, since conservative estimates suggest that 2 million species of plants and animals are dependent on tropical forests. Many other habitats, especially wetlands, are being similarly affected on a world-scale, and current estimates suggest that half to one million species will become extinct by the end of the century (10–20 % of all species on earth!).

Even in Britain the fragments of semi-natural vegetation which have hitherto survived within our lowland agricultural landscape have come under increasing pressure in recent years. Despite our elaborate machinery of planning control, and the increasing effectiveness of conservation agencies during the past 20 or 30 years, the amount of semi-natural habitat has been substantially reduced over this period and many areas of prime importance for nature conservation have either been destroyed or their value much reduced. Figures recently published by the Nature Conservancy Council (Goode, 1981) demonstrate that all kinds of habitat are currently affected, though herb-rich and calcareous grasslands, heathlands, bogs and ancient woodlands appear to be among those which have suffered most.

The reduction of lowland heathland and its effect on populations of species dependent on this habitat were clearly demonstrated by Moore (1962) in his study of the heathlands of Dorset (Fig. 5.7). This showed that not only had the amount of heathland declined by 85% since 1760, but the previously extensive tracts of heathland had been broken up into numerous small fragments, many of which were too small to support populations of the characteristic fauna. Moore estimated that by 1990 the only heathland remaining would be that protected as nature reserves. His prediction has been supported by a recent survey (Webb and Haskins, 1980) which showed that since Moore's survey of 1960 about half of the heathland which he

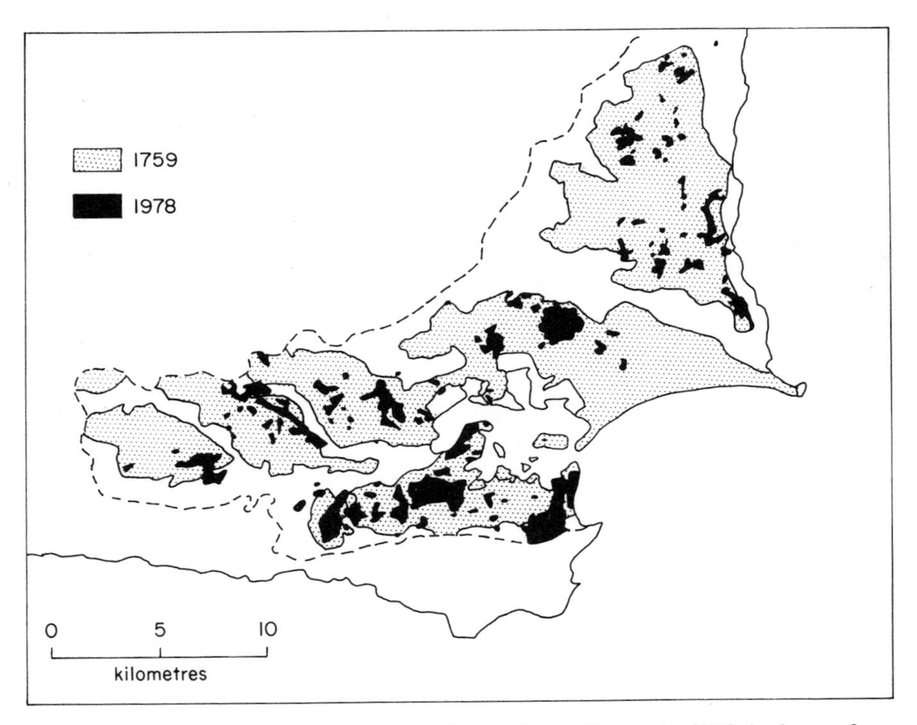

Fig. 5.7 The extent of heathlands in the Poole Basin, Dorset in 1759 (redrawn from the maps of Isaac Taylor) and in 1978 (redrawn from Webb and Haskins, 1980)

recorded had been destroyed and the remaining areas were more fragmented (Fig. 5.8a). Over half of this area is currently designated as SSSI and the NCC plans to designate other areas totalling 80% of the remaining heathland in Dorset.

This is not an unusual example. A recent study of raised bogs (Goode, 1981) has demonstrated a dramatic and continuing decline in the extent of this habitat in Britain over the past 130 years (Fig. 5.8b). The study involved 120 different bogs totalling 13 704 ha. Changes in land-use affecting these bogs have been traced since 1850 using information from the first and second

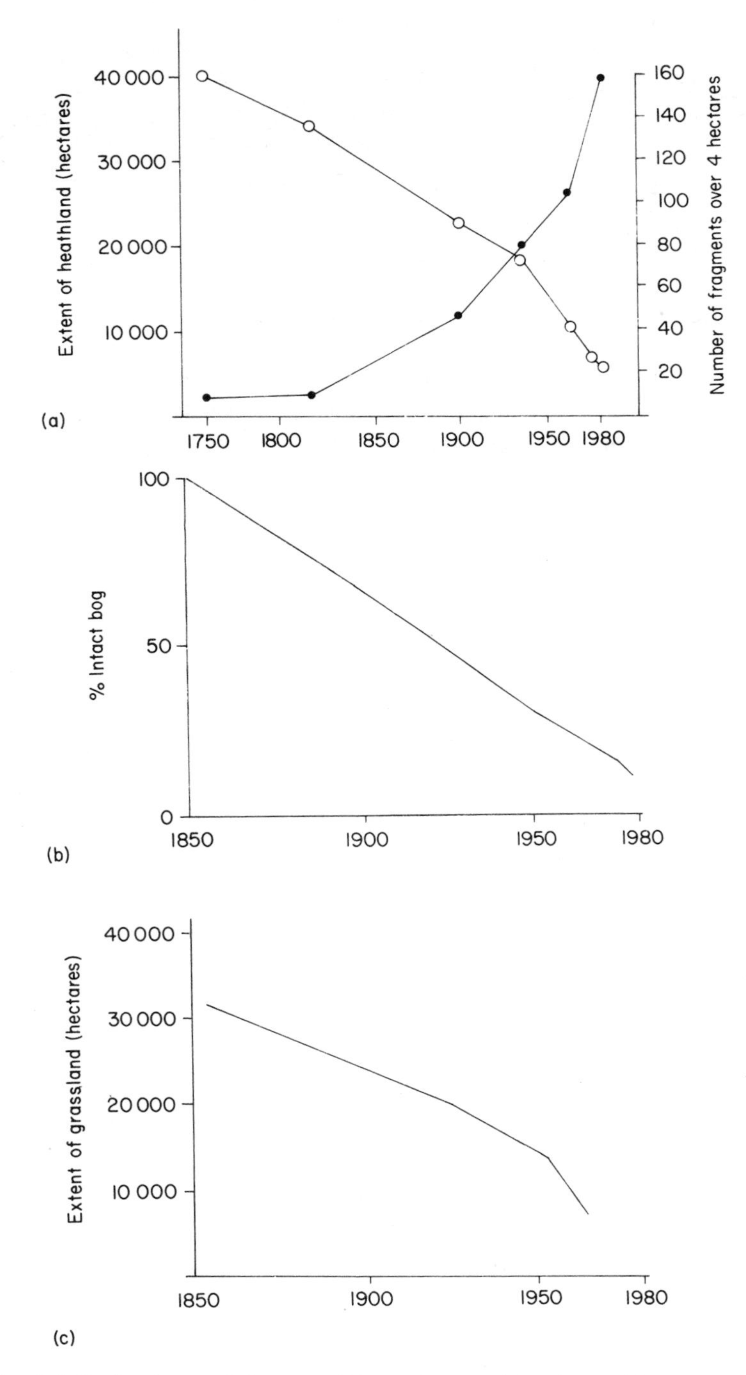

Fig. 5.8 Reduction and fragmentation of natural habitats. (a) Dorset heathland (after Moore, 1962; Webb and Haskins, 1980): ●, number of fragments; ○, extent of heath; (b) Lowland bogs (after Goode, 1981*b*); (c) Chalk Grassland (after Jones, 1973).

edition Ordnance Survey maps and from aerial photographs during the past 40 years. The results show that by 1978, 87% of the area of raised-bog habitat present in 1850 had been utilized in one way or another and only 34 of the original 120 bogs had any bog habitat remaining. Half of those remaining were reduced to fragments of less than 10 ha which no longer support viable raised-bog habitats, and none of the surviving areas of bog was in an entirely natural condition as all had been reduced in area by piecemeal 'reclamation' around their margins. Conversion to agriculture accounted for 36%, forestry 37·5% and peat winning 11% of the total area of raised bog in this study. The area of raised bog has thus been reduced to only 13% of the 1850 figure and even between 1960 and 1978 half of the remaining bog habitat was destroyed. Since there are now so few examples which remain, virtually all the viable areas of this habitat in Britain have been notified as Sites of Special Scientific Interest.

Areas of ancient broad-leaf woodland, especially those which are surviving fragments of the original forest cover of lowland Britain are extremely important in nature conservation terms because of the wealth of species which are dependent on them. The remaining ancient semi-natural woods now cover only about 1·5% of Britain. Although much of the massive clearance of natural woods took place many centuries ago, losses have continued in recent times and have been particularly rapid in the 19th and 20th centuries. The post-war period has seen the destruction of many of the remaining semi-natural woodlands. The NCC has estimated that by 1980 Britain had lost between 30 and 50% of the ancient semi-natural woodland that survived to 1946, and that all the remaining areas of such woodland outside nature reserves or other specially protected areas are likely to be eliminated by the year 2025 (NCC, 1980). The main reasons for such losses over the past 30–40 years have been clearance to agriculture and conversion into plantations.

As this gradual attrition and loss of semi-natural habitats continues so the value of those places which remain in a relatively natural condition increases. This is particularly true of the 'islands' of semi-natural habitat within our lowland agricultural landscape. It is a biological fact that most species can only exist in particular ecological conditions to which they are adapted. The vast majority of our native species of plants and animals cannot survive in areas which are intensively farmed. They are entirely dependent on the remaining fragments of semi-natural habitat within the agricultural landscape and because of this they tend to be restricted to scattered localities. Such places, with their rich assemblage of species, contrast sharply with intensively farmed land, which now supports only those relatively few species which are capable of adapting to the artificial environment, and which tend to be both widespread and abundant.

So it is that in terms of importance for nature conservation certain places and areas have especially high wildlife value and because of this it is possible, in allocating resources and effort, to concentrate on safeguarding the best of these places. This approach forms the now well-established basis of habitat conservation whereby important examples are protected in the long-term either as nature reserves or by means of other legislative mechanisms.

5.4.3 Nature reserves and sites of special scientific interest

Three statutory categories exist in Britain, the National Nature Reserve (NNR) (Fig. 5.9), Site of Special Scientific Interest (SSSI), and Local Nature Reserve (LNR). The first two are the responsibility of the Nature Conservancy Council whilst the Local Nature Reserves fall under the jurisdiction of local authorities. Nearly 1400 other nature reserves are managed by voluntary conservation organizations notably the Royal Society for the Protection of Birds and the many County Conservation Trusts, which are affiliated to the Royal Society for Nature Conservation. The number of areas and total extent of land protected under these various categories are shown in Table 5.7.

Table 5.7 Nature Reserves and Sites of Special Scientific Interest in Britain

	Total number	*Area*
National Nature Reserves	181	138 918 ha
Local Nature Reserves	84	12 234 ha
Sites of Special Scientific Interest	3877	1 361 404 ha
County Conservation Trust Nature Reserves	1300	44 515 ha
RSPB Nature Reserves	80	37 398 ha

N.B. Some of the LNRs and non-statutory nature reserves are also designated as Sites of Special Scientific Interest. Figures are for 1981.

The main difference between SSSIs and the various categories of nature reserve is one of control. National Nature Reserves are areas which the NCC considers to be outstandingly important in the national context and in most cases the primary land use is nature conservation. Control over management of the land in question is obtained by purchase, lease or through a management agreement. Most other nature reserves are similarly managed with the aim of maintaining and enhancing their value as wildlife habitats. However, the effectiveness of nature reserves very often depends on the degree of control vested in the conservation body concerned. In general, conservation objectives are most readily achieved where land is owned, and experience has shown that short-term management agreements are least effective as they

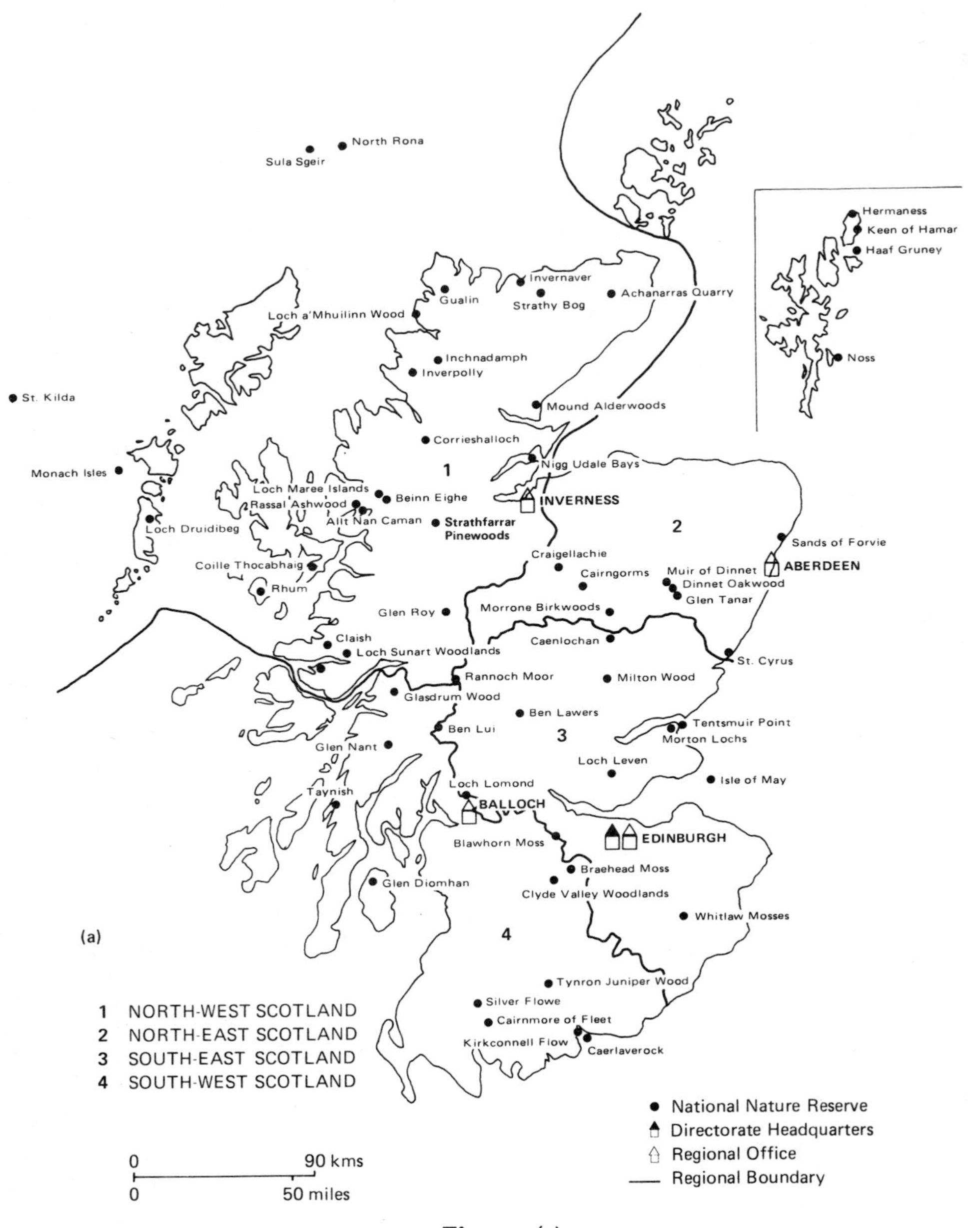

Fig. 5.9 (a)

Fig. 5.9 Distribution of the main National Nature Reserves in (a) Scotland and (b) England (from Nature Conservancy Council).

Fig. 5·9 (b)

are often subject to pressure from other, economically more rewarding, activities in direct conflict with the conservation objectives.

The SSSI designation, as defined in the *Wildlife and Countryside Act* of 1981, requires the NCC to notify any areas of land which it considers to be of special scientific interest, to the relevant planning authority and to owners and occupiers. In so doing the NCC must specify the features which constitute the special interest of the site and also state any operations (except those which are covered by normal planning control) which are likely to damage the scientific interest of the site. Local authorities are then required to consult the NCC about any proposed change of use that requires planning consent, and owners are required to notify the NCC in advance if they propose to carry out any of the specific operations which may be damaging. These may include agricultural or forestry operations which were not subject to any form of statutory control within SSSIs prior to the 1981 Act.

Over the past thirty years a considerable number of sites have been totally destroyed and the value of many others substantially reduced as a result of agricultural or forestry activities such as land drainage clearance of ancient woodland, afforestation of heathland, moorland and bog, and agricultural improvement of permanent pasture. The new requirement, under the 1981 Act, for owners to notify the Nature Conservancy Council of their intention to carry out certain operations which may be damaging, is a very significant change which could make the SSSI system far more effective. It ensures that consultation will take place and provides the opportunity for a positive approach to management of these sites, which in total cover 5·5 % of the land area of Britain (Table 5·9). On the other hand much will depend on the availability of the necessary funds to compensate land owners as required by the 1981 Act. Early indications are that this compensation mechanism will be a very costly way of safeguarding important wildlife areas, far exceeding the funds available for conservation prior to the 1981 Act. In the absence of sufficient funds for compensation there is a real danger that sites will be automatically developed for other purposes, the decision having been taken purely on financial grounds, even though the SSSI series as a whole is crucial to the conservation of wildlife in Britain. Although not fully protected as nature reserves, SSSIs perform a vital function in safeguarding Britain's flora and fauna. Moore (1982) states that:

> 'SSSIs together with NNRs should represent the basic minimum of habitat necessary to conserve the flora and fauna of Britain in something approaching its present diversity and distribution. This minimum should be supported as far as possible by conservation in countryside outside the notified areas. Officially and unofficially protected wildlife habitats support each other. Most individuals of the commoner species are found in habitats outside SSSIs and NNRs, so their continued abundance and availability to the public will depend primarily on conservation in the wider countryside. On the other hand rarer species living outside the notified areas will depend increasingly on the secure bases they have inside the SSSIs'.

He also pointed out that populations of numerous species are already virtually confined to the notified sites and that those species would become extinct in Britain if the sites concerned were destroyed.

The past 20 years has seen the NCC increasingly involved as a major agency concerned with rural planning. Substantial contributions have been made to the development of long-term strategies for conservation through submissions to structure plans, and NCC is continuously involved in appraising the environmental consequences of major development proposals. Voluntary conservation bodies, especially the county conservation trusts, also play a significant part in the planning process by submitting their views on development proposals. In some counties the trusts have recommended non-statutory sites of conservation value (additional to SSSIs) which are considered to be important at county level and in these cases there is automatic consultation between the planning departments and voluntary trusts. Assessments for selection of nature conservation sites at both national and county level, both require comprehensive knowledge of wildlife and habitats throughout the countryside. The 1970s saw a considerable effort on the part of conservation organizations to obtain this comprehensive knowledge, which is essential for the development of conservation strategies at any level. In parallel with this has been the development of more sophisticated methods for assessing conservation priorities.

5.4.4 The selection of conservation priorities

The way in which areas of importance for conservation have been selected has changed dramatically over the past 60–70 years. Early suggestions were based on the personal knowledge of a few eminent naturalists (such as the Rothschild List of 1912) and this was still the case when more comprehensive lists were produced in 1947 (Min. of Town and Country Planning, 1947). Whilst many of the chosen sites were outstandingly important and remain so today, the lists produced at that time were still far from being representative of the full range of British ecosystems. The past 15 years has seen the development of a much more systematic approach involving comprehensive knowledge of the whole wildlife resource and the development of rigorous quantitative methods for assessing the relative conservation value of particular areas (Goode, 1981 *b*). An essential step in the development of this more critical approach was the definition of criteria for assessing conservation value as described by Ratcliffe in the *Nature Conservation Review* (1977). Although these criteria were developed specifically for selecting potential National Nature Reserves the same range of criteria (including, for example, naturalness, diversity and rarity) has subsequently been used in the development of more objective systems of evaluation, which are increasingly being used in the identification of important wildlife sites including SSSIs. The aims of such a refined system of 'conservation evaluation' are to discover all the examples of a particular habitat type within a given geographical region and to select the best examples using methods which are as objective as possible. The process is best described in several stages.

The first requirement is to carry out a comprehensive survey of natural and semi-natural habitats throughout a particular geographical region, so that all examples of each habitat are located. This kind of mapping has now been completed for many parts of Britain. The next stage is the assessment of conservation value which requires comparison of like-with-like. Individual habitat types are therefore dealt with separately in order to decide which are the best examples of each type. For this to be done on a quantitative basis it must be restricted to clearly defined homogeneous habitats such as raised-bog, calcareous grassland or limestone pavement. For each of these the next, and most fundamental step, is to decide what makes a good example of the habitat concerned. It is at this stage that value judgements are made, the choice of criteria being dependent on the particular attributes deemed to be important in each type of habitat. In the case of bogs, 'naturalness' and the degree to which the hydrological regime has been modified, will be crucial factors. In coastal areas it may be the integrity of the whole ecosystem as in the case of salt marshes and sand-dunes. In contrast the value of many grassland habitats lies not in naturalness, but in the continuity of traditional forms of management which have perpetuated the desired structure and species composition.

Having established the basis on which the evaluation is to be made, all the relevant sites within the particular geographical region are examined in a consistent manner. The data collected at this stage are restricted to certain selected parameters which best reflect the criteria already established. For example, in the case of bogs naturalness has been measured by the number of bog plant species set against the number of 'invasive' species, which increase as the degree of modification becomes more pronounced. Similarly Peterken (1974*b*) has developed a technique for assessing the conservation value of woodland by means of 'indicator species'. These are plants which are characteristic of ancient woodland. The method is based on the premise that in the English lowlands such woodlands (i.e. areas with a record of continuous woodland cover throughout historic time) are the most important woodlands for nature conservation. The number of indicator species present in any individual wood can thus be regarded as a measure of naturalness. Ward and Evans (1976) used a combination of two criteria, species-richness and rarity in their assessment of limestone pavements. In all these cases the information collected at this stage is entirely factual. The final stage in this process is simply the ordering of the sites concerned in terms of their 'nature conservation value'. Once the earlier stages have been completed this can be done entirely on a quantitative basis, the value judgements having been built in at the beginning. Although the criteria differ, because they have to suit slightly different conservation objectives in each case, the final result is a series of hierarchies which together provide a firm basis for deciding priorities in nature conservation.

5.4.5 Value judgements and assessment criteria in impact analysis

The use of criteria such as rarity, diversity and naturalness has now become well established in the selection of important nature conservation sites. In recent years such criteria have also been used increasingly in the appraisal of environmental impacts, especially to assess the relative conservation value of alternative sites. A notable example is the environmental appraisal of four alternative water resource schemes in north-west England by Land Use Consultants (Chapter 6). In this case eight criteria were used, some of which were given greater weight than others. Species diversity and species rarity were regarded as particularly significant, together with rarity of each vegetation type, both within the particular sites concerned and in Cumbria as a whole.

The direct application of these criteria to impact analysis can pose practical and conceptual problems because impact studies tend to focus on individual developments, or at best on a few alternatives. Whilst the criteria are still relevant there is often no 'yard-stick' for comparison and the results of a survey using these criteria could be quite meaningless if produced in isolation. For this approach to be fully effective the sites need to be considered in a wider context, as in the case of the water resources study which was fortunate in having ecological data available for the whole county. Setting the development sites in a wider context allows their intrinsic 'conservation value' to be compared with that of other sites where this has already been established in either the national or regional context. It also allows greater significance to be attached to the predicted changes or losses if factors such as rarity and distribution of habitats and species have been quantified.

A good example is provided by Helliwell (1974) who has developed an evaluation method which simply utilizes the regional and national status of individual plant species. It can be applied to any individual habitat or mixture of habitats and as such has considerable potential for use in impact analysis. In this case the 'conservation value' for a species is a factor of its relative frequency on the site together with values for its national and regional rarity. The score for each species on a site can be added to give a total conservation value so allowing different sites to be compared. The method also provides a basis for determining the relative value of different habitats. Helliwell used a farm to illustrate the method and demonstrated that the re-seeded pastures make little contribution to the conservation value of the site compared with their total area. Hedgerows and stream banks had comparable conservation value to re-seeded pastures yet were one-hundredth the area. Woodland and permanent grasslands emerged as the most important habitats. In this case no value judgements were involved in the method except the initial premise that the relative abundance of rare species (in both a regional and national context) provides a sound basis for assessing the conservation value of an area. The basis of the evaluation is quite explicit and unambiguous.

There are inherent difficulties in using more than two or three different criteria within a single evaluation process, especially if they are all quantified. A note of caution is required here, because if several independent criteria are used, which are subsequently amalgamated in some way to produce a final assessment of quality, it can introduce severe problems of interpretation. The relationship between the criteria is complex, and their relative importance varies between habitats and between sites. By amalgamating the criteria the many different value judgements become subsumed within the method and it is virtually impossible for them to remain explicit. The decision-makers may thus be faced with an apparently objective appraisal, the subjective elements of which are obscure.

This is one of the problems inherent in any evaluation method which involves aggregation of parameters to express the environmental impact by a single value. Such 'aggregation methods' have been used extensively in EIA and this kind of approach is perhaps best illustrated by the Environmental Evaluation System developed by the Battelle Laboratories of Columbus, Ohio. The method was developed for assessment of water resource developments but could be used in any impact assessment. The argument in favour of aggregation is stated by Whitman *et al.* (1971)...

> 'the process of choosing between alternatives can be improved by relating all environmental impacts to a single set of units. By expressing parameters in common units the net environmental impact of any project is stated as a single value. This value represents, in relative terms, the nature of a project's impact on an area and the importance of an impact. Because net environmental impact is expressed as a single value, it is easily compared to other alternatives to determine the most environmentally sound approach to development of a particular resource'.

The EES method is essentially a highly developed checklist which also incorporated an evaluation process for each individual parameter. It is based on a checklist of 78 environmental and socio-economic parameters of significance in water resource development. To overcome the problem of comparing and summing impacts, parameters are weighted so that they can be related to each other in terms of their relative importance. This was done by a Delphi technique using a panel of experts. Each parameter was also examined to establish a relationship between the potential range of the parameter and an environmental quality scale from 0 (very bad) to 1 (very good). Any change in the condition of each parameter can then be reflected on the environmental quality scale. These scores are multiplied by the appropriate weights and the resulting scores are summed for each alternative. The final numerical index provides a preference ranking for the alternative projects.

So in this technique separate value judgements are made for each individual parameter firstly to decide how important each parameter is, and secondly to decide what conditions are desirable or otherwise as defined by each parameter. Examples of the weightings used are dissolved oxygen (31), pH (18), waterfowl (14), housing (13), employment opportunities (13), inorganic phosphate (28). There is a clear bias towards the physical and chemical aspects of water quality at the expense of other parameters identified. Clearly the composition of the panel of experts is a crucial factor in such a system. There are difficulties too at the next stage, i.e. relating each parameter to a scale of environmental quality. How, for instance, does one deal with pH which has a natural range of variation? The degree of change which is acceptable may well vary according to different limnological conditions. The quantification of every factor in this way is bound to result in a gross simplification of the complex ecological and social systems which the method is intended to assess.

This kind of approach has some grave disadvantages not least because a complex analytical method is allowed to replace the normal processes involving value judgements, which are after all concerned with human feelings and preferences. But there are other objections to this method. Sorensen and Moss (1973) argue that assigning numerical values to individual impacts and incorporating these in a single value disguises all the individual impacts. Thus aggregation inhibits rather than encourages public discussion of specific impacts. Indeed there is a danger that the analytical method could totally replace the normal process of decision-making. Clark *et al.* (1979) point out that 'Impacts on different environmental components are reduced to a seemingly neutral, non-controversial number, untainted by special interest, bias or inaccuracy. Laymen, including decision-makers, may believe that social and environmental scientists can predict accurately the impacts of a development. Consequently they may be inhibited from questioning results obtained from the use of this method'. Under these circumstances it is quite possible that the decisions will be made by those carrying out the investigation and not by those responsible for deciding whether a project should proceed or not (Dee *et al.*, 1973). It is, however, heartening to find that more recent studies by the Battelle Laboratories have provided disaggregated information, and guidelines for evaluating the significance of measured environmental effects. In fact Duke *et al.* (1977) recognize that methods which provide composite indices of environmental impacts are unsuitable for presentation of environmental impact data to decision-makers.

5.4.6 Value to people

Much of what is written about values and priorities in nature conservation is concerned with the intrinsic features of individual areas rather than the value which is placed upon such areas by society. Whilst conservation bodies have developed sound systems for evaluation which allow different places to be put on a scale of importance in nature conservation terms, nevertheless the kind of value judgements involved are not directly related to the basic underlying reasons for conservation. For instance the criteria for selection of National Nature Reserves, or SSSIs, are specifically concerned with selecting a representative series of habitats and ensuring the survival of our native species of flora and fauna. In applying these criteria the emphasis is placed on the characteristics of individual sites such as naturalness, diversity and rarity. But these do not in themselves explain why such areas should be conserved. The more fundamental reasons for nature conservation may be stated explicitly, as in Cmd 7122 (Min. of Town and Country Planning, 1947) and in the Nature Conservation Review (Ratcliffe, 1977), but very often there is a tendency for such reasoning to be assumed or disregarded in making decisions on individual cases. The arguments for protection of important areas as nature resources were discussed at length in Cmd 7122 and also by Tansley (1945). At that time the main emphasis was placed on the need to ensure the protection of an adequate series of representative habitats for scientific study and education, though Tansley also argued strongly that the beauty of the landscape depended largely on areas of natural and semi-natural vegetation and for that reason 'we should preserve it as far as we can'.

As man's impacts on the natural world have increased in post-war years so the arguments for conservation have gradually shifted. Apart from the moral arguments, that we have a responsibility towards the natural world, or towards future generations of mankind, there are strong arguments for nature conservation on the grounds that it is in our own interest either culturally or materially. Reasons for conservation include the simple enjoyment of nature, the satisfaction of scientific discovery, the potential utility to man of all species of plants and animals as a natural resource, and the need to maintain a healthy 'biosphere' for mankind's own survival!

The relative weight given to these different arguments varies according to circumstances, particularly the scale at which conservation is being considered. On a world scale emphasis has been placed by international agencies on the need for conservation of vital ecological systems and the prudent management of natural resources as a basis for sustained development. In contrast, aesthetic and cultural arguments have gained prominence in recent years in Britain and several other European countries as the fundamental basis of conservation. This rapidly developing ethos, which is reflected in a wide range of human activities, seems to stem from a deep desire amongst many people for greater contact with nature. Mabey (1980) pointed out that preservation of a series of 'special' sites as nature reserves or SSSIs does little to provide for the continued existence of ordinary wildlife for people to enjoy in their own local surroundings. Since then, and particularly since the publication in 1980 of Shoard's analysis of recent changes in the British countryside, there has been a dramatic upsurge of concern and interest in wildlife and habitats. This shift in social attitudes may well be of great significance and radically affect the practice of nature conservation. Already there are signs that many local communities are taking positive action to develop conservation programmes, as in the case of 'Parish Action Plans'. The emphasis is on the needs of people and their local environment, rather than on the 'scientific' justification for conservation of specific examples of the natural world. The distinction is perhaps most pronounced in the case of urban environments, where the importance of habitats depends on their value to local communities and the potential of such areas for creative conservation rather than on the intrinsic merits of individual sites.

So there is now a pressing need for the development of an additional set of criteria, based on values for people, which can apply to both the urban environment and the ordinary features of our countryside. Such a shift in emphasis does not, however, bring into the question the need for nature reserves or other specially important sites, as defined on scientific criteria; such areas have a vital role to play in the effective conservation of wildlife in Britain, and will continue to do so whatsovever happens at the 'local level'. What this shift in emphasis does mean is that there are other values stemming from the needs of local communities, which have not been catered for to any great extent in the past and which need to find expression in the development of future land-use policies.

Recent years have also seen strong arguments advanced by conservation bodies for a national land-use strategy. It is argued that such a strategy, including the major interests of agriculture, forestry, water resources and nature conservation, could be achieved by an appropriate framework of fiscal incentives. The NCC has argued in evidence to Parliament (House of Lords Select Comm., 1980) that – 'Britain has entered on an intensified phase of resource exploitation which requires a fundamental rethinking of national policy if nature conservation is to survive as a meaningful concern and activity into the 21st century; and that the sectional approach to land and resource use which has proved acceptable in the past is no longer valid'. Moore (1980) has indicated the essential ingredients of such a strategy, which would require guidelines on national priorities to be accepted by all government departments and agencies. These guidelines on priorities would then form the basis of grant aid, for agriculture, forestry or conservation. Agreed priority use would involve presumptions that, for instance, agriculture would be the primary land use on all grade 1 and 2 land and in the same way Nature Conservation would have priority on all grade 1 and 2 sites as defined in the Nature Conservation Review (Ratcliffe, 1977). Moore points out that 92 % (by area) of such conservation sites is poor quality agricultural land and there would be few serious clashes with grade 1 or 2 land, or even grade 3 land as previously defined. This kind of principle has already been accepted in Scotland, in the context of development control. The National Planning Guidelines for Scotland indicate that any development proposal affecting grade 1 and grade 2 conservation sites, which the planning authority propose to approve against NCC advice, is likely to raise a national issue. This must then be notified to the Secretary of State. The guidelines suggest that 'it will be in the interests of both the developer and the planning authority, and a more rapid determination will be achieved, if it is accompanied by a project appraisal statement describing the project and its implications for the environment'.

In effect such a strategy, agreed at Government level, would provide an opportunity for values other than those of financial profitability to be accommodated within the planning of our countryside. Although applied at the national level through fiscal policies, such a strategy could provide a meaningful approach even at the regional or local level since it would operate through financial incentives rather than through a form of planning control.

The importance of environmental values in such a strategy is illustrated by the fact that conservation bodies are increasingly questioning the economic arguments advanced by other land users, especially in cases of marginal agriculture and forestry. Those concerned with conservation argue that other values need to be taken into account in assessing the best use for society as a whole. In 1980 a proposal to convert 81 ha (200 acres) of intertidal saltmarsh on the Wash into arable farmland was opposed by the NCC along with several voluntary conservation bodies. In its evidence to the public inquiry the NCC argued that the area formed part of an internationally important area for wildlife conservation which should have precedence over the proposed agricultural use since there was no national need for the extra agricultural output which would result from the reclamation. NCC provided a detailed cost-benefit analysis of the agricultural proposals, from which it concluded that the overall scheme would have a negative value to society. The application was subsequently refused.

In parallel with these arguments for an agreed land use strategy the NCC has advocated (NCC, 1981) that the principle of environmental impact analysis should be applied to major policies affecting the countryside, such as the EEC Common Agricultural Policy. It has pointed out that the environmental effects of changes in such policies are given virtually no consideration, yet even minor changes may have profound environmental consequences.

In all these cases, whether they deal with local communities or national strategies, the one common aim is the attempt to ensure that environmental values form part of the normal process of decision making. This was the very basis of the *National Environmental Policy Act* of the USA (US Senate, 1969). Although this Act is probably best known for introducing EIA procedures, it is worth recalling that the Act embodied a broad statement of environmental policy which identified the need for proper consideration of environmental *values*. The Act required all Federal agencies 'to identify and develop methods and procedures which will ensure that presently unquantified environmental amenities and values may be given appropriate consideration in decision making along with technical and economic considerations'.

Ten years later Ashby (1979) pointed out that there has been a remarkable trend in social evolution. 'Just over a century ago it was still arguable whether or not you might sell a man, compel a child to work in the mines, be cruel to a dog, dig up a rare orchid. All these are now forbidden by law in many nations. The laws made explicit what was implicit in the values held by society. They did more than this; they paved the way for the development of more refined values in the attitude of man toward his fellow man and toward other living things.' Legislation for the protection of endangered species now means that the presence of single species, like the Furbish Lousewort or Snaildarter fish, may preclude major industrial or water resource schemes. Ashby believes that the extent to which social values continue to flow along this channel of widening empathy, will depend on the support which exists for an environmental ethic. All the signs are that such an ethic is rapidly developing. The environmental values which I have discussed are manifestations of this underlying ethos – man's concern for the natural world.

5.5 PLANNING A NEW COUNTRYSIDE

B. H. Green

5.5.1 Introduction

Before the impact of modern intensive agricultural and forestry practices described in Section 5.2 these industries created and maintained diverse ecosystems and an attractive countryside. Such ecosystems, for example, coppiced woodlands and permanent pastures, were mainly composed of unsown, indigenous species and some were analogous with quite natural communities long disappeared from most of Britain. If one wishes to maintain similar ecosystems today one can moderate agricultural and forestry practice to reduce their damaging environmental impacts, or one can set aside areas which are managed in the traditional way.

5.5.2 Some ecological principles relevant to countryside planning

The fact that agriculture and traditional forestry practice once maintained a rural environment rich in wildlife and recreational opportunity, and that these land uses determine the nature of nearly all the countryside (agriculture more than 80%, forestry perhaps 10%), has led most people to believe that most provision for the protection of wildlife, landscape and access must be made by integrating these land-uses with modern agriculture and forestry (e.g. Barber, 1970; Leonard and Cobham, 1977; Steele, 1972). Attempts have indeed been made to demonstrate that the retention of hedgerows and the adoption of more diverse crop ecosystems might also be advantageous to agriculture and forestry (Allen, 1974). There is some evidence that diversity in crops can reduce pest infestation (e.g. Southwood, 1981). Unfortunately ecological principles seem to dictate that this can only be bought by a fall in production and room for real compromise seems limited.

The overriding objectives of both agricultural and forestry policy are to maximize production of food and timber. High primary production is typically a feature of fertile ecosystems populated by a few species with rapid relative growth rates, such as, for example, reedswamps. Agricultural ecosystems are analogues of such ecosystems with vigorous species responsive to fertile conditions – 'dung heap superweeds' (or crops) (Pirie, 1969) – being cultivated in monocultures. All their water and nutrient needs are supplied and competitors and predators eliminated with herbicides and pesticides.

The overriding objectives of conservationists, however, are quite the opposite. A list of attributes thought desirable in the selection of sites worthy of protection for wildlife reasons has been published by the Nature Conservancy Council (Ratcliffe, 1977). I have argued (Green, 1981) that the two most important of these ecosystems attributes are diversity and rarity. If one wishes to protect as many species and habitats as possible, protecting a diverse ecosystem will clearly be more effective than protecting a simple one. Rare species and habitats merit protection simply because their small populations or extent render them more vulnerable than common species and habitats. Protecting rarity and diversity are also arguably the main objectives of landscape and access conservation. Uniformity is the burden of modern society, but 'variety is the spice of life', diversity means options, the possibility of choice. Rarity, or the unusual, is something which is of universal appeal whether it be a species, view or a postage stamp.

It is by no means certain as to what generates diversity in ecosystems. An essential prerequisite is obviously a diverse fauna and flora which can provide a rich pool of species to draw upon. Environmental heterogeneity, and its stability over sufficient time for species to evolve into a variety of different situations seems to be important for the evolution of a lot of species. Tropical regions are thought to have such diverse biota because the relative stability of their climates over long periods of time has allowed many species to evolve and survive compared with more temperate areas where glaciations have caused major upheavals and loss of species. Ease of colonization is also obviously an important factor controlling the species richness of

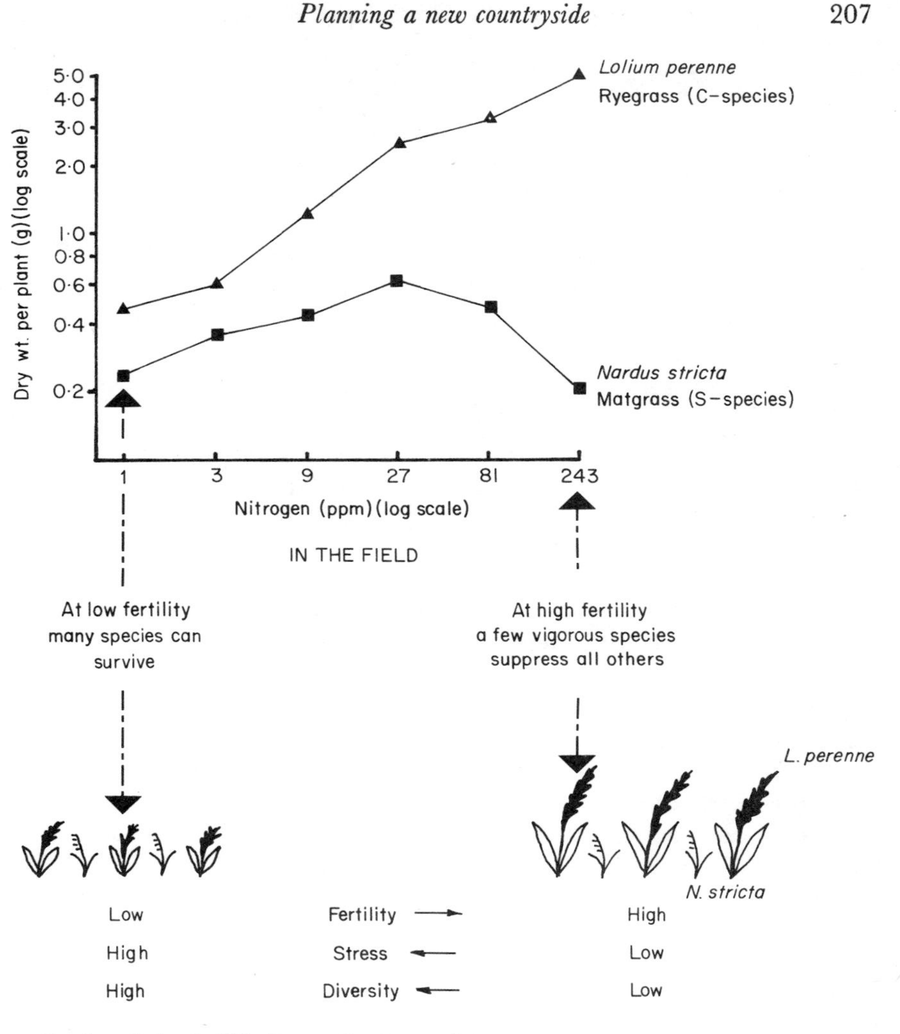

Fig. 5.10 Variation in yield of grass species in and culture with different levels of nitrate nitrogen (adapted from Bradshaw *et al.*, 1964).

an area; there seems to be considerable empirical evidence for larger, less isolated islands, or island habitats, having richer flora and faunas, even though the mechanisms proposed to explain this in island biogeographic theory now seem to be in doubt (MacArthur and Wilson, 1967; Sullivan and Shaffer, 1975; Simberloff and Abele, 1976; Gilbert, 1980). The conditions under which species available from the biota come together into diverse ecosystems are perhaps better understood than those which lead to the evolution of a diversity of species. The rather obvious argument once propounded that abundant resources and high production provide opportunities for a lot of species is now known to be quite wrong. Indeed, at least insofar as herbaceous communities are concerned, it is clear that the least productive are the most diverse and that infertility and other forms of environmental stress like predation grazing, or fire, are key factors in generating diverse ecosystems (Paine, 1966; Harper, 1968; Grime, 1973). All these forms of stress generate diversity by constraining the growth or population sizes of vigorous species thus allowing many weaker competitors to survive (Fig. 5.10).

If this mechanism of competitive exclusion is to account for ecosystem diversity there must obviously be only relatively few strongly competitive species, and this indeed seems to be the case. When the population sizes of all species in an ecosystem are sampled there are typically a few species with large populations and many species with small populations (Fig. 5.11). Common species are rare and rare species are common. The common species with large populations are usually those of wide tolerance which are vigorous and competitive under productive conditions. The rare species are the weaker competitors under these conditions, specialized to occupy narrow niches in extreme environments. It seems that two of the main adaptive strategies in plants and animals are either to become a good competitor at the cost of being able to utilize extreme environments, or to evolve the means of doing this, but sacrificing the ability to compete under more productive conditions (Drury, 1974). These correspond roughly to the adaptable, competitive (C–) and adapted, stress (S–) strategy species of Grime (1977).

The maintenance of diversity and rarity as important conservation objectives is given added weight by the fact that most species are rare species and thus that diverse ecosystems are constituted mainly of them. Common species are adapted to the very same moderate and productive conditions that man is constantly creating to favour his crop species. Such conditions, exemplified by river silt banks or animal dunging areas, are scarce or transient in nature and being unpredictable environments are probably difficult to speciate into. This perhaps is why common, adaptable species are relatively few in number. But they are now favoured in man-made environments and are therefore in little need of protection. If you make environments suitable for nettles and cow parsley, sparrows and starlings, rats and mice, then these are the species which will predominate in them. If you want a world in which there is the rare and the diverse then marshes must remain undrained, pastures unfertilized and deserts unirrigated. The objectives of conservationists and farmers and foresters are thus diametrically opposed. The difference between their two strategies is beautifully illustrated by the Park Grass fertilizer trials at Rothamsted (Fig. 5.12). The 60 species recorded

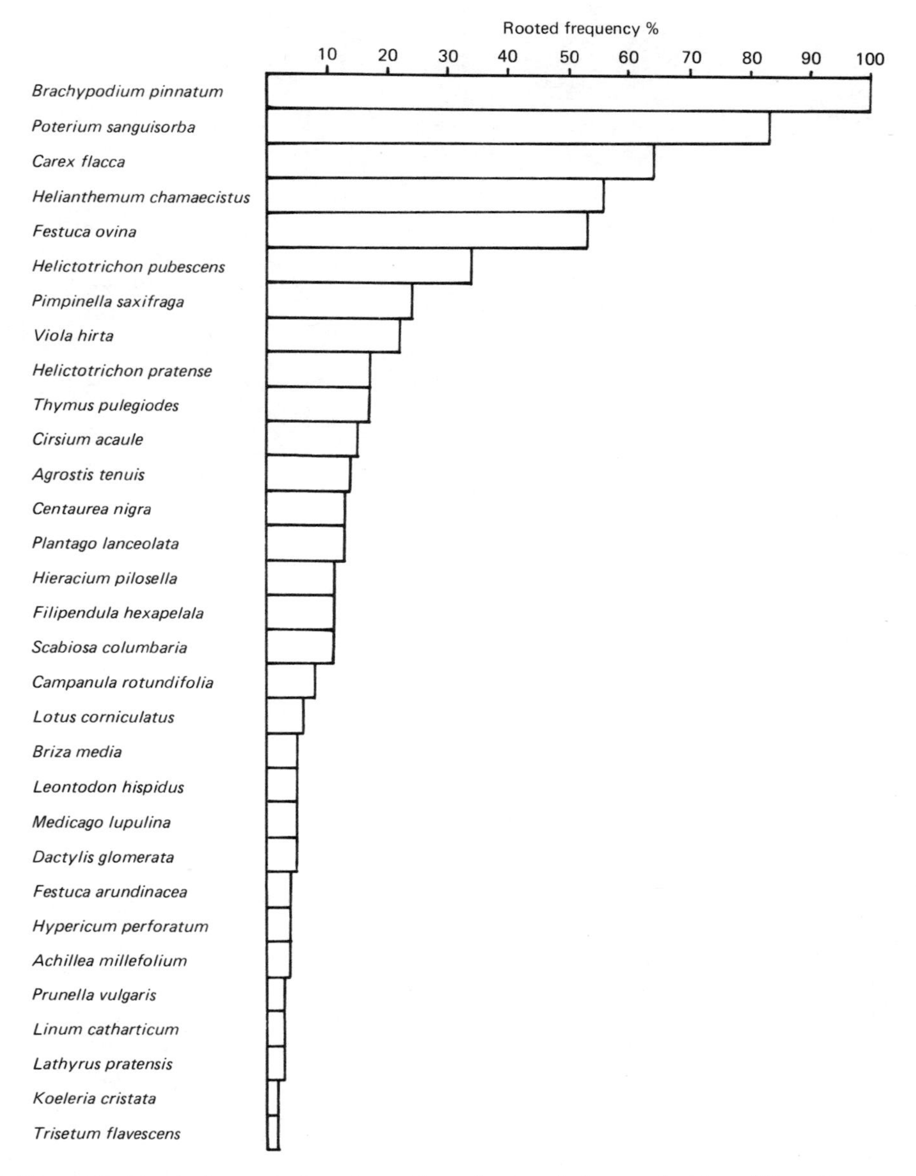

Fig. 5.11 Ranked population sizes of species in chalk grassland at Wye, Kent. Similar patterns of species abundance, with few common species and many rare species, are typical of many groups of species in many ecosystems, e.g. moths caught in light traps (author's data).

Fig. 5.12 The Park Grass Trials, Rothamsted Experimental Station, Harpenden, England. The plots in the foreground have received a complete fertilizer treatment and are much more productive, but less diverse, than the unmanured plots to the left (photograph: B. H. Green).

in the untreated unproductive control plots have been reduced to 2 or 3 in some of the productive fertilized plots. There is no doubt about which state the conservationist would like to see most of our pastures, nor about the state to which farmers have made them.

5.5.3 Planning and the countryside

The essence of the present Town and Country Planning legislation in Britain stems from the principles developed during World War II and embodied in the 1947 Act. The 1947 and subsequent *Town and Country Planning Acts* were designed in a large part to keep the town, both urban sprawl and urbanites, off the countryside. The main mechanism for doing this was development control, but there was no provision for control of agriculture and forestry for, far from being seen as threats, they were regarded as the main forces maintaining the amenities of the countryside. In 1942, the influential Scott Committee on Land Utilization in Rural Areas reported 'Farmers and foresters are unconsciously the nation's landscape gardeners . . . even were there no economic, social or strategic reasons for the maintenance of agriculture, the cheapest way, indeed the only way of preserving the countryside in anything like its traditional aspect would still be to farm it'.

Nor at that time was it foreseen that agricultural practice would change so dramatically 'In our opinion a radical alteration of the types of farming is not probable and no striking change in the pattern of the open countryside is to be expected.' Despite the publication of excellent reports by the state conservation agencies outlining the severe environmental impacts of modern agricultural practice (Westmacott and Worthington, 1974; Nature Conservancy Council, 1977) it has taken a long time for the full significance of the changes in agriculture to be appreciated. In 1978, the Advisory Council for Agriculture and Horticulture reported: 'There is an evident concern about the harmful effects of many current farming practices upon both landscape and nature conservation, coupled with a widespread feeling that agriculture can no longer be counted the prime architect of conservation, nor farmers accepted as the "natural custodians of the countryside"' (Strutt, 1978). Yet there is still evidence in its reports that even the interdepartmental Countryside Review Committee set up to examine changes in the countryside has not fully grasped the magnitude of the problem (Countryside Review Committee, 1976–1979).

This failure fully to recognize what has been happening has bedevilled the development of measures adequately to protect wildlife, landscape, access and the other cultural and amenity values of the countryside. Much faith has been placed in our superficially impressive state system of National Parks (NPs), Areas of Outstanding Natural Beauty (AONBs), National Nature Reserves (NNRs) and Sites of Special Scientific Interest (SSSIs) (Table 5.8). Yet although National Parks cover some 9% of England and Wales the protection that designation confers against the changes in agriculture and forestry which are the main threats to them is minimal. The huge loss of moorland in National Parks has been documented in an earlier section. The protection conferred by AONBs is even less (Anderson, 1980). Likewise only a quarter of the tiny NNR acreage (less than 1% of GB) is owned freehold and under the complete control of the Nature Conservancy Council and there is an enormous shortfall of some 0·75 m ha between what is protected and deemed worthy of such protection in the Nature Conservation Review (NCR) (Ratcliffe, 1977). SSSIs confer some protection to NCR sites against development controlled by the Town and Country Planning legislation, but this does not include any control over agriculture and forestry which are the main threats to them. We are lucky that the National Trusts, Royal Society for the Protection of Birds, County Naturalists Trusts and other voluntary bodies help protect some of this land, but apart from that held inalienably by the National Trusts the security of tenure of much of this land is even weaker than much of that held as NNR.

It has always been recognized that specially designated reserve areas can

Table 5.8

Statutory	*Nos.*	*ha*	
	Protected land in Great Britain		
Local Nature Reserves (LA)	56	7 129	(1977)
National Nature Reserves (NCC)	161	132 853[a]	(1979)
National Park access lands	?	?	
County Parks (LA mainly)	153	20 324	(1977)
Other public open spaces	?	?	
Statutory access areas (LA)	105	36 907[h]	(1975)
Ministry of Defence land	?	287 679[b]	
Voluntary			
Country Naturalists Trusts	1080	36 470[c]	(1978)
Royal Society for the Protection of Birds	72	31 696[d]	(1978)
National Trusts	?	246 300[e]	(1977)
	Land worthy of protection in Great Britain		
NCR sites G1	395	627 100	
NCR sites G2	207	286 300	
NCR sites G1 and G2	702	913 400[f]	
SSSIs	3900	1 300 000[g]	
National Parks	10	1 360 100	(1978)
AONBs	33	1 448 000	(1978)
Commons	?	607 050[h]	(1958)

Notes

(a) 25% owned by NCC, 60% agreements, 15% leased
(b) Much of this land is of considerable value for nature conservation
(c) Nearly 50% (65% area) CNT reserves are SSSIs and 17% (47% area) of these are G1/2 sites. CNTs own 28%, private landowners 37%, local authorities 11%, Forestry Commission 8%
(d) UK not GB
(e) Includes gardens as well as open spaces
(f) Estimated that over 90% of low agricultural value, 9% already NNR, some also protected by other conservation bodies e.g. NTs 5% and other public agencies e.g. MOD, LA
(g) Includes all key (G1 and 2) NCR sites
(h) England and Wales not GB
(i) There is, of course, substantial overlap between all these designations

Main Source: Digest of countryside recreation statistics (Countryside Commission, 1978).

never realistically be expected to protect more than the very most important sites and it has always been an additional objective of conservation organization to persuade land owners to manage their land in a way whereby it retains its scientific and amenity value. There are provisions in the 1949 *National Parks and Access to the Countryside Act* whereby National Park Authorities and Local Authorities can enter into agreements with landowners to provide public access to open country, and in the 1968 *Countryside Act* whereby the Nature Conservancy Council can do so to ensure that the management of an SSSI is carried out in such a way as to protect its flora, fauna or geological interest. Where the land is of little agricultural potential such agreements can provide some small, but welcome, additional income to the landowner from the state and happily add to the extent of protected land. Unfortunately, with modern technology there is now very little land which is not of agricultural potential and landowners solicited to enter into such agreements naturally expect to be reimbursed to the full opportunity cost they would realize were they to go ahead with their agricultural improvements. The difference in farm income between rough grazing on a marsh or moor and enclosed arable or improved pasture can be more than £100 ha^{-1} an^{-1}, and the capital value of the land can increase from hundreds to thousands of pounds per hectare. With a million hectares of land regarded as worthy of protection for nature conservation purposes alone it is clear that vastly larger sums than are presently allocated for conservation would be needed if land is to be protected by compensation agreements of this kind. And this does not include the large areas in the matrix of the countryside where conservation management is desirable for landscape, access and additional wildlife protection. Some landowners are prepared to make small concessions to scientific and amenity interests by foregoing some intensification by leaving copses, hedges, ponds and planting trees for their own pleasure and altruistic reasons without any reimbursement. Such measures can make a big difference at the local level but do only little to balance the great losses incurred in ploughing meadows, draining marshes and grubbing out woods.

Despite their active promotion as a main plank of conservation policy many conservationists now conclude that voluntary agreements with landowners have failed to stem adequately the loss of countryside amenities. Furthermore, at a time when many stable foodstuffs are in surplus in the European Economic Community there has been a fundamental questioning of the enormous state subsidy for the agricultural industry through commodity price support and agricultural improvement grants, without which it is certain that much of the loss of countryside and wildlife would not take place. In particular it has been asked why agricultural developments should be immune from planning control, and why there should be compensation for the control of agricultural, but not other kinds of industrial development (Shoard, 1980). There have been calls from a government-appointed committee for the planning control of forestry in National Parks (Sandford, 1974) but not yet for the control of agriculture. But moves in this direction have begun. The study for the Department of the Environment and Ministry of Agriculture to investigate moorland loss in Exmoor National Park (Porchester, 1977) concluded that management arrangements to protect moorland should be orders rather than agreements (i.e. mandatory, rather than voluntarily entered into) and that compensation should be on a once and for all, rather than annual basis. Conservation organizations have pressed strongly for similar measures to be introduced into the current Wildlife and Countryside Bill. These proposals have been equally strongly resisted by the powerful agricultural lobby, principally the National Farmers Union and County Landowners Association. The Department of the Environment and the Ministry of Agriculture seem to have aligned themselves with their respective interests and there has been a remarkable amount of parliamentary time spent on such a measure. It seems likely that the government will make some concessions whereby landowners proposing agricultural or forestry developments in at least SSSIs will, for the first time, be statutorily bound to consult the state conservation agencies and that there will be some arbitration system over management agreements.

Whether this measure proves to be successful in stemming the loss of countryside amenities would seem to depend on a number of factors. Even if enough money is made available for compensation there is some evidence, both in this country and on the continent, that many farmers are just not temperamentally inclined to take money to farm in a way which is inefficient by present standards and become 'park managers' (Newby *et al.*, 1977; van der Weijden *et al.*, 1978). Nor is it clear how compensation can be calculated, or abuse of the system avoided – what is to prevent a landowner seeking compensation for not felling a wood which he has no real intention of touching? It is difficulties such as these which have led to the calls for control of agriculture and forestry by normal planning procedures. But this also presents considerable problems, especially in defining developments to be controlled. Should planning permission, with all its consequent delay, be necessary for fertilizing a meadow, or cutting a hedge? It seems unlikely that any extension of planning control, except perhaps to cover the more major improvement schemes, is going to be entertained at a time when the planning system is already under attack for its delaying of developments and large bureaucracy. But there are indications that a change in government could reverse this situation.

For so long as it is agricultural policy to maximize production and publicly support farm incomes principally through subsidizing product prices there can be only limited opportunity of persuading farmers to forego making the most profitable use of their land. In a period of rising costs it may be necessary for a farmer to bring more and more land into production just to maintain his living standards. The Common Agricultural Policy, is, however, now under debate for its enormous costs (£6000 million/an, or 60% of the entire EEC budget), surplus foodstuffs, and inequitability within the farming community. Its flat rate price support per unit yield, set to guarantee smaller farmers a reasonable livelihood, means the larger farmers reap huge incomes, and it is these large enterprises which are responsible for generating much of the unwanted food surpluses and environmental damage. A change to a system encompassing differential income support and product quotas seem possible. Some economists, including an erstwhile EEC Agricultural Commissioner, Professor Mansholt, have argued that for agricultural reasons millions of hectares of agricultural land should be released from production (Boddington, 1973). Conservationists are now making their voices heard on this debate to ensure that any changes made in agricultural policy relieve the current imperative to intensify farming and incidentally destroy countryside amenities (e.g. Royal Society for Nature Conservation, 1981).

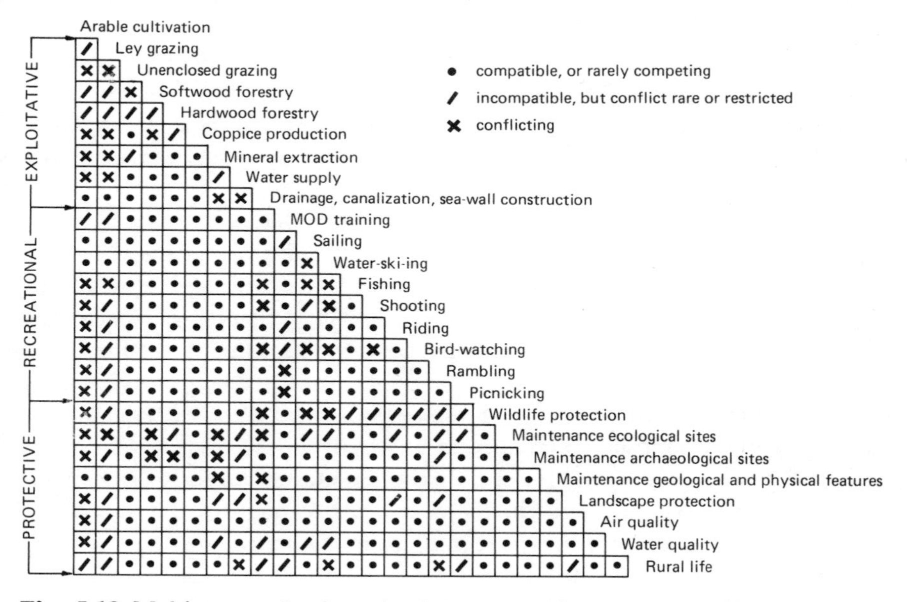

Fig. 5.13 Multipurpose land-use in the countryside – a compatability matrix of competing activities (from Green, 1977).

In the short term I have argued that it would seem better to concentrate conservation resources into protecting as much land as possible in large designated conservation areas with compatible land uses (Fig. 5.13) rather than using the money to try and achieve dubious agreements with landowners to integrate some very much secondary conservation objectives into the primary agricultural use of land (Green, 1975, 1977). It is only by this means that the large tracts of natural and semi-natural ecosystems required by many species can be protected. To protect such 'regional conservation areas' there will have to be much closer co-operation between conservation bodies concerned with wildlife protection and those concerned with landscape protection and public access; both in site acquisition policies, and in the use of zoning systems like those of conservation areas in other parts of the world employed to avoid conflicts between people and environment. Management of such areas where plagioclimax communities such as downland or heathland are present can also be difficult in the absence of rangeland grazing systems or other forms of extensive agriculture. The threat to such areas by seral development of rank swards and scrub is, however, often exaggerated and the ability to maintain them readily, effectively and safely by mowing or prescribed burning, underestimated.

Forestry practice perhaps offers more opportunity for the integration of wildlife, landscape and access conservation into woodland management than does modern agricultural practice. Peterken (1977) has outlined some general principles by which this might be achieved. A review of government forestry policy in 1972 recognized this as an important future objective. The new policy and grant schemes announced in 1974 provided for additional grants to be made to woodland owners for planting hardwoods and introduced measures for closer consultation between the Forestry Commission and local planning authorities and the conservation agencies.

There is still concern amongst conservationists that afforestation in the uplands especially is a potentially serious threat to areas of scientific and amenity value. The issues are much less clear cut than with agriculture and a review paper promised by the Nature Conservancy Council has been long delayed. The main conservation requirements are that the protection of the best woodland and moorland sites should be regarded as an over-riding commitment, and that afforestation elsewhere is undertaken with the protection of wildlife, landscape and access as important secondary objectives. As the Forestry Commission is not only the state forest authority, but by far the biggest forestry enterprise in the country, this ought to be able to be achieved much more readily by consultation without resort to planning control than is the case with agriculture, where a much larger number of private farming enterprises are involved. But this advantage may disappear. A new review of forestry policy is currently being undertaken and the extent of public forestry and the release of Forestry Commission holdings are important topics under consideration.

5.5.4 Conclusions

It is arguable that more has been achieved by designating various kinds of protected areas for conservation and recreation outside the agricultural and forestry ecosystems, i.e. by segregating land uses, than by attempting to moderate the environmental impacts of the new technologies of these industries. But such protected areas can only ever be expected to cover a very small proportion of the land surface and there is plenty of evidence in the advancing tide of scrub over many protected heaths and downs that the necessary kind of traditional management is often very difficult to implement.

If we want a countryside rich in wildlife, landscape and recreational opportunity it is clear that both approaches must be pursued in tandem.

REFERENCES

Aldhous, J.R. (1972) in *Lowland Forestry and Wildlife Conservation* (ed. R.C. Steele), Monks Wood Experimental Station. Symp. No. 6.

Allen, R. (1974) *New Scientist*, **63**, 528–9.

Anderson, M. (1979) in *Ecology and Design in Amenity Land Management* (eds S.E. Wright & G.P. Buckley), Wye College, London.

Anderson, M.A. (1980) *Landscape Planning*, **7**, 1–22.

Ashby, E. (1979) *The Search for an Environmental Ethic*, The Tanner Lecture on Human Values, University of Utah.

Avery, B.W., Findlay, D.C. & Mackney, D. (1975) *Soil Map of England and Wales*, 1:1,000,000 Ordnance Survey.

Ball, D.F., Dale, J., Sheail, J. & Heal, O.W. (1982) *Vegetation Change in Upland Landscapes*, Institute of Terrestrial Ecology, Cambridge.

Barber, D. (ed.) (1970) *Farming and Wildlife: a Study in Compromise*, Royal Society for the Protection of Birds, Tring, Herts.

Beresford, T. (1975) *We Plough the Fields*, Penguin, Harmondsworth.

Best, R.H. (1976) *Town and Country Planning*, **44**, 171–6.

Bibby, J.S. (1980) *Land Assessment in Scotland* (ed. M.F. Thomas & J.T. Coppock), Aberdeen University Press, Aberdeen, pp. 25–36.

Bishop, I.D. (1978) *Land Use in Rural Cumbria – a Linear Programming Model*, Institute of Terrestrial Ecology, Grange-over-Sands.

Blacksell, M. & Gilg, A.W. (1981) *The Countryside: Planning and Change*. Allen & Unwin, London.

Boddington, M. (1973) *Built Environment*, **2**, 443–5.

Bradshaw, A.D., Chadwick, M.J., Jowett, D. & Snaydon, R.W. (1964) *J. Ecol.*, **52**, 665–70.

Bunce, R.G.H., Morrell, S.K. & Stel, H.E. (1975) *J. Environ. Mgmt.*, **8**, 151–65.

Bunce, R.G.H. & Smith, R.S. (1978) *An Ecological Survey of Cumbria*, Structure Plan Working Paper, No. 4, Cumbria County Council, Kendal.

Bunce, R.G.H., Barr, C.J. & Whittaker, H.A. (1981*a*) *Land Classes in Britain: Preliminary Descriptions of the Merlewood Method of Land Classification*, Institute of Terrestrial Ecology, Grange-over-Sands.

Bunce, R.G.H., Barr, C.J. & Whittaker, H.A. (1981*b*) Annual Report Institute of Terrestrial Ecology 1980, 28–33.

Bunce, R.G.H., Barr, C.J. & Whittaker, H.A. (in press) in *Ecological Mapping from Land, Air and Space*, Institute of Terrestrial Ecology, Cambridge.

Buse, A. (1974) *J. Appl. Ecol.*, **11**, 517–28.

Centre for Agricultural Strategy (1980) *Strategy for the UK Forest Industry*, (CAS report No. 6), Centre for Agricultural Strategy, University of Reading, Reading.

Chandler, T.J. & Gregory, S. (1976) *The Climate of the British Isles*, Longmans, London.

Clark, B.D., Chapman, K., Bisset, R. & Wathern, P. (1978) *Environmental Impact Assessment in the USA: A Critical Review*, DOE Research Report, No. 26, Department of the Environment, HMSO, London.

Clark, B.D., Chapman, K., Bisset, R. & Wathern, P. (1979) in *Land Use and Landscape Planning* (ed. D. Lovejoy), Leonard Hill.

Coleman, A. (1961) *Geog. J.*, **127**, 168–86.

Coppock, J.T. & Gebbett, L.F. (1978) *Land Use and Town and Country Planning*, Reviews of UK Statistical Sources, Vol. 8, Pergamon, Oxford.

Countryside Commission (1978) *Digest of Countryside Recreation Statistics*, CCP86, Countryside Commission, Cheltenham.

Countryside Review Committee (1976) *The Countryside – Problems and Policies*, HMSO, London.

Countryside Review Committee (1977) *Leisure and the Countryside*, HMSO, London.

Countryside Review Committee (1978) *Food Production in the Countryside*, HMSO, London.

Countryside Review Committee (1979) *Conservation and the Countryside Heritage*, HMSO, London.

Cunningham, J.M.M., Eadie, J., Maxwell, J.J. & Sibbald, A.R. (1978) *Scottish Forestry*, **32**, 182–93.

Dane, C.W., Meador, N.C. & White, J.B. (1977) *J. Forestry*, **75**, 325–9.

Dee, N., Baker, J.K., Drobny, N.L., Duke, K.M., Whitman, I. & Fahringer, D.C. (1973) *Water Resources Res.*, **9**, 523–35.

Dent, A. (1974) *Lost Beasts of Britain*, Harrap, London.

Dimbleby, G.W. (1952) *J. Ecol.*, **40**, 331–41.

Drury, W.H. (1974) *Biol. Conserv.*, **6**, 162–9.

Duffield, B.S. & Owen, M.L. (1970) *Leisure + Countryside = a Geographical Appraisal of Countryside Recreation in Lanarkshire* (ed. J.T. Coppock), Department of Geography, University of Edinburgh, Edinburgh.

Duke, K.M., Dee, N., Fahringer, D.C., Maiden, B.G., Moody, C.W., Pomeroy, S.E. & Walkins, G.A. (1977) *Environmental Quality Assessment in Multiobjective Planning*, Battelle Columbus Laboratories, Columbus, Ohio.

Elton, C.S. (1966) *The Pattern of Animal Communities*, Methuen, London.

Fanstone, G.H. & Himsworth, K. (1976) *Choices for Cumbria. Report of Survey*, Cumbria County Council and Lake District Special Planning Board, Kendal.

Ford, E.D., Malcolm, D.C. & Atterson, J. (1979) *The Ecology of Even-aged Forest Plantations*, (eds E.D. Ford, D.C. Malcom & J. Atterson), Institute of Terrestrial Ecology, Cambridge.

Forestry Commission (1952) *Census of Woodlands* 1947–1949, Census Report No. 1, HMSO, London.

Forestry Commission (1977) *The Wood Production Outlook in Britain: a Review*, Edinburgh.

Fryer, J.D. & Chancellor, R.J. (1974) in *The Flora of a Changing Britain* (ed. F. Perring), Classey, Faringdon.

Gilbert, F.S. (1980) *J. Biogeog.*, **7**, 209–35.

Goode, D.A. (1981*a*) *New Scientist*, **89**, 219–223.

Goode, D.A. (1981*b*) in: *Values and Evaluation* (ed. C.I. Rose), Discussion Paper in Conservation No. 36, University College, London.

Green, B.H. (1975) *Landscape Planning*, **2**, 179–95.

Green, B.H. (1977) *The Planner*, **63**, 67–9.

Green, B.H. (1978) *Biologist*, **25**, 75–9.

Green, B.H. (1981) *Countryside Conservation.* George Allen and Unwin, London.

Greene, L.A. & Walker, P. (1970) *Water Treatment and Examination*, **19**, 169–82.

Grime, J.P. (1973) *J. Environ. Mgmt.*, **1**, 151–67.

Grime, J.P. (1977) *Am. Nat.*, **111**, 1169–94.

Harper, J.L. (1968) in *Diversity and Stability in Ecological Systems* (eds G.M. Woodwell & H.H. Smith), *Brookhaven Symp. Biol.*, **22**, 48–61.

Hawskworth, D.L. (ed.) (1974) *The Changing Flora and Fauna of Britain*, Systematics Association, Special Volume 6.

Helliwell, D.R. (1974) Bellside Farm: a botanical survey and evaluation, Institute of Terrestrial Ecology, Grange-over-Sands.

Hill, M.O., Bunce, R.G.H. & Shaw, M.W. (1975) *J. Ecol.*, **63**, 597–613.

Hill, M.O. & Jones, E.W. (1978) *J. Ecol.*, **66**, 433–56.

Horn, H.S. (1975) in *Ecology and Evolution of Communities* (ed. M.L. Cody & J.M. Diamond) Harvard University Press, Cambridge, Mass., pp. 196–211.

House of Lords Select Committee on Science and Technology (Sess. 1979/80) (1980) Second report: Scientific aspects of forestry: Vol. II: Minutes of evidence, HMSO, London.

Huband, P. (1969) in *Hedges and Hedgerow Trees* (ed. M.D. Hooper & M.W. Holdgate), Monk's Wood Exp. St. Symp. No. 4, Nature Conservancy, London.

Institute of Terrestrial Ecology (1978) *Upland Land Use in England and Wales* (Countryside Commission publication 111), Countryside Commission, Cheltenham.

Jenkinson, D.S. (1971) *Rothamsted Experimental Station Report for 1970*, Part 2, 113–37.

Leonard, P.L. & Cobham, R.O. (1977) *Landscape Planning*, **4**, 205–36.

Linton, D.L. (1968) *Scottish Geog. Mag.*, **2**, 219–38.

Locke, G.M.L. (1970) *Forestry Commission Census of Woodlands* 1965–67, HMSO, London.

McVean, D.N. & Ratcliffe, D.A. (1962) *Plant Communities in the Scottish Highlands*, HMSO, London.

Mabey, R. (1980) *The Common Ground*, Hutchinson, London.

MacArthur, R.H. & Wilson, E.O. (1967) *The Theory of Island Biogeography*, Princetown University Press, Princeton, N.J.

Maltby, E. (1975) *J. Biogeog.*, **2**, 117–36.

Marquiss, M., Newton, I. & Ratcliffe, D.A. (1978) *J. Appl. Ecol.*, **15**, 129–44.

Maxwell, J.J., Sibbald, A.R. & Eadie, J. (1979) *Agric. Systems*, **4**, 161–88.

Miles, J. & Young, W.F. (1980) *Bull. d'Ecologie*, **11**, 233–42.

Miller, H.G. (1979) in *The Ecology of Even-aged Forest Plantations* (eds E.D. Ford, D.C. Malcolm & J. Atterson), Institute of Terrestrial Ecology, Cambridge, pp. 221–56.

Ministry of Agriculture, Fisheries and Food; Department of Agriculture and Fisheries for Scotland (1968) *A Century of Agricultural Statistics, Great Britain 1866–1966*. HMSO, London.

Ministry of Agriculture, Fisheries and Food (1974) *Agriculture Land Classification of England and Wales*, Agricultural Development Advisory Service, Land Service, MAFF, London.

Ministry of Agriculture, Fisheries and Food (1979*a*) *Farming and the Nation*, Cmnd. 7458, HMSO, London.

Ministry of Agriculture, Fisheries and Food (1979*b*) *Possible Patterns of Agricultural Production in the United Kingdom by 1983*, HMSO, London.

Ministry of Agriculture, Fisheries and Food; Department of Agriculture and Fisheries for Scotland (1981) *Annual Review of Agriculture*, HMSO, London.

Ministry of Town and Country Planning (1947) *Conservation of Nature in England and Wales*: Report of the Wildlife Conservation Special Committee (England and Wales), Cmd. 7122, HMSO, London.

Miron, J.R. (1976) *Regional Development and Land Use Models: an Overview of Optimisation Methodology*, International Institute for Applied Systems Analysis, Research Memorandum RM-76-27, Laxenburg.

Moore, N.W. (1962) *J. Ecol.*, **50**, 369–91.

Moore, N.W. (1980) in *Conflicts in Rural Land Use – the Need for a National Policy* (ed. K. Mellanby), Conference Rpt. Instit. of Biol., London.

Moore, N.W. (1982) *New Scientist*, **93**, 147–9.

Myers, N. (1979) *The Sinking Ark*, Pergamon, Oxford.

Natural Environment Research Council (1976) *Report of the Council for the period 1 April 1975–31 March 1976*, HMSO, London.

Nature Conservancy Council (1977) *Nature Conservation and Agriculture*, NCC, London.
House of Lords Select Committee on the European Communities, Session 1980/81 (1981) Eleventh Report, Environmental Assessment of Projects: Minutes of evidence, Submission from the NCC, HMSO, London, p. 156.
Newby, H., Bell, C., Saunders, P. & Rose, D. (1977) *Countryside Recreation Review*, **2**, 23–30.
Newton, I. (1974) *J. Appl. Ecol.*, **11**, 95–101.
Ovington, J.D. (1958) *J. Ecol.*, **46**, 391–405.
Ovington, J.D. (1965) *Woodlands*, English University Press, London.
Ozenda, P. (1979) *Vegetation Map of the Council of Europe Member States*, European Committee for the Conservation of Nature and Natural Resources, Strasbourg.
Paine, R.T. (1966) *Am. Nat.*, **100**, 65–75.
Parry, M.L. (1976) *Geog. J.*, **142**, 101–10.
Perring, F. (1974) *The Flora of a Changing Britain*, Classey, Faringdon.
Perring, F.H. & Walters, S.M. (1962) *Atlas of the British Flora*, Nelson, London.
Peterken, G.F. (1974*a*) *Q.J. For.*, **68**, 141–9.
Peterken, G.F. (1974*b*) *Biol. Conserv.*, **6**, 239–45.
Peterken, G.F. (1977) *Forestry*, **50**, 27–48.
Pirie, N.W. (1969) *Food Resources: Conventional and Novel*, Penguin, Harmondsworth.
Pollard, E. (1969) in *Hedges and Hedgerow Trees* (ed. M.D. Hooper and M.W. Holdgate), Monks Wood Experimental Station Symp. No. 4.
Pollard, E. (1979) *Proc. Trans. Br. Entomolog. Natural History Soc.*, **12**, 77–90.
Pollard, E., Hooper, M.D. & Moore, N.W. (1974) *Hedges*, Collins, London.
Porchester, Lord (1977) *A Study of Exmoor*, Department of Environment/Ministry of Agriculture, Fisheries and Food, HMSO, London.
Potts, G.R. (1971) *Outlook Agric.*, **6**, 267–71.
Price, C. (1976) in *Future Landscapes* (ed. M. MacEwen), Chatto & Windus.
Rackham, O. (1976) *Trees and Woodland in the British Landscape*, Dent, London.
Ratcliffe, D.A. (1972) *Bird Study*, **19**, 117–56.
Ratcliffe, D.A. (ed.) (1977) *A Nature Conservation Review*, Cambridge University Press, Cambridge.
Rawes, M. (1981) *J. Ecol.*, **69**, 651–69.
Richardson, S.D. (1970) *Advancement of Science*, **27**.
Robinson, D.G., Laurie, I.C., Wager, J.F. & Traill, A.L. (eds) (1976) *Landscape Evaluation: the Landscape Evaluation Research Project 1970–1975*, Centre for Urban and Regional Research, Manchester.
Royal Commission on Environmental Pollution (1979) *Seventh Report: Agriculture and Pollution*, Cmnd. 7644, HMSO, London.
Royal Society for Nature Conservation (1981) *Towards* 2000 *A Place for Wildlife in a Land-use Strategy*, RSNC, Lincoln.
Sandford, Lord (1974) *Report of the National Parks Policies Review Committee*, Department of the Environment, HMSO, London.
Selman, T.H. (1981) *Ecology and Planning*, Goodwin, London.
Sharrock, J.T.R. (1976) *The Atlas of Breeding Birds in Britain and Ireland*, British Trust for Ornithology, Tring, Herts.
Sheail, J. (1980) *Historical Ecology: The Documentary Evidence*, Institute of Terrestrial Ecology, Cambridge.
Shoard, M. (1976) *Forma*, **4**, 128–35.
Shoard, M. (1980) *The Theft of the Countryside*, Temple Smith, London.
Simberloff, D.S. & Abele, L.G. (1976) *Science*, **191**, 285–6.
Smith, R. (1900) *Scottish Geog. Mag.*, **16**, 385–416.

Smith, R.S. & Budd, R.E. (1982) *Land Use in Upland Cumbria: A Model for Forestry/Farming Strategies in the Sedbergh Area* (Research Monographs in Technological Economics no. 4), University of Stirling, Stirling.

Sorensen, J.C. & Moss, M.L. (1973) *Procedures and Programmes to Assist in the Environmental Impact Statement Process*, Institute of Urban Regional Development, University of California.

Southwood, T.R.E. (1981) in *Tall Timbers Conference on Ecological Animal Control*, Proc. No. 3, 29–51.

Stamp, L.D. (1937–47) *The Land of Britain: The Final Report of the Land Utilisation Survey of Britain*, Geographical Publications, London.

Steele, R.C. (1972) *Wildlife Conservation in Woodlands*, HMSO, London.

Stone, E.L. (1975) *Phil. Trans. R. Soc. London*, **B271**, 149–62.

Strutt, N. (1978) *Agriculture and the Countryside*, Advisory Council for Agriculture and Horticulture in England and Wales, London.

Stuttard, P. & Williamson, K. (1971) *Bird Study*, **18**, 9–14.

Sullivan, A.L. & Shaffer, M.L. (1975) *Science*, **189**, 13–7.

Tansley, A.G. (1939) *The British Islands and Their Vegetation*, Cambridge University Press, Cambridge.

Tansley, A.G. (1945) *Our Heritage of Wild Nature*. Cambridge University Press, Cambridge.

Thomas, A.S. (1960) *J. Ecol.*, **48**, 278–306.

Toleman, R.D.L. & Pyatt, D.G. (1974) *Proc. Tenth Commonwealth Forestry Conf.*, Forestry Commission, London, pp. 1–21.

Tomlinson, T.E. (1971) *Outlook Agric.*, **6**, 272–8.

Turner, J.R. (1980) in *Land Assessment in Scotland* (ed. M.F. Thomas & J.T. Coppock) Aberdeen University Press, Aberdeen, pp. 79–87.

United States Senate (1969) *National Environmental Policy Act of 1969*, Public Law 91–190, 91st Congress, 2nd Session, December 1969.

Usher, M.B. (1980) *Field Studies*, **5**, 323–48.

van der Weijden, W.J., ter Keurs, W.J. & van der Zande, A.N. (1978) *Ecol. Q.*, Winter, 317–35.

Ward, S.D. & Evans, D.F. (1976) *Biol. Conserv.*, **9**, 217–33.

Watt, A.S. (1960) *J. Ecol.*, **48**, 605–29.

Webb, N.R. & Haskins, L.E. (1980) *Biol. Conserv.*, **17**, 281–96.

Wells, T.C.E. (1968) *Biol. Conserv.*, **1**, 37–44.

Westmacott, R. & Worthington, T. (1974) *New Agricultural Landscapes*, Countryside Commission, Cheltenham.

Whitman, I.L., Dee, N., McGinnis, J.T., Fahringer, D.C. & Baker, J.K. (1971) *Design of an Environmental Evaluation System*. Battelle Columbus Laboratories, Columbus, Ohio.

Williams, R. (1973) *The Country and the City*, Chatto & Windus, London.

CHAPTER SIX

Policy Planning

6.1 OVERVIEW

Policy planning is concerned with stating goals, classifying options, and broad allocation of land-use priorities. Planning processes in the UK, for example, developed in the post-war era when agriculture and forestry were automatically assumed to enhance the countryside. The main thrust of planning decisions has therefore been concerned with encouraging urban development and limiting proliferation into the rural setting. As a consequence planning policy is concerned as much with improving or maintaining the urban environment in a physical condition to encourage industrial development and provide residential sites with a comprehensive array of community and social facilities, as it is with assessing the impact of individual projects on its immediate site.

The formulation of policy, be it a decision in favour of nuclear, rather than coal-fired power stations, or the reconciliation of conflicting considerations in the form of stated goals, has a marked determinate effect on the more detailed project planning processes. As the decision process moves to lower tiers, from national policies to local projects, the viable alternatives to the proposed action become progressively excluded and the level of willingness to consider alternatives may decline. The incorporation of ecological information into the lower tiers only will inevitably be of limited usefulness. It is for this reason that some authorities have proposed a 'two-tier' arrangement for appraising major schemes. Under this system a general investigation of the broad issues relevant to the project setting would be carried out at a generic level prior to a specific impact analysis at the local planning level.

The papers in this section examine the requirements for ecological information in such generic investigations and assess how they may be carried out. Four specific examples are considered relating to (a) policy development for the cumulative assessment of numerous medium-scale developments (Handley, Section 6.2), (b) for regional nature conservation (Currie and MacLennan, Section 6.3), (c) strategic studies of water supply options (Nelson *et al.*, Section 6.4) and (d) major highway developments (Colwill and Thompson, Section 6.5).

It has often been assumed that unacceptable environmental changes will be prevented if some form of mandatory impact analysis were required for major developments above a critical size. As emphasized by Handley (Section 6.2) the number of projects which fall into this category is very small, the vast majority being excluded on the basis of size alone. It is, however, the cumulative effects of small- and medium-scale developments which may result in significant impacts, such as land loss and structural changes. These are, however, too small to merit detailed individual impact analysis and too numerous to permit such investigations without causing undue delay. The solution to the problem adopted by Merseyside County Council has been to survey the status of environmental factors sensitive to these developments – such as the occurrence of rare habitats and environmental pollution – as a prelude to policy formulations. The result has been a Structure Plan based on a promotion of urban regeneration balanced with a maintained capacity to meet the leisure, recreation, education and conservation requirements of the general public.

The requirement for rapid assessment of a growing number of proposals, at the time of rapid growth of Scottish oil and petrochemical development, was also the main motivating force behind the preparation of the 'Conservation Prospectus' for the Moray Firth as a conservation policy document (Currie and MacLennon, Section 6.3). Preparation of the Prospectus involved an assessment by ground-survey techniques of all potential conservation sites and their classification into five categories by integrating criteria of scientific merit with geographic and other factors although only grade 1 and 2 (nationally or internationally important) sites were involved in the published document. Despite the large number of man-hours required to produce and update it, Currie and MacLennon argue that the Prospectus has served many valuable functions in resolving and avoiding, land-use conflicts. Whether the comparable national Nature Conservation Review (Ratcliffe, 1977) proves to be as productive remains to be seen.

All the papers in this section emphasize the advantages of considering ecological parameters at an early stage in the planning process. Particular advantages include sounder environmental decisions, minimum delay in the consideration phases and reduction in costs associated with planning delays and process designs necessary to meet environmental standards. This is especially the case in assessments of strategic options such as those for water resource sites (Nelson *et al.*, Section 6.4) and major roadway developments (Colwill and Thompson, Section 6.5) which may involve complex impact factors in complex environments. The resolution of these complexities may require detailed, time-consuming investigations if potential impacts are to be assessed in a quantitative way so environmental considerations must clearly be taken into account at an early stage of policy as well as project planning.

6.2 ECOLOGICAL REQUIREMENTS FOR DECISION MAKING REGARDING MEDIUM-SCALE DEVELOPMENTS IN THE URBAN ENVIRONMENT

J. F. Handley

6.2.1 Introduction

Land-use planning in the UK has traditionally been concerned with the problem of urban growth. A major consideration has been to control urban encroachment into the countryside and maintenance of the agricultural and forestry, and to a lesser extent, recreational and amenity components of the rural setting. A period of sustained urban growth may be seen as an expression of population expansion and inadequate social and recreational facilities, but population growth is now virtually static and many older conurbations are experiencing a net loss of population. Here control of further encroachment may be achieved through a strategy of urban regeneration designed to improve employment prospects and resident standards with comprehensive community and environmental facilities. In the major conurbations, development proposals need to be assessed as much for their overall effects on the physical fabric of the urban environment as for specific biological impacts.

The development of methodologies for environmental impact assessment has of necessity focused on major development proposals such as oil refineries, power stations and aluminium smelters (Clark *et al.*, 1976). Catlow and Thirlwall (1976) suggested that a formal environmental impact analysis is required when a development would, in its particular setting, be likely to cause large-scale and complex environmental impacts. These authors asked Government planning directorates in the UK to estimate (on the basis of criteria set out in Appendix V of their report) how many projects dealt with in the three years up to March 1974 would have required environmental impact assessment. Their replies suggested that the number would average between 25 and 50 per year.

It is salutary to compare this figure with the total number of planning applications determined annually in England and Wales which averaged 464 000 per annum over the period 1965–1975. It is clear that the handful of major projects which are likely to justify a formal EIA represents the tip of a rather large iceberg! The localized impact of smaller-scale development may be quite severe and the accumulative effect of incremental development may lead to structual change of great ecological significance. This paper examines the requirements for rapid decision-making on medium-scale development which may contribute to fundamental change over a longer time scale.

It is not proposed that an elaborate Environmental Impact Statement should be prepared for more than the few exceptional development proposals identified by Catlow and Thirlwall. But it is suggested that environmental impact analysis, which is understood to be a structured approach to project appraisal, can assist decision-making at all levels in the planning system.

6.2.2 The development control spectrum

The vast majority of planning applications are locally determined by the planning committee of a County or District Council. Over the period of 1965–1975 less than 2% of all applications were decided on appeal by central Government in the person of the Secretary of State for the Environment or his appointed Inspector. The Secretary of State also has powers to require any planning application to be referred to him for a decision by himself (*Town and Country Planning Act* 1971, s35 and General Development Order 1977, art. 19) but these powers are rarely exercised.

The majority of planning decisions refer to small-scale development proposals which, considered in isolation, would have only a minor environmental impact. Clark *et al.* (1976) suggest that developments requiring the use of detailed assessment procedures would fall within one or more of the following categories:

1. Have a large land take
2. Be of a contentious nature
3. Have a significant impact on the physical environment, as well as on local employment structure and levels of service provision
4. Be of 'national' or 'regional' significance
5. Be major departures from approved development plans

Criteria 2–5 are not easy to quantify but it is possible to assess the spectrum of development opportunities in terms of land take for industry and housing. The frequency distribution of potential development sites for industry and housing in Merseyside, a Metropolitan County in north-west England, is shown in Figs 6.1(a) and 6.1(b) respectively. These diagrams are based on data held in a computer file in the Merseyside Land Resources System which includes land parcels with planning consent or which have been identified by the Local Authorities as having potential for a particular land-use (Merseyside County Council, 1979).

It can be seen from Fig. 6.1 that 54% of industrial development sites in Merseyside and 78% of housing sites are less than 0·5 ha in area. On the other hand opportunities for major industrial development, in excess of 100 ha are apparently absent. It is development at an intermediate scale between these two extremes which accounts for the greatest land take and is

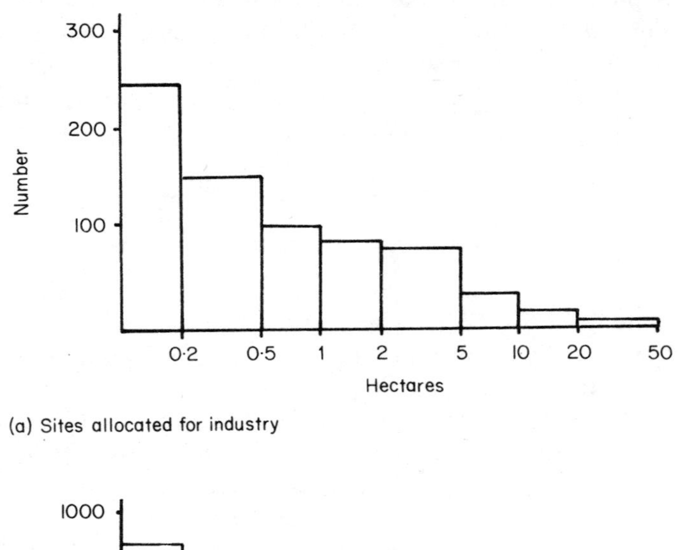

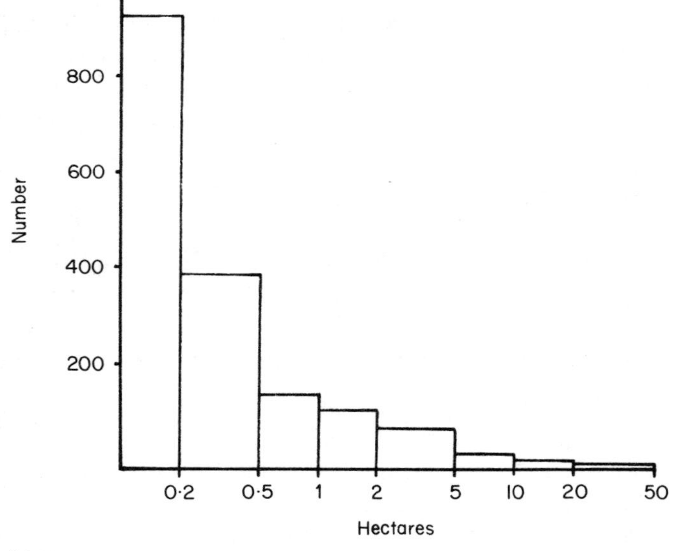

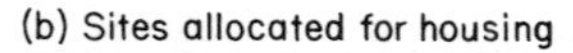

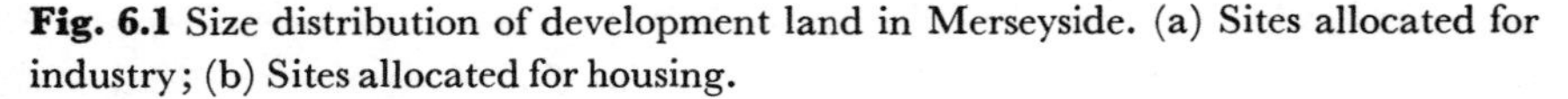
Fig. 6.1 Size distribution of development land in Merseyside. (a) Sites allocated for industry; (b) Sites allocated for housing.

likely to contribute towards structural change. Arrangements are made through the Development Control Scheme for County Councils to give a view and in certain instances, such as mineral workings, to determine planning application submitted to the District Councils (DOE Circular 74173). By way of illustration the degree of involvement of Merseyside County

Council in decision-making is shown by the fact that during the period 1977–1979, the District Councils received 31 789 planning applications of which only 5% had a County Council involvement.

It is clear that this involvement in the planning process need not lead to unacceptable delay for a view was given on 80% of all county-involved applications within 6 week cycles. The problem is rather that of being equipped to respond effectively to a development proposal which raises complex environmental issues, within the statutory period of eight weeks.

The time scale for determining a planning application is strongly dependent on a consultation process which increases in complexity according to the sensitivity of the proposed development. But the Planning Authorities can be under considerable pressure to determine applications quickly and this may increase the risk that necessary consultations are either overlooked or overtaken by events.

The environmental consequences can then be serious as in 1968 when, without consulting HM Alkali Inspectorate, St Helens Borough Council granted outline planning consent for a 'double contact' sulphuric acid plant in a densely built up area with residential development within 200 m (Miller, 1979). The acid plant, as originally installed, had design faults. Major shortcomings were in the acid-cooling system and in the system of gas flow. There resulted 45 pollution incidents resulting mainly from high SO_2 ground level concentrations coinciding with plant start ups, acid spray and gas leaks from faulty equipment, and gas or visible fume escapes from leakages and spillages, many of which led to public complaint. The localized impact of atmospheric pollution on the natural environment may also be considerable as documented by Vick and Handley (1977) in the vicinity of a phosphoric acid factory at Kirkby, Merseyside.

6.2.3 The cumulative impact of development on environmental quality

The localized ecological damage associated with medium-scale development may be less severe in the long term than the cumulative effect of land-use decisions. This is particularly evident in an area such as Merseyside where rapid urban development has occurred on a natural resource base of high quality (Merseyside County Council, 1976*a*, *b*). The principal ecological consequences of this are habitat loss and a reduction in the complexity and productivity of natural systems due to environmental pollution.

(a) Habitat loss

The magnitude of land-use change over a ten year period (1966–1976) is indicated by a comparison of the 2nd Land Utilization Survey and a more recent survey of Merseyside carried out for the County Council by Environmental Information Services Ltd in a six month period in 1976 (Fig. 6.2 (a)) The loss of farmland, in absolute or relative terms, is in fact much greater than the loss of semi-natural habitat. The removal of about 50 km² of farmland from productive agriculture is a matter of more than local significance in a County when 54% of farmland is classified by the Ministry of Agriculture, Fisheries and Food as of Grade 1 and 2 quality, with a national average of 17% for England and Wales (Handley, 1982).

The loss of semi-natural habitat is partly due to development and partly to reclamation for agriculture, which is effectively outside the scope of planning control. In southern England, Moore (1962) has carefully documented the changes in land-use which led to the reduction of the heathland area in Dorset to its present figure of one third of the 1811 acreage. Losses of habitat were attributed to mineral extraction afforestation, agricultural reclamation and military training. Recent observations have shown that this picture is typical of southern heaths in general and that the process of attrition continues (Shoard, 1980). The heathlands, moorlands and sand dunes of Merseyside have suffered a similar fate (Merseyside County Council, 1976*a*). The dune system of the Sefton coast is still the fourth largest in Britain but recent mapping indicated that 46% of the sand dune area was lost to urban development and afforestation between 1801 and 1974 (Fig. 6.2(b)). Estimates based on air-photo interpretation indicate that 67% of the remaining dune habitat is in a semi-natural condition, though in places severely degraded by visitor pressure (Merseyside County Council, 1976*a*).

The consequences of habitat loss for species populations are not well understood. Batten (1972) has examined the relationship between urban development and the status of breeding birds in the vicinity of Brent in north-west London, where detailed ornithological records are available from 1833 onwards. This was a rural area in 1833 with 72 regular breeding species and by 1970 (with 65% urbanization) this total had declined almost linearly to 43. A study of breeding bird populations in fully urbanized areas suggested that the number would fall to around 20 if urbanization proceeded to completion.

From an ecological standpoint, habitat fragmentation may be at least as important as net loss of habitat and rare species are especially vulnerable. Moore (1962) has linked the decline of the Dartford warbler (*Sylvia undata*) in southern England to heathland fragmentation and the decline of the sand lizard (*Lacerta agilis*) in the Merseyside dunes has been attributed to this among other factors. The species diversity of woodlands in Merseyside must be severely limited by the fact that 67% are less than 2 ha in area and that almost all are of recent origin. Unfortunately the only woodlands of real antiquity in river valleys on the Wirral peninsula, have already been

(a) Summary of land-use change in Merseyside, 1960s–1976

Category of use	1960s (km²)	1976 (km²)	Net change (km²)
Residential/commercial	144.9	172.0	+27.1
Industry	18.3	20.5	+2.2
Mineral extraction	1.2	1.1	−0.1
Waste disposal	4.7	3.1	−1.6
Derelict land	3.7	6.3	+2.6
Transport	82.4	88.2	+5.8
Open space	57.9	74.6	+16.7
Total settlement	313.1	365.8	+52.7
Arable	96.1	96.7	+0.6
Horticulture	47.8	30.6	−17.2
Improved grass	110.8	77.1	−33.7
Total improved farmland	254.7	204.4	−50.3
Coastal	23.5	18.6	−4.9
Heath, Moss and Marsh	4.4	4.2	−0.2
Rough Grass and Scrub	23.5	29.0	+5.5
Woodland	27.2	25.0	−2.2
Water	6.4	5.8	−0.6
Semi-natural habitat	85.0	82.6	−2.4

(b)

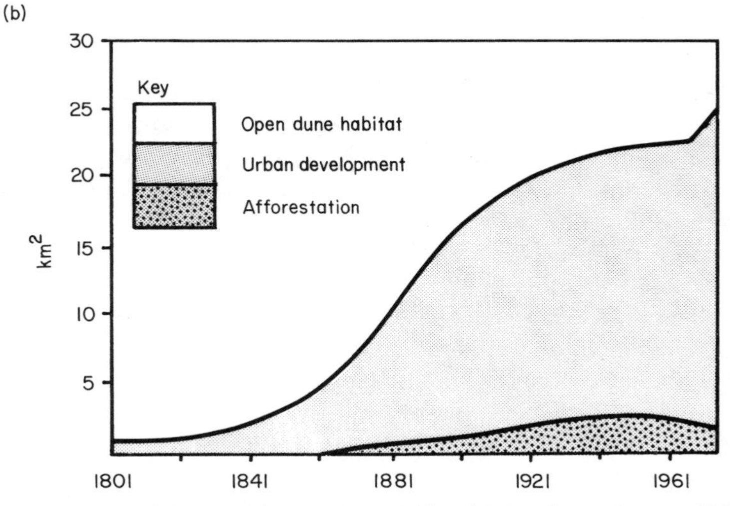

Fig. 6.2 Cumulative habitat loss in Merseyside: (a) Land-use change, 1960s–1976 (Handley, 1982); (b) Urbanization of sand dunes, Sefton (Jackson, 1979).

adversely affected by fragmentation and adjacent urban development (Roberts, 1975).

From a positive standpoint the data shows a substantial increase in the area of rough grassland and derelict land. Similar observations have been made from Land Utilization Survey data for the Thames Estuary (Coleman 1975). Bradshaw (1979) has shown the ecological potential of damaged and neglected land and opportunities may exist for creative conservation in these areas. For example an ecological survey in the Sankey Valley of Merseyside provided the basis for a subsequent programme of landscape restoration and countryside management (Anderson, 1975).

(b) Environmental pollution

In 1973 a planning strategy for the north west of England highlighted the poor condition of the physical environment in the Mersey Valley giving particular emphasis to the inherited problems of atmospheric pollution, industrial dereliction and water pollution (Strategic Plan for the North West, 1973). It was made clear that these are fundamental structural problems which can only be solved by close working between Local Planning Authorities and the agencies responsible for pollution control, together with major investment in smoke control, land reclamation and the sewerage system.

The condition of the Mersey estuary, which is perhaps the most intractable pollution problem in the region, has developed gradually over a long period:

> 'Merseyside has for centuries been one of Britains foremost trading and industrial regions. There has been no meteroric rise to prominence, but rather a continuous expansion of population and industry over a long period. The estuary's pollution problem has similarly built up rather gradually, although there has been an accelerated decline in water quality in the last few decades (Porter, 1973).'

The result is that today the Mersey Estuary is one of the most heavily polluted tidal waters in Britain. The principal source of pollution is the discharge of untreated sewage from a surrounding population of nearly a million people. This is made worse by industrial discharges and the poor condition of the rivers flowing into the estuary including the Mersey itself and the Manchester Ship Canal which has a heavily industrialized catchment with a surrounding population of about five million (Gouge *e tal.*, 1977).

The biochemical oxygen demand associated with the heavy pollution load causes oxygen depletion in the upper estuary leading to odour problems, especially at low tide in warm summer weather. It seems that the effect would be more severe but for the high concentrations of nitrate in inflowing water which act as an oxygen source at 10% water saturation.

Fortunately the problem of estuarine pollution by a heavy organic load is reversible given adequate capital investment, as experience in the Thames Estuary has shown. Through the mathematical model developed by the North West Water Authority it is possible to test the likely effectiveness of capital investment strategies in reducing the oxygen-depletion effect. Already substantial progress has been made at the head of the estuary and proposals

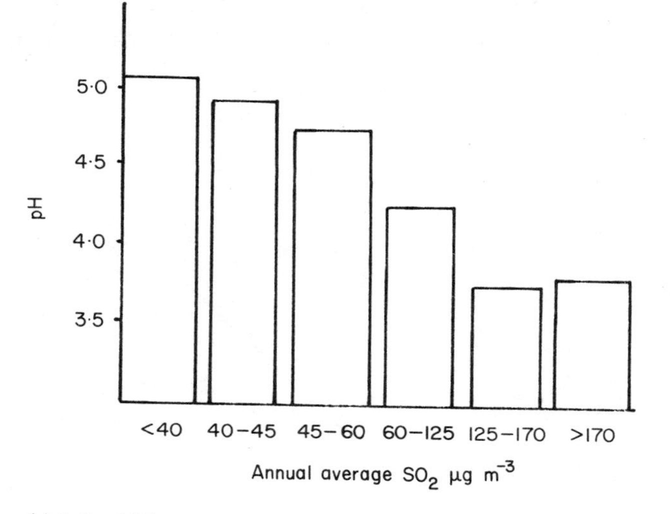

(a) Soil acidification

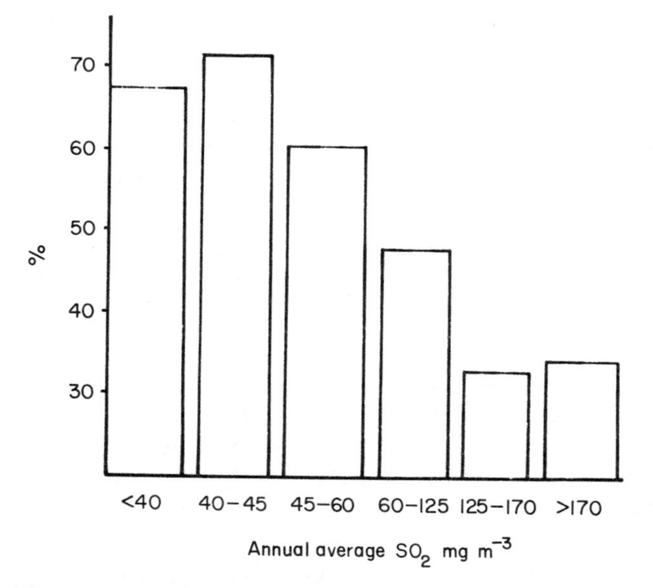

(b) % Base saturation

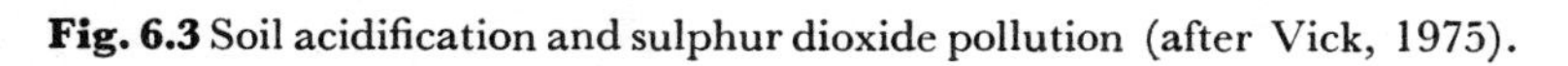
Fig. 6.3 Soil acidification and sulphur dioxide pollution (after Vick, 1975).

are being brought forward by the North West Water Authority for interception and primary treatment of crude sewage flows to the lower estuary. The accumulation of pollutants with long residence times, such as heavy metals, may in the long term be of greater concern within the estuary and beyond into Liverpool Bay.

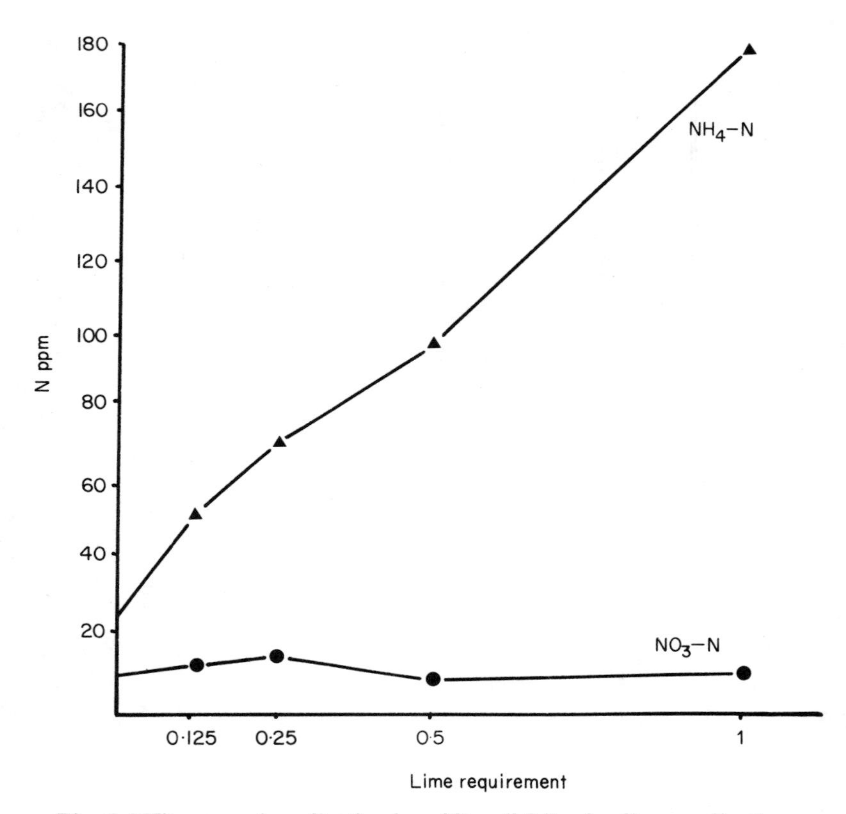

Fig. 6.4 Nitrogen mineralization in acidic soil following lime application.

Despite a sustained improvement over recent decades atmospheric pollution by smoke and sulphur dioxide remains high by UK standards in the older part of the Merseyside conurbation. Here too we can anticipate a cumulative impact on terrestrial ecosystems as described by Bradshaw in Section 2.2. In parkland soils dry deposition and washout of sulphur compounds has led to progressive acidification (Fig. 6.3) giving an impoverished sward with a low carrying capacity (Vick and Handley, 1975). In practice this may be overcome by a surface application of ground limestone without resorting to reseeding (Fig. 6.4).

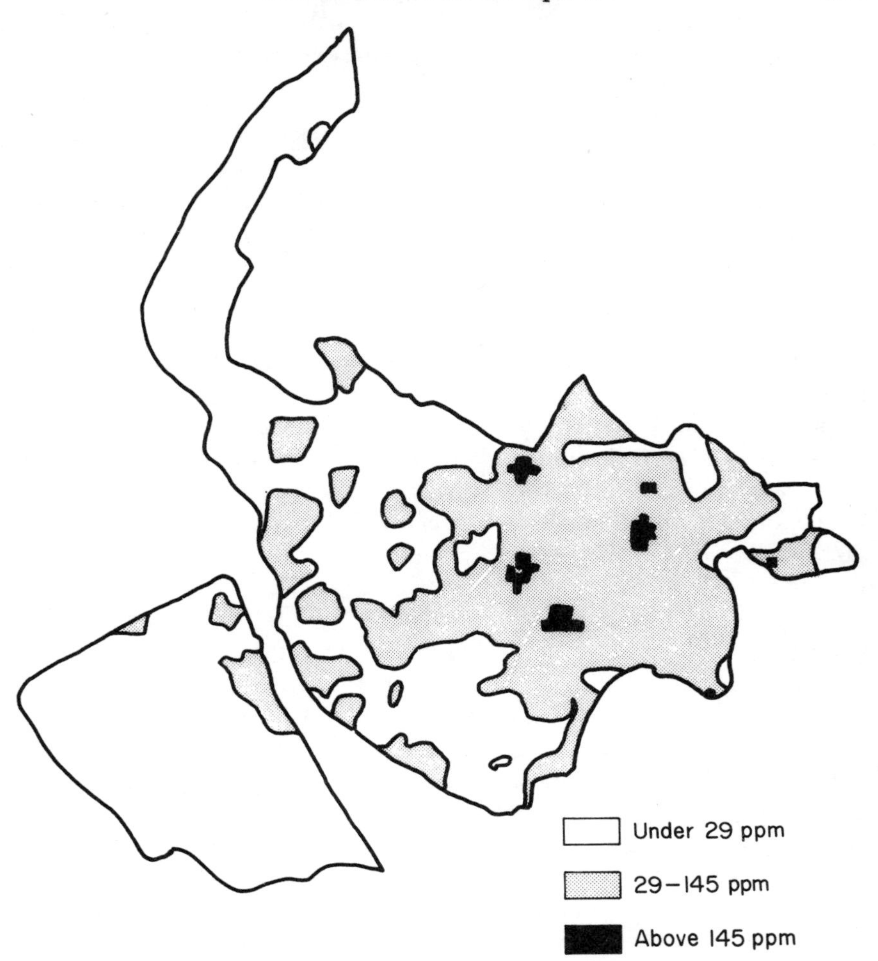

Fig. 6.5 Available copper in Merseyside soils.

However, as in the estuarine system, pollutants with a more lasting effect, such as heavy metals, may accumulate in parkland soils (Purves, 1972). A systematic survey of heavy metal concentrations in Merseyside soils showed that in some instances, notably copper and cadmium, elevated metal concentrations were associated with specific industrial complexes (Fig. 6.5). In fact in the present area the evolution of metal tolerance is a pre-requisite for the survival of grasses in the lawns of a large factory complex (Bradshaw, 1976).

6.2.4 The County Structure Plan

From the foregoing it is clear that the assessment of medium-scale development has two important characteristics: (1) Decisions are required in a short time-scale; and (2) Contextual information is required for the consequences of development to be properly assessed. It is unlikely that these requirements can be met unless a framework is already available within which decisions can be taken. In England and Wales this framework is provided by the County Structure Plan (Booth, Section 2.1). The County Structure Plan provides the planning authorities and the public with an opportunity to examine, in a rational framework, the physical problems of an area and to decide, within the limits of what is attainable, how our physical environment should be ordered and what priorities should be selected.

From the ecological standpoint the base-line for the County Structure Plan was prepared in the form of two Report of Survey Documents dealing with Natural Resources and Environmental Pollution respectively (Merseyside County Council, 1976*a*, *b*). For the most part this represented a collection of existing information and this information was assessed, and gaps in our knowledge identified, by consultative groups. The Environmental Control Consultation Group included representatives from all pollution control agencies operating within the sub-region and their guidance was particularly valuable in identifying key subject areas about which information was required.

It was possible to fill some of the gaps in our knowledge quite rapidly by commissioning survey work through a modest specialist advice budget (£50 000 per annum) and by making full use of labour-saving techniques such as remote sensing (Handley, 1980). Thermatic surveys based on air photograph-interpretation were commissioned to provide a map record at 1:10 000 scale of derelict and despoiled land, the distribution and condition of natural environments and the incidence of archaeological artefacts. It is recognized that aerial surveys cannot replace ground survey work but they can greatly reduce the effort involved and increase the precision of field work. One advantage of the method of working is that the air photo record, and the imagery obtained from scanning systems, provides a permanent and objective source of information about the physical environment. This type of survey may be repeated at low cost so that trends can be identified and the effectiveness of policies and programmes can be tested. This monitoring function is an essential feature of the structure planning process and the access to a time dimension can also be extremely useful in project appraisal.

The County Council compared three alternative structure plan strategies for Merseyside by assessing in broad terms their likely cost and the impact of each on employment, housing, transport and the environment. These alternatives were:

(a) Passive decline–continuation of present levels of urban depopulation particularly from the more recent out-of-town estates.
(b) Managed dispersal – a concentration of new development on green field sites on the urban periphery with an accelerated clearence programme in older urban areas.
(c) Urban regeneration – the promotion of new developments within older urban areas and strict limitation on development at the periphery.

An Environmental Impact Statement as such was not prepared but the method of working bore a strong resemblance to EIA procedures for project appraisal. The adopted strategy (alternative (c)) seeks to promote the urban regeneration of Merseyside by encouraging investment and development within the urban county and particularly in those areas with the most acute problems, enhancing the environment and assisting housing and industrial development on derelict and disused sites. There is a reciprocal effort to enhance and conserve the County's open land and its agriculture whilst ensuring that its capacity to meet the needs of leisure, recreation and informal education is exploited.

The Department has gone to unusual lengths to ensure that the detailed exposition of the strategy is readily understood both by the general public and elected members. There has been very full consultation with interested parties at all stages in plan preparation including mandatory public consultation on the Draft Document prior to submission to the Secretary of State for the Environment and an examination in public held in June 1978.

The structure plan provides a basis for the assessment of development proposals and also for the preparation of detailed local plans by the District Council. The local plans are submitted to the County Council for their approval. As a matter of routine, the Environmental Quality Division of the County Planning Department sets out these proposals in an impact matrix against the relevant Structure Plan Policies. In this way significant impacts are quickly identified and conflict is kept to a minimum because it can be seen that the local plan is being assessed in the light of adopted policy rather than the arbitrary judgement of another professional at a different level in the planning hierarchy.

6.2.5 Development control

The decision on a planning application is made not by a professional adviser but by a Planning Committee consisting of elected representatives. O'Riordan (1971) has provided a general model of the decision-making process in resource management which is readily applicable in this context (Fig. 6.6). The model visualizes decision-making as a learning process consisting of four principal stages each of which is interconnected by feed-back loops. O'Riordan points out that when the decision-maker is an elected representative and hence when his technical knowledge is limited, the influence of technical advice, the clarity with which he sees the problem and his role in the decision hierarchy are important features affecting his final choice.

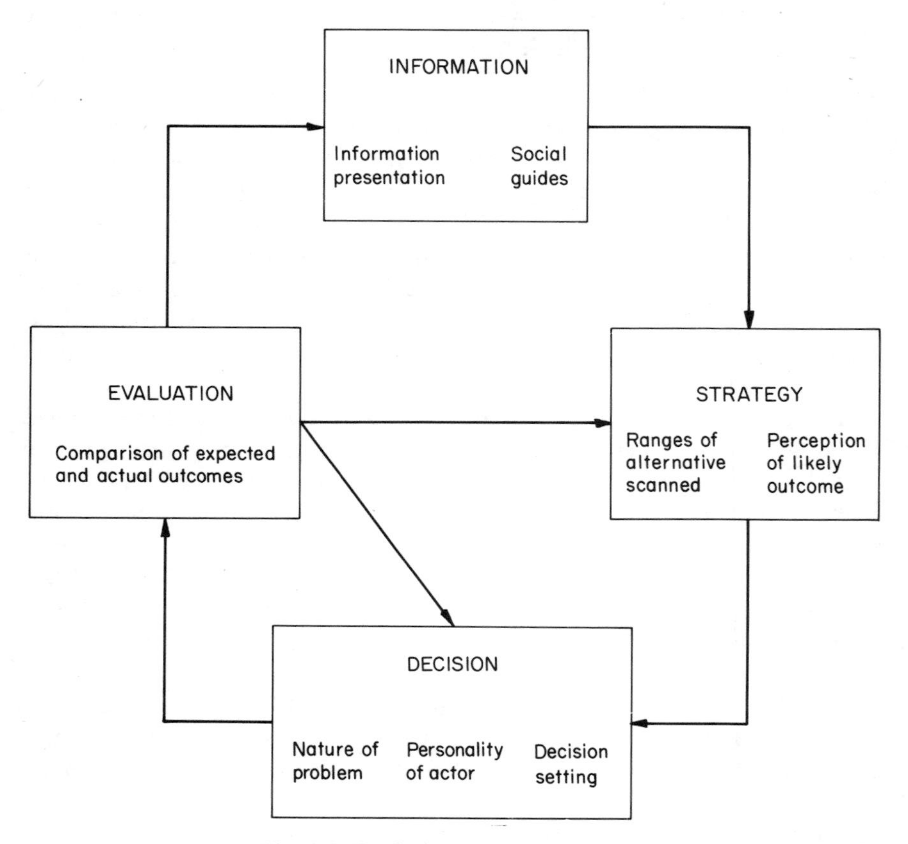

Fig. 6.6 The decision-making process.

In forming a view of a planning application the Committee will be assisted by information in the form of a report which summarizes the development, the views expressed by those consulted, and the view of the Planning Officer. The principal social guides are the recommendation of the Planning Officer and adopted policy in the form of the County Structure Plan. In Merseyside the committee report is rarely more than two sides, otherwise it will not be read, and the development site is usually illustrated on a contemporary air photo base because this is more meaningful than an Ordnance Survey base map (though such maps are made available at Committee to meet the statutoryobligations of the Planning Authority).

The crucial role for environmental impact analysis is in decision-setting. If there is a dialogue between the developer and the Planning Department in the pre-application phase then potential conflict can be identified and hopefully resolved. The impact analysis is usually undertaken as a prelude to the committee report and is rarely, if ever presented at Committee.

Perhaps the most difficult applications for a committee to handle are those which bring considerable social benefit to the community at large but may result in a severe loss of amenity, and therefore a high political cost if in their immediate locality. Developments of this type could include highway proposals, solid-waste disposal, substantial industrial development (25–100 ha) and Special Category Industry. In each of these instances sites or routes which are potentially suitable, or will minimize environmental disbenefits, have been identified within my own authority in advance of an application being received. In an area which continues to suffer from high levels of unemployment this is essential if elected members are not to be faced with an impossible choice between employment opportunity and environmental disruption.

The most demanding role for the professional adviser concerns the prediction of the likely outcome of development. The problem here is twofold: (a) to quantify the interaction effects within the impact matrix; and (b) to assess whether the development will in fact proceed along the lines indicated in the planning application.

The difficulties inherent in quantification are described by other contributors to this publication. The course which a development will take in practice may be constrained by administrative devices such as a conditional planning consent. But planning conditions are notoriously difficult to enforce in practice and they are effectively considered inappropriate where environmental control is available through alternative legislation (for example the registration and licencing procedures of the Waste Disposal Authority, the Regional Water Authority and the Health and Safety Executive, incorporating HM Alkali Inspectorate. There is a growing tendency with potentially damaging developments, such as mineral working, to cover the planning consent by a legal agreement (Section 52 *Town and Country Planning Act* 1971) which may include a financial commitment in the form of a bond. The outstanding example of the positive use of such a measure, thought in this instance by a specific Act of Parliament rather than through planning legislation, is the operation of the Ironstone Restoration Fund in Northamptonshire.

6.2.6 Conclusions

Decision-making on medium-scale development is usually rapid but the consequences of such a decision may have significant localized effects or contribute to a longer term trend with important ecological consequences. A framework is needed to assist decision-making and, in England and Wales, this is provided by the County Structure Plan. The professional advisor should provide elected representatives with good quality information in a succinct and well-presented format. Where possible, difficult areas of decision making should be anticipated and impact analysis may assist a dialogue between developer and officials in the pre-planning application stage.

6.3 A PROSPECTUS FOR NATURE CONSERVATION WITHIN THE MORAY FIRTH: IN RETROSPECT

A. Currie and A. S. MacLennan

6.3.1 Introduction

During the late 1960s and early 1970s the advent of large-scale industry coupled with proposals for further developments exerted considerable change upon the social, political, economic and planning framework of the hitherto predominantly rural Moray Firth area in north-east Scotland. Within this setting the Nature Conservancy (NC) and its successor, the Nature Conservancy Council (NCC) had a statutory remit to advise central government, local authorities, other statutory bodies and developers on matters relating to the conservation of wildlife and the environment. The Nature Conservancy also had a duty to identify areas of nature conservation importance, to notify the most important of these as Sites of Special Scientific Interest (SSSIs) and to consider appropriate safeguards for these.

One of the tools employed by the Nature Conservancy in order to carry out this remit was the preparation and publication of a report entitled '*A Prospectus for Nature Conservation within the Moray Firth*'. This report was addressed primarily to planners, developers and statutory bodies. Its purpose was to draw attention to the changes in land use which were taking place, to assess the effects of these changes on natural habitats and wildlife, and to suggest means by which adverse affects could be avoided or mitigated. The Prospectus may be considered as a form of environmental impact statement. It had a secondary objective of influencing the thinking of both planners and the general public, so as to enlighten them as to the rationale behind nature conservation policies. In this respect, the Prospectus was also a promotional document.

This paper examines the Prospectus published in 1972 and its successor in 1978, as a case study, relating their aims to the changing economic and planning situation of the period. It considers in retrospect their value to nature conservation, as a means of creating awareness of the impact of developments, and of gaining support for nature conservation policies. The mechanics of production and circulation are also discussed, as well as the applicability of the method to other situations.

6.3.2 Industrial development and the need for a 'prospectus'

(a) The Moray Firth

The Moray Firth area (Fig. 6.7) consists of a narrow coastal strip, focusing on the Dornoch Firth, the Cromarty Firth and the Beauly Firth, along with the intervening peninsulas of Easter Ross and Black Isle. Between Brora to the north and Burghead to the south and east there are over 350 kilometres of varied coastline. Inland there is an upland interior rising to over 1000 metres, and penetrated by narrow tongues of low land along the major river valleys.

A number of administrative divisions are encompassed including parts of the Districts of Sutherland, Ross and Cromarty, Inverness, Nairn and Badenoch and Strathspey, all within Highland Region; and Moray District in Grampian Region. Prior to local government reorganization in 1975, the area was administered by the four County Councils of Sutherland, Ross and Cromarty, Inverness and Moray and Nairn.

The main town is Inverness and there are smaller concentrations of population in small towns like Golspie, Tain, Invergordon, Dingwall, Nairn and Forres. The Moray Firth is essentially rural, with its economy based traditionally upon agriculture, forestry, fishing and tourism along with the service industries. Some of the farmland of Easter Ross and Moray and Nairn is of very high quality.

Throughout the 1950s and 1960s in common with the rest of the Highlands, the Moray Firth suffered from unemployment at a level considerably higher than the national average. In addition to this, the area experienced depopulation through migration, which though less than that experienced in other parts of the Highlands, was nevertheless a major source of concern (Jack Holmes Planning Group, 1968).

(b) The advent of large-scale industrial development

By the mid 1960s there was strong pressure to create greater employment opportunity, and to redress population loss. A plan to create such conditions was produced in 1968 by the Jack Holmes Planning Group (The Holmes Report). This report was sponsored by the Highlands and Islands Development Board (HIDB) which was itself established in 1965. This plan has acted as a guide for many of the subsequent planning and development decisions. A central theme was the need to capitalize on the resources of flat land and deep-water facilities of the Cromarty Firth.

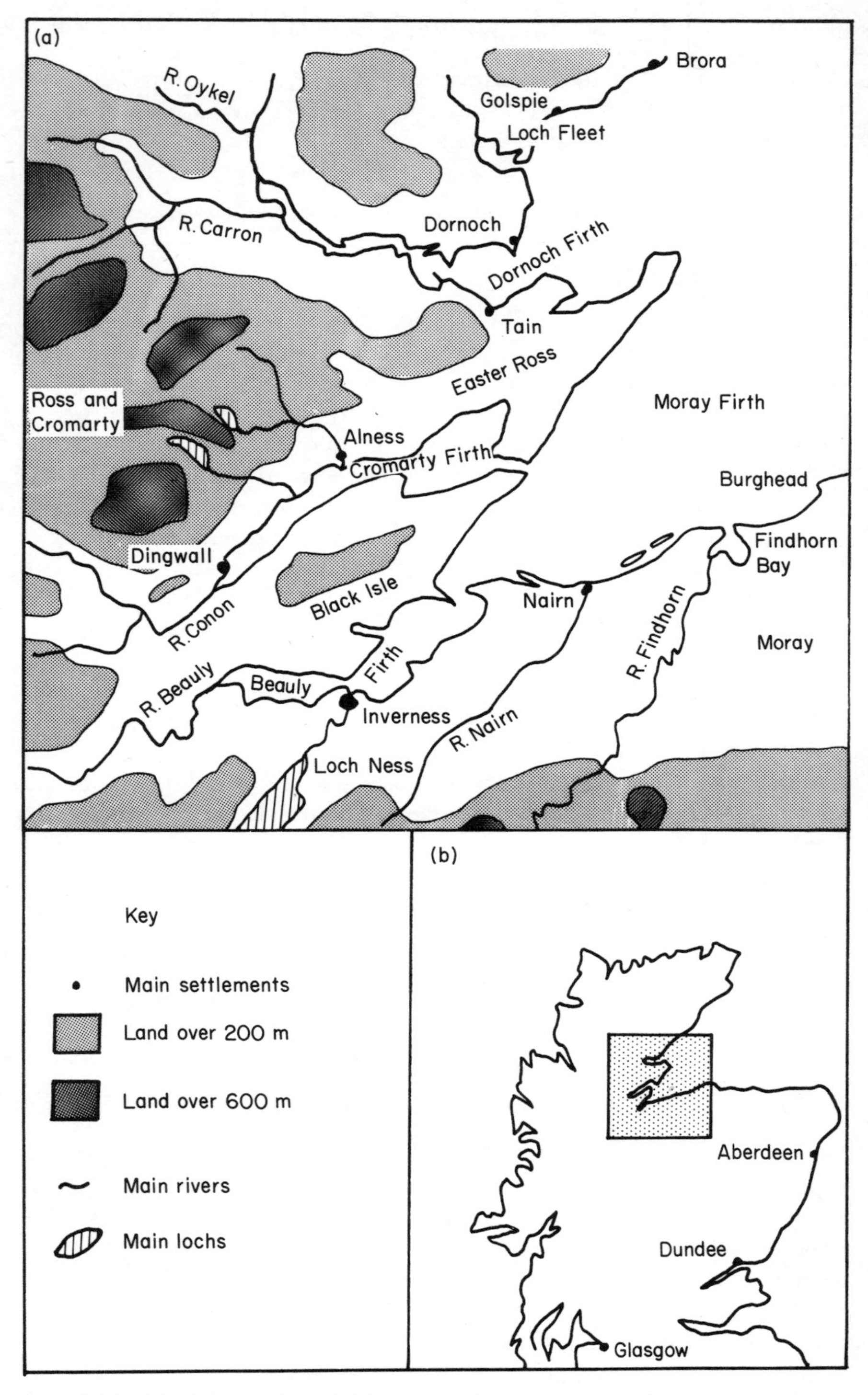

Fig. 6.7 (a) The Moray Firth; (b) Location of area covered by Prospectus editions.

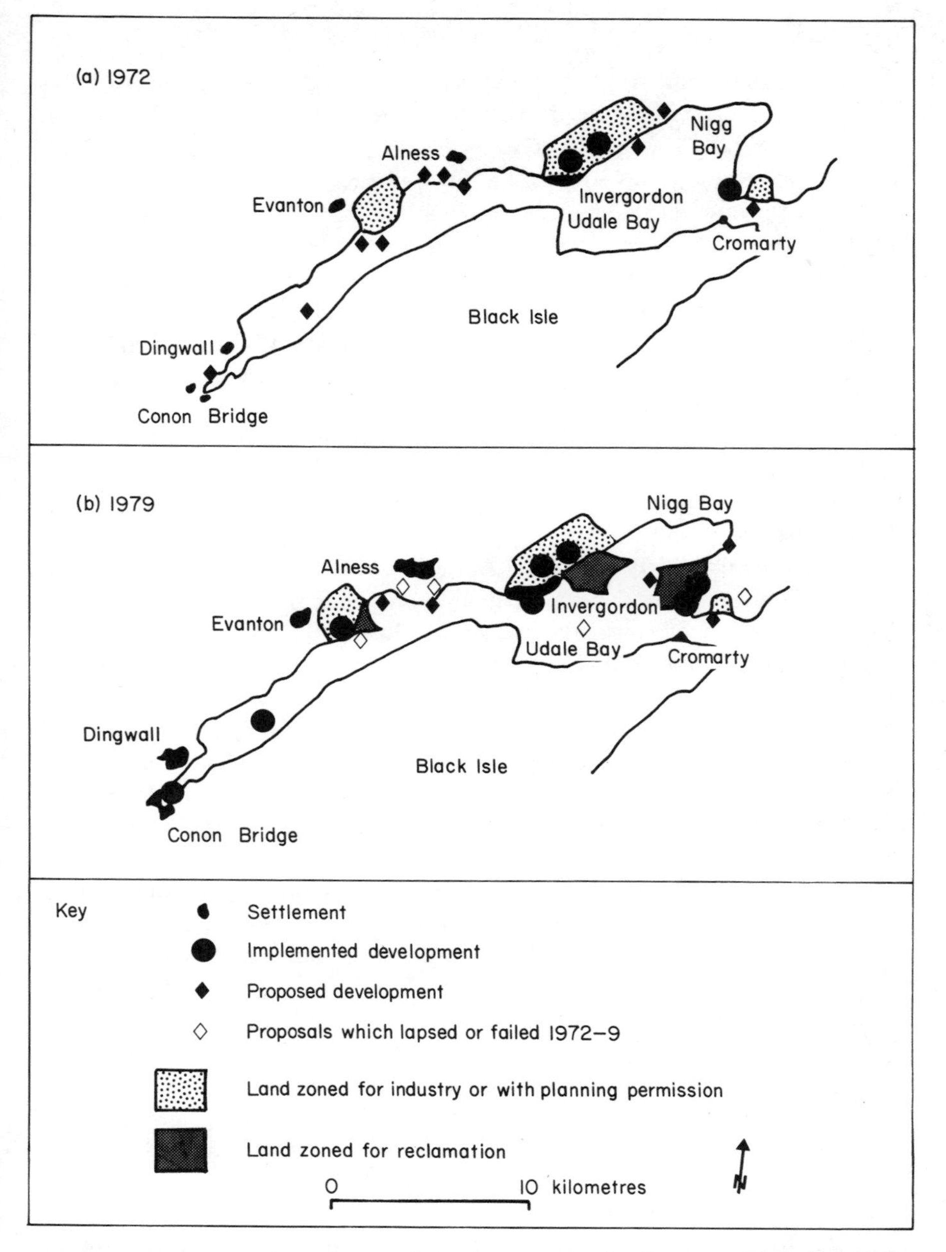

Fig. 6.8 Proposed and implemented development in the Cromarty Firth ((a) 1972, (b) 1979).

An early development was the establishment of a distillery at Invergordon in 1964, and this was followed in 1971 by the completion of an aluminium smelter nearby. By 1972 exploration for North Sea oil was rapidly expanding and this triggered off a period of land acquisition and speculation, coupled with a large number of planning applications for construction and service activities related to the oil industry. By the end of 1972, such applications in the Cromarty Firth included five for oil rig construction while some 4000 hectares of land adjacent to the firth had been zoned for industry (Fig. 6.8(a)). Other applications were lodged for several sites between Brora and Whitness Head while a considerable number of proposals, speculations and ideas did not reach the planning stage. The period between 1971 and 1974 was a peak, 'with industry putting in massive applications and then fading away, and population estimates swinging from 200 000 back to 50 000 . . .' (Nairn, Sunday Times, 28 July 1974). It was during the period that the first Prospectus was produced.

In 1973 the Cromarty Firth Port Authority (CFPA) was set up in order to cope with snowballing development proposals, but by 1974, the peak had passed. A period of consolidation set in, during which almost all of the planning applications which had not been carried to fulfilment fell into abeyance. In 1976, however, planning permission was granted for the construction of an oil refinery at Nigg Point. Commercially exploitable quantities of oil were discovered in 1976 some 15 miles offshore from East Sutherland. Oil from this, the Beatrice Field now operated by the British National Oil Corporation (BNOC), was planned to be piped ashore to Shandwick and thence to east Nigg Bay in Easter Ross.

Subsequent development proposals have been predominantly for the east Nigg area, involving terminals and petrochemicals. In 1979, a group headed by Dow Chemical Company submitted plans for major petrochemical development on land to be reclaimed in Nigg Bay, while the British Gas Council and Highland Hydrocarbons also produced similar proposals for the same area. In 1980, BNOC oil terminal and transhipment facility was under construction, and a service base for the same company at Invergordon was given planning approval. It was during this continuing and planned development period (Fig. 6.8(b)) that the revised Prospectus was introduced in 1978.

Industrial growth led to an increase in population, and almost every settlement within the area expanded, particularly in Easter Ross and around Inverness. By 1980, these areas had experienced an 8% population increase over the 1961 census figures, while elsewhere the population had been stabilized. Settlement expansion increased the need for the provision of sewage and refuse-disposal facilities.

In addition to the expansion of settlements, a number of new roads were constructed and many others upgraded. Ideas for a new road crossing the three firths crystallized with the publication of a promotional pamphlet (Clarke *et al.*, 1969) and in 1980, a new bridge crossing the Cromarty Firth was opened. At the same time, one crossing the Inverness Firth was under construction, and engineering and environmental studies for the Dornoch Firth Bridge had been undertaken.

All the constructional activity, whether for industry, settlement or road-works, created a demand for aggregates, especially sand and gravel. This has been met by considerably expanding existing major quarries especially in the Inverness and Alness areas, as well as opening a number of new ones. Plentiful resources of aggregates exist (Harris and Peacock, 1969) but increased demand put considerable pressure upon many of the landform features of the Moray Firth.

Growth of industrial activity greatly increased shipping traffic in the Cromarty and Inverness Firths.

(c) Nature conservancy involvement and strategy

The advent of major industry, with the resultant expanding settlement, communications and leisure requirements, represented a major impact on nature conservation interests, and in many cases a threat to them. Other important land interests were also affected, including agriculture, forestry and tourism. When this became apparent, a number of bodies began to point out the social, environmental and land-use implications. One early concern was at the loss to industry of high-quality arable land. An Easter Ross Land-use Committee was formed which gave evidence at three public inquiries. The National Farmers' Union of Scotland (NFU) lodged objections with the Scottish Development Department (SDD) regarding the zoning of land at Invergordon for industrial development, and it was at the request of the NFU that the HIDB undertook a survey of the agricultural interest of the area (Ormiston, 1973). More recent research by Smith (1979) has substantiated this concern.

Throughout this period, the Nature Conservancy was required, as statutory advisors to the planning authorities, to provide advice and comment on a range of development proposals and land zoning, and also to give advice on certain planning documents such as the Holmes Report (see above). One sample case concerned the proposed new road crossing the Cromarty Firth. Early designs indicated a structure with only 10% open channel. However, as a result of advice given to the SDD by the NC, a causeway was built with 90% open structure, unlikely to have adverse effects upon the ecology of the upper firth. This was the sort of co-operation which the NC sought.

Based upon early experience, it became apparent that there was a need to set out a broad case for nature conservation throughout the area, and to emphasize the need for nature conservation interests to be considered at as early a stage as possible in development proposals. Furthermore, it was necessary to create an atmosphere of concern regarding environmental impacts in terms of pollution, loss of habitat and changes in hydrology and sediment patterns as the result of land and estuarine reclamation. One typical area of concern was the effect which expanding oil-related activities and increased shipping would have on the incidence of oil pollution. It rapidly became evident that the large bird populations within the firths were susceptible even to small spills (Currie, 1974).

There was also the need to provide planners, developers and the general public with information on the location and importance of specific areas of nature conservation value. Politically, it was necessary to identify these areas in advance of planning proposals, rather than seemingly in response to them.

The major problem which existed for the NC in fulfilling these roles was that by 1967 little basic work had been carried out in the Moray Firth area. There was a National Nature Reserve (NNR) at Mound Alderwoods, in Sutherland, and a handful of SSSIs were notified to local authorities. However, by 1968 a preliminary survey of wildlife was prepared by Dr I. D. Pennie, NC Warden-Naturalist, and this was included in the Holmes Report. It contained a compact statement of the area's potential as well as suggesting specific sites of value. It also highlighted the need for more data, especially for bird numbers (Appendix 4, Holmes Report).

Further work was carried out by Currie, who in an internal paper to NC (1971) set out proposals relating to wildfowl conservation in the Cromarty Firth. In response to the rapidly changing requirements of the situation, this was subsequently expanded into a total nature conservation policy and promotional statement for the whole Moray Firth. The resulting report, *A Prospectus for Nature Conservation within the Moray Firth* was published in 1972.

6.3.3 A prospectus for nature conservation within the Moray Firth

(a) Introduction

The report, which comprised 81 pages with 13 maps and 16 photographs, was in three parts. Parts 1 and 2 were concise policy and promotional sections (20 pages), which highlighted the development of industry along with some of its actual and potential impacts, stated the case for nature conservation and suggested the means by which conservation measures might be brought about. A total of 36 sites of SSSI standard were identified and proposed, and 23 other areas of high nature-conservation value were recognized. Part 3 contained appendices in which factual information on the natural environment and detailed site descriptions were set out.

(b) Survey and data collection

Considerable time and resources were spent in carrying out basic survey between the year 1969 and 1972, in order to identify areas of conservation value, and to assess their relative importance. The work was carried out largely by NC regional staff. For example, a survey of intertidal wildfowl feeding plants such as *Zostera*, *Ruppia* and *Salicornia*, was instigated, and this survey still continues (Currie and MacLennan, in preparation). Another input of effort was by NC specialists in different habitats and disciplines. An example was a survey of intertidal invertebrates carried out by Miss Sheila Anderson, then of the NC Coastal Ecology Research Unit. This survey was continued by the Department of Zoology, University of Aberdeen.

Ross and Cromarty County Council, with financial assistance for the HIDB, instigated a survey of changes in the framework of the total environment, and information from this source became available to the NC (University of Aberdeen, Department of Geography, 1969–81). Yet more information was provided by a number of organizations which included the Royal Society for the Protection of Birds (RSBP), the Scottish Wildlife Trust (SWT) and the Botanical Society of the British Isles (BSBI) as well as a number of individuals.

One outstanding requirement was the need to obtain accurate figures for bird numbers in the Firths. Regular wildfowl counts, to augment the figures of Atkinson-Willes (1963), had begun in 1965 as the result of local initiative, and initial reports were prepared for the SWT by C.G. Headlam (SWT, 1969). In the same year, the British Trust for Ornithology, the RSPB and subsequently the Wildfowl Trust, launched the Birds and Estuaries Enquiry, which provided a framework within which local counters could work. Since the enquiry was a National one, information from the Moray Firth could be set in the national context (Prater, 1981).

In all, some 20 Nature Conservancy scientific staff were employed in obtaining data, requiring an estimated total of at least 450 man-days. The Birds of Estuaries Enquiry alone involved an average of 30 voluntary man-days per month, and while no estimates are available for the total relevant work carried out by other bodies and individuals which was incorporated into the Prospectus, this was considerable. Once sufficient data had been obtained, and the important sites for nature conservation were identified, their boundaries were defined and statements of their interest prepared.

(c) Revised prospectus

The first edition of the Prospectus had been produced hurriedly, during the peak period of industrial activity, and to fit the needs of the moment. Although it incorporated all available research and survey material readily available at the time, there were many omissions, since there was no time to await much essential data. By 1975 a considerable amount of additional ecological work had been carried out, much of it instigated and inspired by the early Prospectus. While the revised Prospectus leaned heavily upon the survey work carried out for the original edition, every effort was made to cover as many additional aspects of the ecology of the Moray Firth as was possible. A much wider range of NCC specialists were invited to participate, and other surveys were carried out under contract by universities and research organizations which included the Institute of Terrestrial Ecology. This additional survey work resulted in the identification and notification of 22 additional SSSIs.

A number of other factors also combined to suggest the need for an updated and modified report. One of the main requirements was to widen the scope beyond one aimed simply at industrial development, in order to include the whole spectrum of land use in the area. Thus, the relationships, between nature conservation on the one hand, and agriculture, forestry, settlement, recreation and other land uses on the other, were considered in greater detail. Certain developments which either had been given little coverage, or had not been anticipated during the production of the early Prospectus, were also treated. These included the effects of offshore oil exploration and transportation upon important seaduck concentrations (Mudge, 1978; Allen, 1979; Mudge and Allen 1980), and the effects of greatly increased demand for aggregates upon the sand and gravel landform resources of the area.

The revised Prospectus also took into account the fact that in May 1975 the old County Councils had ceased to exist, to be replaced by the Highland Regional Council which had an executive planning role over the most of the area. In 1973 the Nature Conservancy had become the Nature Conservancy Council with a new remit and it had by 1978 produced a number of policy statements such as 'Nature Conservation and Agriculture' (NCC, 1977) information from which was incorporated into the new Prospectus along with that from the major national survey report 'A Nature Conservation Review' (Ratcliffe, 1977).

By 1978 the Scottish Development Department had sponsored reports on the production of environmental impact statements (for example Clark *et al.*, 1976). Their National Planning Guidelines series provided a further important set of policy documents incorporating major changes in planning practice. As a result development proposals affecting nationally important SSSIs had to be referred to the Secretary of State for Scotland where NCC had registered an objection.

Perhaps the greatest need by 1978 for a modified document stemmed from the fact that a number of the main pressures which had led to the original report had been modified or had ceased to exist. There was by that time a wider acceptance of the case for nature conservation and the need for protection of sites as well as for more detailed evaluation of the effects of proposed developments upon the environment. An example of this was the production of amendment 10 of the Ross and Cromarty Development Plan

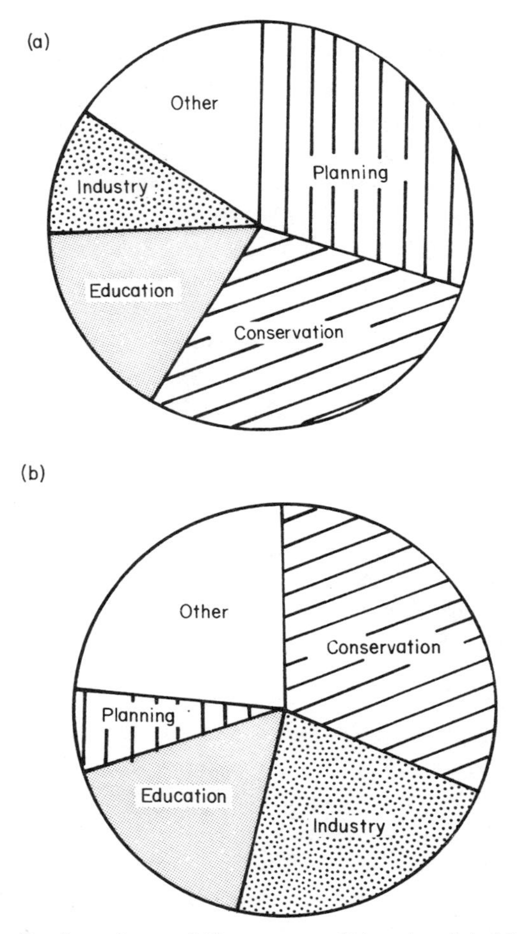

Fig. 6.9 (a) Distribution of second Prospectus edition (total is 180); (b) sources of requests for a Prospectus copy (total is 80).

(1973) which incorporated nature conservation zonings. Thus while the revised Prospectus was based very strongly upon the original, it represented an advance in terms of knowledge, philosophy and advice from the NCC. Its content was orientated more towards means of effecting nature conservation rather than attempting to create a case for it.

(d) Circulation and promotion

In order to fulfil two of the main functions of the Prospectus, both editions were circulated to as wide a chosen audience as possible. Publicity was angled towards the case for nature conservation and the value of the report as a planning document, rather than towards the publicizing of individual sites. The mailing list included all planning authorities, all central government agencies with a remit in the Moray Firth, university departments, industrialists known to have an interest in the area, major landowners and a range of independent conservation bodies. Local libraries were also provided with copies to allow the general public to consult.

Further circulation was to organizations and individuals from a variety of backgrounds who requested copies (Fig. 6.9(a)). In order to reach a wider public, an eight page folding leaflet was produced and given away free at lectures, local agricultural shows and exhibitions. In total, between the date of publication and the end of 1975, some 350 copies of the initial Prospectus had been distributed and around 3000 leaflets given out. The revised Prospectus was distributed by NCC to a similar range of interested parties as the original (Fig. 6.9(b)) while a major attempt was made to inform the media of its production as a planning tool. By 1980 some 230 copies had been distributed and a steady demand was being maintained.

6.3.4 Impact of the prospectus

It is impossible to assess comprehensively the impact of the two editions of the Prospectus on planners, public and developers, since these people were also subject to other external pressures and other contacts with NC and NCC. At the time of publication the media, particularly local newspapers, gave good coverage to the documents so it may be assumed that there was fairly broad public knowledge of the subject. One sample review stated 'This Prospectus is a positive-minded document, and if it helps to bring planners and industrialists into awareness of the environment at the very early stage of their thinking, and not as an exasperating afterthought, it may mark the start of real sense over oil, rather than wild emotion.' (Leigh, The Scotsman, 14 April 1973).

Incoming industry was given a clear indication of likely NC and NCC reaction to certain proposals, and sensitive sites were pinpointed. The reports allowed ecological and nature conservation views to be seen by industry as part of the total environmental impact analysis, and not merely as unrelated fringe matters. Knetch and Freeman (1979) have pinpointed the damage caused by the separation environmental accounting from economic evaluation of projects, thus discouraging the consideration of impacts during the early stages of project formulation. With the Prospectus to hand, some of the environmental factors which would of necessity affect certain sites could be considered by designers at the very beginning. The number of organizations which requested a copy (Fig. 6.9(a)) indicated that the publication was fulfilling a demand in this sector.

Responses from professional planners were wholly favourable. In particular, they welcomed the overall perspective regarding priority areas for nature conservation, and they welcomed the advance indication as to how NC and NCC might react to certain types of development. The documents also permitted a reasoned approach to nature conservation in the broader countryside outwith the statutory SSSIs and NNRs. More information is contained within the Prospectus than was included in the then standard site-notification schedules, or could be incorporated into planning documents such as local plans.

The main opposition to nature conservation and the main criticism of the Prospectus editions came from an influential nucleus of local politicians. They believed that their task was to attract industry, and therefore jobs, by all means, and regarded nature conservation as a nuisance. This view was not universally held, and the NC and NCC found it a great advantage to use the Prospectus as a means of countering the mythology which grew up regarding the role and views of NC during the main phase of development in the Moray Firth. Elected representatives, many of whom were taking major planning decisions, had access to the facts regarding the aims and objectives of nature conservation, and misgivings could be dispelled by having clear statements on paper. Sufficient impact was made at national level to cause the then Secretary of State for Scotland, Mr Gordon Campbell, to ask Ross and Cromarty County Council in late 1972 'to take full account' of the first Prospectus.

One further impact of the Prospectus was that it listed surveys which had been carried out, and pinpointed areas where there was a lack of knowledge. This acted as a stimulus to encourage research by other bodies or individuals. The independent nature conservation bodies were provided with a 'shopping list' of sites which they could take into account when planning their own programmes of survey and reserve acquisition. With the exception of Loch Fleet which became a SWT reserve in 1970, all the reserves now managed by the SWT within the Moray Firth were on a list in the original Prospectus denoting sites suitable for such action.

Perhaps the only real assessment of achievement can be made by looking at the results in real nature conservation terms on the ground. Since the publication of the first Prospectus, over 50 additional SSSIs have been designated as well as an NNR in Nigg and Udale Bays. Though two areas of Nigg Bay are zoned for reclamation, the remaining two-thirds of the bay and the whole of the Udale Bay are zoned for nature conservation under the *Cromarty Firth Port Authority Act* 1973. Other nature conservation designations include Local Nature Reserves, one at Munlochy Bay administered by the Highland Region and one in Findhorn Bay administered by Moray District. The RSPB have established a reserve at Nairn Bar off Culbin while the SWT have added one reserve in Easter Ross and two on the Black Isle to the previously established Loch Fleet in Sutherland. Loch Eye in Easter Ross is a statutory Bird Sanctuary under the 1954 *Protection of Birds Act*.

All these designations plus other advice on nature conservation have now been incorporated into the appropriate local plans produced by the Highland Regional Council. Such opposition as there was has been allayed following reasoned discussion, and there was broad support from the public, landowners, planning authorities and other interested parties. Most of the designations were proposed in the first Prospectus, and it is doubtful if such progress could have been achieved without first of all preparing the way through the publication of that report.

One other major area in which progress may be seen is the fact that environmental impact assessments of major developments are now considerably more an accepted feature of planning permission. The development of the Beatrice oil field off the coast of East Sutherland and the proposals to establish an oil refinery at Nigg Point were both subjects of such studies. Furthermore, following the initiation of recent major projects such as the Nigg Point refinery and the BNOC Beatrice Field pipeline, environmental monitoring has commenced in order to establish baselines and to ensure that adverse effects from these developments are detected and minimized. Though this has come about largely as a result of central government direction, it epitomizes the kind of consideration for the natural environment which the Prospectus sought to promote.

However, many of the aims and objectives set out in the two documents have not been fulfilled. One major example is the lack so far of a comprehensive hydrological model to facilitate the assessment of large-scale reclamation in the Cromarty Firth. There are still few NNRs in an area with a number of nationally important sites and with many actual or possible threats upon them. One can only hope that increased awareness by owners and planners will give them some measure of protection.

6.3.5 Problems and disadvantages of the prospectus approach

There are difficulties associated with the production of a Prospectus which merit discussion.

It is very important to consider the readership to whom the report is being presented. In the present case, the main readers were planners, developers and landowners, most of whom were expert within their own chosen fields of experience, but few of whom had any ecological training or skill. Though the text of the Prospectus was kept simple, without being simplistic, there were quite contrary reactions. Some readers considered the reports to be too scientific, while others complained that they were not scientific enough. This illustrates the problems of achieving the correct balance for such a knowledgeable, diverse public.

Probably the greatest disadvantage is the time involved in producing a thorough document (some 3 to 4 years in the cases in question) and the specialist manpower which is required to achieve an authoritative text. As time goes on these requirements can only increase. During the production period of a Prospectus, major developments can arise which might seriously impair the impact of the document when it is published. Even after publication, the length of time which a Prospectus can last without becoming dated is limited. This is particularly the case in a rapidly changing situation, and especially so when an ageing document is used without checking in order to ascertain more current NCC site assessments and policies (see postscript).

One further disadvantage is that once produced, such documents may be used as evidence in Planning Inquiries, where by quoting excerpts out of context, opposing council have derived meanings the opposite of what was intended. Such possibilities may never be fully avoided, but it does impose a further time factor on production to ensure that all ambiguity of text is excised and that no major errors exist which might be exploited at a later date.

Because of the secrecy which for commercial reasons often surrounds industrial activity, it may often be difficult to ensure that such a report as the Prospectus falls into the proper hands in good time. There is also the possibility that a company that intends to set up an industry which potentially may cause pollution may be encouraged to keep silent regarding its plans for even longer than at present on the strength of an expected response to their proposals derived from a Prospectus document.

There can be, especially in areas of high unemployment, a reluctance on the part of the planning authorities to draw the attention of developers to a Prospectus document lest it frighten them away. A further problem occurs where a Prospectus covers the geographical areas of several planning authorities as the first edition of the Prospectus did. If the bulk of the important sites fall within the territory of one authority there may be a reluctance to accept the importance of these sites since it may seem to give unfair advantage to neighbouring areas. There is no doubt that in the Moray Firth, some developers played on this, in order to gain as favourable planning terms as possible. Such problems occur to a lesser extent since regional reorganization and since the automatic referral of planning proposals affecting nationally important sites to the Secretary of State for Scotland, in cases where the NCC had objected.

Though the grading of sites was specifically welcomed by planners and others, we have found that sites given lower individual gradings have occasionally been dismissed as being of lesser account, and priority has only been given to those sites of major importance.

A Prospectus may also be used by a variety of fringe individuals or groups, either 'pro-environmentalist' or 'anti-development', for their own ends, often reducing the reputability of the report. While such a document represents the NCC views of nature conservation, it does not and cannot represent the views of independent conservation bodies such as the Scottish Wildlife Trust and the Royal Society for the Protection of Birds, who represent many thousands of private individuals who quite rightly expect to be able to express independent views.

6.3.6 The future use of the prospectus approach

There is no doubt that the economic development of the Moray Firth, and perhaps other areas in Northern Scotland, will continue. There is, therefore, a need for continued liaison with appropriate bodies, in order to ensure that nature conservation is given due consideration. In this context, we feel certain that the Prospectus method is assured of a future. Its advantages exceed its disadvantages, especially if NCC staff maintain close regular contact with local authority planners. There is no reason either why the method may not be used to promote similar aims in other parts of the country.

The first Moray Firth Prospectus was produced rapidly in response to, and in the face of, rapid development. It is intended that documents for other areas will be produced in advance of major land-use changes, so that incoming industries, and expanding existing land-uses such as forestry, agriculture and settlement, will have before them a blueprint to ensure the proper conservation of the natural environment. Though local plans may partly fulfil this requirement, they cannot provide the promotional function which we have found so valuable.

It may seem that major land-use changes in areas like Skye or North-West Sutherland are unlikely. But we have found that during the three years of the compilation of Caithness and East Sutherland Prospectus, major proposals hitherto unconsidered, have arisen including large-scale afforestation, onshore oil exploration, uranium exploration and large-scale peat extraction. As these develop, some at least will have a major effect upon the environment and upon nature conservation interests.

There still remains in many such parts of the Highlands and Islands of Scotland, the need to put over the case for nature conservation and a reluctance in many local communities to accept it. This may be partly because of concern that in high unemployment areas, the imposition of nature conservation designations will discourage future developments with their associated jobs. Another reason may be the fact that wildlife resources in these areas are not seen to be under threat; extensive peatlands, lochs, uplands and coastline have always seemed to be a limitless resource to the layman in which the choice of specific areas for conservation and protections looks a pointless exercise. Landowners may see themselves as losing money as the result of nature conservation measures. We therefore see a continuing requirement for promotional publications, pointing out the environmental impact of projected developments.

Postscript

Since this paper was written, the dynamic development situation in the Moray Firth has altered considerably. Some major changes affecting the area include:

(a) Government decision not to fund gas-gathering venture in North Sea has shelved major petrochemical plans for Nigg Bay.
(b) Application from Highland Hydrocarbons for MTBE Plant.
(c) Proposals for mountain railway and leisure complex, Ben Wyvis.
(d) New set of oil exploration licences for Inner Moray Firth.
(e) Proposals passed to reclaim whole of upper Beauly Firth for agriculture, then subsequently shelved.
(f) Closure of British Aluminium smelter with loss of 900 jobs.
(g) Completion of Inverness bridge and A9 road link.
(h) Planning permission granted for construction of 3 km pipelines in Alness Bay.
(i) Proposals for declarations of Easter Ross as an industrial 'Enterprise Zone'.

Acknowledgements

We would like to acknowledge the help of R. N. Campbell, Miss L. V. Cranna, P. E. Durham, G. Pease, Dr J. S. Smith and Dr P. J. Tilbrook.

6.4 ENVIRONMENTAL IMPACT ASSESSMENT PROCEDURES USED IN A STRATEGIC STUDY OF WATER RESOURCE DEVELOPMENT OPTIONS

P. J. Nelson, J. Corlett and C. Swanwick

6.4.1 Introduction

Many of the basic analytical techniques in Environmental Impact Assessment have been used in Britain for a number of years especially in the area of mineral resource development, but the use of EIA as a management tool, as a component of project research and design, and as an aid in the process of plan-making and decision-taking is still in its infancy.

The recent Regional Water Resource Studies conducted by the North

West Water Authority (NWWA) have been described as a pioneering attempt to encompass the wider aspects of EIA. In this paper, it is the authors' intention first to describe the regional studies concentrating particularly on those characteristics which represent innovations in approach, and secondly to seek to illustrate these features by special reference to the ecological surveys and analysis.

6.4.2 Reasons for the study

Following reorganization of the water industry in 1974, NWWA was confronted with a wide range of potential supply options which might be developed to meet any further increase in water demand in its region. Those options had been examined in outline by the Water Resources Board in conjunction with the three former River Authorities in the North West during the 1960s. Between 1974 and 1976 an internal review of more than 20 schemes was carried out by NWWA culminating in a shortlist of four alternatives, each of which would be capable on its own, of meeting a substantial growth in the region's demand for water (Fig. 6.10). The schemes in question were:

(a) Enlargement of Haweswater Reservoir in the Lake District.
(b) Construction of a new reservoir in Cumbria on the Borrow Beck, a tributary of the River Lune.
(c) Impoundment of the River Ribble at Hellifield in North Yorkshire.
(d) Storage in Morecambe Bay.

At this stage, the Authority decided that before a final choice could be made, comparative data would be required not only on the technical feasibility and probable order of costs of the short-listed alternatives but on the relative effects of the engineering works on the environment. Informal discussions were held with the Chief Planning Officers for the seven Structure Planning Authorities in the region about the most appropriate means of achieving the latter objective. It was concluded that a joint study should be undertaken using Consultants, and staff of both the Water Authority and the Local Authorities.

In addition to the environmental studies, the Water Authority engaged the services of three firms of Consulting Engineers: Binnie and Partners, Babtie Shaw and Morton, and Rofe Kennard and Lapworth, to review the findings of earlier work and to complete feasibility studies for each of the four schemes in conjunction with the environmental assessment.

The studies were intended to be strategic in nature in that they were not based on any specific forecast of growth in demand for water supplies within the region. Indeed, the Authority chose deliberately not to use any fixed estimate of future demand, but to work on a series of projected levels of

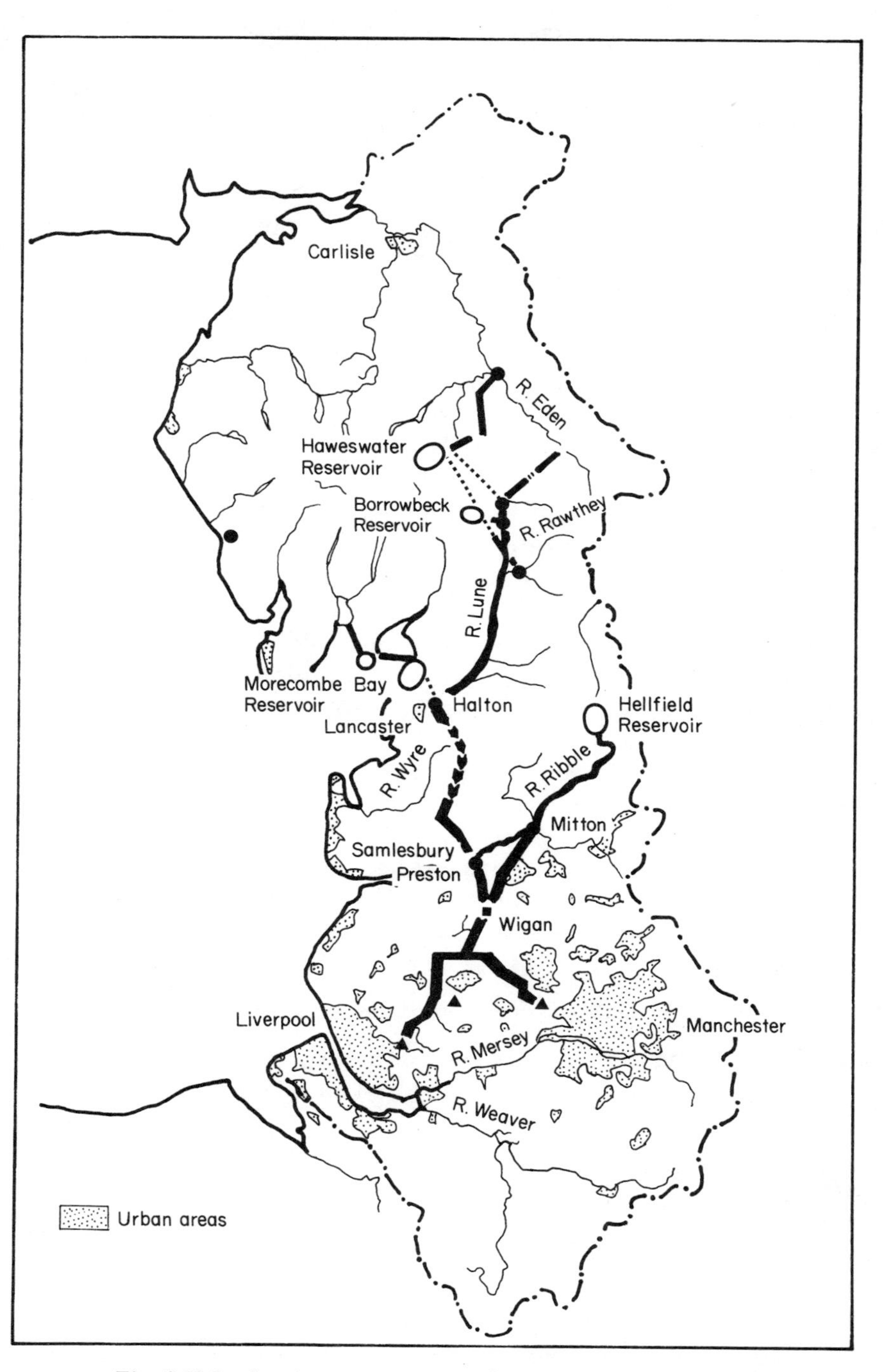

Fig. 6.10 Regional water resource studies – schemes considered.

deficits in supply based upon standardized growth patterns. SDGPs are revised regularly by the Authority and consist of a range of possible future demands comprising a 'preferred forecast' complemented by upper and lower bounds. For the purpose of the Study, these levels were fixed in 1976 at a potential deficit in supply of 300, 600 and 900 Ml day^{-1} in the year 2011. The thinking underlying this approach recognized the uncertainty of long-term forecasting, especially in the area of water consumption where upward trends which remained relatively constant during the 1960s and early 1970s have become increasingly unreliable in the past five years.

Midway through the study a decision was taken to eliminate the upper bound of 900 Ml day^{-1}, in response to updated SDGPs which reflected a marked falling in demand, especially in the industrial sector. However, while the remaining SDGPs of 300 and 600 Ml day^{-1} were designed to focus attention on the potential order of yield required from the new sources, they were not intended to act as a constraint on the optimum level of development to be considered for any scheme, nor as a set of target yields to be met precisely by each and every variant. From the Authority's standpoint, the purpose of the studies was to identify the relative merits of the alternative sources should a requirement arise to provide new supplies within the time horizon of 2011. The studies were not intended – as some commentators have since suggested they might have been – to answer the question of whether or not new sources of this type should be developed at all.

The environmental component of the Regional Water Resource Studies was thus designed to be project-orientated, but by the very nature of the four schemes and the size and scale of the regional water supply systems considered, the study findings can be expected to influence policy decisions on wider aspects of regional water resource management and to help to focus future areas of research, whether or not future levels of water demand necessitate the development of any one of the schemes in question.

6.4.3 Description of the schemes

Each of the four schemes have a common element in the form of the aqueduct supply system south of Wheelton in Mid-Lancashire but in most other respects they differ widely.

Haweswater is an existing supply reservoir serving the Manchester area by direct gravity. Its further development would involve introducing new sources of water to augment the supplies fed into the reservoir from its existing and indirect catchments, and the release of the additional stored water through a new aqueduct to regulate the River Lune in periods of low flow. In this way increased abstractions could be made from the Lune a little above its tidal limits near Lancaster. The Lune is already tapped as part of the Lancashire Conjunctive Use Scheme, and the intake at Halton would be developed to serve both systems. A new aqueduct would be laid from the intake down to Wheelton.

Borrow Beck is one of the few major streams draining the eastern fells of the Lake District which has not been tapped as a water supply. A dam, built across a remote section of the valley near its confluence with the Lune, could create a reservoir with a substantial volume of storage which would be filled by direct run-off from the catchment. This source could be supplemented either by pumping water direct from the River Lune and/or the River Eden, or by linking the reservoir with Haweswater. Water would be released for storage to regulate the Lune and recovered at Halton.

Hellifield differs significantly in character from the other inland sites in that the reservoir would occupy the bed of a former glacial lake in the broad fertile plain of the River Ribble below Settle. As such, it would be the only scheme directly impounding a major river. Stored water would be released to augment low flows and permit continuous abstraction from the Ribble in its lower reaches near Preston. From here water would be transmitted into the regional distribution network through a new aqueduct to Wheelton.

Morecambe Bay was initially seen as offering opportunities for tapping rivers which discharge into it, and storing water either by constructing barriers across the estuaries of the rivers themselves, or by transferring it to bunded reservoirs constructed on the intertidal sand flats. The range of options has been narrowed down during the course of the studies to two basic possibilities which will be described in greater detail below. In essence they rely on abstracting water from the River Lune at Halton at times when the level of discharge is above the prescribed flow, and feeding it directly to supply. Under conditions of low river flow in the Lune, water would be taken from a bunded storage reservoir located either in the East Bay, or off the Cartmel Peninsula. These reservoirs would be refilled by pumping from the River Lune and River Leven respectively, whenever sufficient water was available to run the pumps efficiently and economically.

6.4.4 Organization and framework of the study

Land Use Consultants (LUC) were appointed to advise the Authority on environmental matters lying beyond its own statutory responsibilities and to co-ordinate the study. The direction of the study was controlled by an Environmental Impact Panel (EIP), comprising the Water Authority's Director of Resource Planning and his assistants, each of the Chief Planning Officers for the seven Structure Planning Authorities in the region, and Principals of Land Use Consultants. This group met at two monthly intervals throughout the eighteen-month study. The Consultants were responsible for the study's conduct and management, and their own skills and resources were supplemented by those of staff of the local authorities, and in some cases by outside government agencies.

During the first four months of the study, four specialist working groups were established to undertake the field work and analysis required at each of the reservoir sites. These groups covered the topics of agriculture and local economy, recreation, landscape and ecology.

Members of groups were drawn from amongst the specialist staff of the local authorities and, in the case of the agriculture and local economy team, from the Ministry of Agriculture Fisheries and Food's Regional Land Service.

Each group was given terms of reference approved by the EIP and its method of working was coordinated by the Consultants to ensure consistency of approach. The content and scope, and methods of data collection and analysis for each survey were, however, proposed jointly by the Consultants and individual groups before being ratified by the EIP. It was firmly established that members of the groups were free to work as professionals in their own field, without allowing the potential interests of their employing authorities to influence their judgement in any way. In addition to the working groups described above, the Water Authority established its own internal teams to cover hydrology, water quality, river ecology, river management, engineering and costing, and treatment plants and aqueducts.

Liaison between working teams was maintained by a separate body referred to as the coordinator's Group which was chaired by LUC's Project Manager, and consisted of senior staff members of the Local Authorities and Water Authority. Members of this group were responsible for ensuring that other officers from their own Authorities were kept informed of study progress and that representatives of the individual working groups were fully briefed on the role they were expected to perform.

6.4.5 Method of approach

The method of environmental appraisal adopted for the study constituted a modified version of the approach proposed by Clark *et al.* (1976). It differed from the original only through the introduction of a series of preliminary stages in the process of environmental evaluation, which was necessary to allow continual assessment of the engineering options and feedback to the design teams.

This process consisted of:

(a) Examining existing planning policies relevant to the proposal.
(b) Surveying the environmental characteristics of the areas potentially affected by each scheme variant.
(c) Analysing the characteristics of the engineering proposals.
(d) Defining the probable interactions between the engineering proposals and the environmental character of the affected area, and assessing the relative significance of particular impacts.

(e) Examining the scope for reducing or modifying the effects of adverse environmental impacts by altering the engineering proposals.
(f) Making a final assessment of those environmental impacts which were inherent and unavoidable.

In practice, a series of iterations were required through each component of the method before any confidence could be placed in discrete elements. This point can be illustrated with reference to the problems which were experienced initially in defining the geographical areas to be covered either by physical surveys or literature reviews. In the case of pipeline routing for example, adequate knowledge of planning and land-use policies, existing environmental conditions, and the basic engineering design parameters for pipeline construction had to be assimilated before a decision could be taken on the degree of ecological and other survey detail which would be required and the full extent of the area to be covered.

From a very early stage in the joint studies, a steady flow of information was established between the engineering and environmental teams. Not unnaturally, the working relationships which developed between the different groups varied considerably in style, but the common framework and organization of the studies was most effective in ensuring a unified approach.

Although each of the four potential water-storage sites had been the subject of previous investigations, the amount of data which had been accumulated on each of them varied considerably, and the schemes themselves varied from detailed proposals for new dam construction at Haweswater, to the generalized concept of bunded storage in some part of Morecambe Bay which had emerged in the closing stages of the Water Resource Board's study in 1970. It was therefore necessary to define the range of options to be considered at each site, and to undertake a preliminary appraisal of the environmental impacts which would be incurred. This work occupied the first six months of the joint study, and by October 1977 more than 12 options for water supply, storage and transmission had been defined in Morecambe Bay alone, with a further 18 Variants, as they were named, identified at the three inland sites.

Over the next eight months, these Variants were subjected to continuous refinement and reappraisal, in order to ameliorate or mitigate areas of adverse environmental impact whenever possible, and to produce cost-efficient and effective solutions to technical problems identified by the engineering studies. At pre-determined stages in the course of the study, formal conferences and meetings were held to debate the respective merits of the variants produced for each of the four major schemes. Those which were shown to be deficient from both an engineering and environmental standpoint, when compared with an equivalent variant for the same scheme, were eliminated from further study. At the same time this process of 'paired comparisons' helped to identify features which conferred either engineering and/or environmental advantages of particular scheme Variants. These refinements were then incorporated in existing or new Variants. In total, 54 variations were generated and studied over a period of ten months. By February 1978, the list had been reduced to 20. Further additions and deletions occurred between February and April, and in June 1978 the final shortlist of 14 variants was agreed. Each of the shortlisted Variants was subjected to a final assessment of unavoidable adverse impacts and potential environmental benefits, and the findings were published in a two volume report entitled the 'Environmental Appraisal'.

6.4.6 Programme of consultation

The main report was subsequently circulated to a large number of statutory authorities and other interested bodies and individuals, and reviewed extensively in the local press. A six month consultation programme was initiated by the Water Authority in 1979, and participants were invited to comment on the findings of the study. The response from Consultees was consistent in praising the Authority's approach towards long-term planning for water resource development, although most organizations were equally direct in stating their view that every reasonable alternative to the development of new water sources should be explored before the high cost, and level of adverse environmental impact associated with most schemes, should be entertained.

6.4.7 Preliminary conclusions

In the light of the study findings and the comments it has received, the Authority has made a preliminary statement on its long-term strategy. It has ruled out the prospect of developing a reservoir at Hellifield on the River Ribble, due to a considerable range of adverse impacts, and in particular to the loss of better quality agricultural land which supports livestock farming over a much wider area than the reservoir site itself. The Authority has also concluded that development of a bunded storage reservoir in Morecambe Bay now seems substantially less attractive than it had appeared prior to the commissioning of the studies. This is because the high costs of the engineering works are seen to be matched by the potentially damaging effects of bunded storage reservoirs on the intertidal flats forming part of the Morecambe Bay Grade 1 Site of Special Scientific Interest.

Without ruling out the possibility of development in Morecambe Bay if future growth in water demand were sufficiently high, the Authority has indicated that a scheme to regulate the River Lune is the most favoured option. This would either entail increasing the throughput of water, and possibly the storage capacity of the existing Haweswater Reservoir, or building a new dam on the Borrowbeck tributary of the River Lune.

6.4.8 Ecological content of the studies

Four main areas of ecological impact to the studies were identified from an initial assessment of the different engineering components making up each of the four schemes. These concerned the possible direct or indirect effects of development on: (a) water supply catchments; (b) river systems used for supply or regulation; (c) water storage areas; (d) areas crossed by aqueducts.

(a) Water supply catchments

In the interests of protecting public water supplies, land-use practices in many of the traditional gathering grounds of north-west England have been substantially modified in the past. The recent introduction of new treatment works has largely removed the need for controls over fertilizer application, stocking rates, and public access to such areas. However, concern was expressed by farming interests at the outset of the study that the development of a major new reservoir on the Upper Ribble could lead to constraints on agricultural improvements upstream of the reservoir if problems of water quality control emerged during its operation. This particular fear was not accepted as reasonable by the Authority's officers on the grounds that water stored in the reservoir in question would be released for river regulation purposes, and would be fully treated subsequently before being fed into the public supply network.

Nevertheless, given the long-term horizon for which the schemes were being planned, it was agreed that a rapid assessment should be made of the basic vegetation and land-use characteristics by the four schemes. This would serve a multiple purpose in providing a context for the more detailed site surveys of the reservoir sites themselves, and allowing an assessment to be made of long-term trends which might be expected to have a bearing on land management and the local economy.

The problem of making such a broad assessment within the three or four months available was resolved by commissioning the Institute of Terrestrial Ecology at Merlewood to extend their work on the Ecological Survey of Cumbria to cover the additional areas. Details of the ITE Methodology and Approach are described by Heal and Bunce in Chapter 5.

In the final analysis this overview did not contribute significantly to the elimination of Variants although the problems inherent in developing water-supply systems in areas receiving substantial agricultural run-off were noted.

(b) River systems used for supply or regulation

The Water Authority's internal working team on river ecology under the direction of Dr J. Leeming was responsible for carrying out all studies relating to the effects of water abstraction, transport and storage, on the natural river systems. Much of their work was concerned with the extrapolation of data already collected as part of the Authority's continuous programme of water quality monitoring, and river management. However, a special one-year algal bioassay project was initiated with the primary purpose of enabling some general predictions to be made about the consequences of mixing and storing waters from the different sources suggested in the course of the study. Given the strategic nature of the study and the short time scale of eighteen months in which it was to be completed, the opportunities for new research were extremely limited and the main function of the River Ecology Group was to advise on the elimination of unacceptable scheme Variants on the basis of broad ecological principles.

Issues considered by the Group which proved to have particular significance included the potentially disruptive effects of inter-river transfers from the nutrient-rich waters of the River Eden to the River Lune, or Haweswater Reservoir, both of which have a much lower order of productivity; the establishment of appropriate levels for 'prescribed flows' to regulate abstractions at alternative intake sites; the consequences of changed flow regimes in both supply and regulatory rivers on river ecology and river management; the effects of alternative intakes and discharge sites on migratory fish movements; and the potential for development of fisheries in the alternative reservoirs.

(c) Water storage areas

Responsibility for assessing the ecological consequences of development at each of the reservoir sites was divided between the Terrestrial Ecology team; the Water Authority's River Ecology and Water Chemistry team; and, in the case of bunded storage schemes in Morecambe Bay, associated consultants. The division of labour reflected the widely varying character of the potential storage sites. In the case of the three inland sites, Haweswater, Borrowbeck and Hellifield, the Terrestrial Ecology group carried out all initial field work and made an assessment of the probable effects of each scheme variant. The character and probable impacts of bunded storage sited in Morecambe Bay, was assessed by the associate consultants. The Nature Conservancy Council contributed to the work on all four schemes by making available and discussing its own very detailed information collected during the course of the study.

The Water Authority's internal teams were responsible for reviewing the existing conditions of Haweswater reservoir, and for making predictions about the types of ecosystem which might be developed within each of the potential reservoirs. In the following sections attention is focused on the work of the Terrestrial Ecology Team, the appraisal of bunded storage sites in Morecambe Bay, with particular emphasis being placed on the way in which ecological findings were used to advance the objectives of the study.

The Terrestrial Ecology Group was initially briefed by the Environmental Impact Panel to: (a) identify important habitats within areas liable to be flooded or otherwise affected by construction works; and (b) provide a basic description of habitat types identified in (a) in terms of plant communities, dominant species and any notified rarities. It became apparent, however, as the study proceeded, that more detailed analysis would be appropriate of the degree of ecological variation within each site, and of the relative importance of specific habitats. Both these factors, it was agreed, could have an important influence on the siting of major development works including the dam and reservoir itself and in defining any thresholds to the size of development which might be achieved – in terms of water levels and inundated area – before a major increment in the scale of adverse impact would be incurred.

In developing a survey methodology, the Group agreed that it should be capable of responding to the following criteria:

(a) The need to minimize the effects of observer bias, and preserve objectivity.
(b) The need to ensure a consistent approach between sites.
(c) The need to rely on sampling rather than blanket coverage, due to constraints on time and resource.
(d) The importance of identifying the conservation value of the different vegetation types.

Following discussion, a method of survey and analysis was selected which drew on elements of the objective random sampling method originally used to study British semi-natural woodlands, as modified for use in the joint ITE/Cumbria County Council survey of Cumbria, and the Nature Conservancy Council's Land Use/Habitat survey method (Anon, 1975; Bunce and Shaw 1975). The essential elements of the combined survey were:

(i) Mapping of habitats in which the vegetation appeared to be visually homogeneous.
(ii) Random selection of 200 m^2 quadrants within the mapped habitats, the number being loosely determined in proportion to the area.
(iii) Recording of all plant species present within each quadrat.

In addition, where extensive linear habitats occurred, for example along field boundaries or water-courses, plant species growing in 25×1 m survey strips were also recorded.

The data obtained from the surveys of each reservoir site were subsequently subjected to indicator species analysis which enabled the individual quadrat samples to be reallocated to standardize vegetation types for the area in question, without reference to the mapped habitat in which they were originally located. In this way a check could be exercised over the accuracy with which each surveyor had identified specific types of habitat. Although the analysis indicated that surveyors had difficulty in recognizing the distinctions between certain closely similar vegetation types and that some habitats had been incorrectly described initially, the combination of the two approaches enabled the survey of the three inland sites, covering more than 3000 hectares, to be accomplished in the space of only a few weeks, with a level of accuracy which was more than adequate for the purpose to which the information was put.

The second part of the Working Group's task was to determine the conservation value of the vegetation types thus identified. As the Working Group's own report concludes 'The assessment of ecological and conservation value has been and will continue to be, a subject of debate. Many attempts have been made in recent years to devise methodologies for assigning numerical conservation values to different habitats for use in large-scale vegetation surveys . . . No one method has found widespread acceptance'.

The Group decided to use the largely subjective criteria listed by NCC in the Nature Conservation Review as a starting point for assessing ecological importance and value of each habitat. It was agreed, however, that some quantitative measure of the relative importance of one habitat to another should be developed in order to be able to advise the engineering design team about the merits of alternative Variants (i.e. dam sites and areas of impoundment). This was achieved by linking the criteria with the statistical data collected from the habitat areas and quadrat samples.

To begin with, scores were derived for each habitat using the five NCR criteria judged to be the most amenable to quantification.

In the case of Morecambe Bay, no comparable data existed outside the immediate survey areas, and an informed judgement was therefore made about the rarity of given habitats.

Naturalness and Recreatability were also introduced in the assessment since although no quantitative measure of the criteria was applicable, it was felt that the considered opinion of the Group should be used to provide a relative value of this important guide to conservation value.

Scoring was undertaken for each of the survey areas by running through the six criteria and identifying the habitat which: (a) was largest in extent; (b) had the highest mean number of species per quadrat; (c) had the greatest community diversity; (d) had the largest number of species of limited distribution; (e) was the least common in the study area; (f) was the most natural and least able to be recreated.

The habitats thus identified were given a score of ten, and the remaining habitats were then ranked in relation to them by adjusting, numerically, the measured parameters to give scores in direct proportion for each criterion.

It will be appreciated that the objectivity of the analysis described so far suffers from certain inherent weaknesses. The most important of these is the fact that the criteria are not necessarily of equal importance in assessing the conservation value of a particular habitat. With this in mind and recognizing, also, that greater confidence could be placed in some of the measured data than in others, the scores for each criteria were weighted as follows:

(a) Criteria considered to be of special conservation significance in relation to the types of impact envisaged (i.e. construction damage and inundation), and for which reliable data were available – weighting ×3.

    Species Rarity
    Habitat Rarity (within the site)

(b) Criteria of significance in assessing conservation value but where data were unreliable or subjective – weighting ×2.

    Habitat Rarity (within the wider study area)
    Naturalness and Recreatability

(c) Criteria of limited importance in assessing conservation value, or dependent on unreliable and subjective data (see footnote) – weighting ×1.

    Extent
    Community Diversity

Having weighted the scores for each habitat, the totals were summed. Various sensitivity tests were then applied to assess the effects of ignoring particular criteria or altering the relative weightings. The method of scoring proved to be very robust and the hierarchy of habitats was seldom affected. The assessment of conservation value was completed by dividing the habitats into three classes within the total range of scores from 50 to 155. Habitats with scores of 120 or more were graded of high relative value. Those between 85 and 120 were classed as medium and those below 85 were regarded as of low relative value.

Information obtained from the ecological surveys and analysis was related directly to the different engineering options, by overlaying tracings of dams, reservoir sites, and other construction works, on the habitat maps. This permitted measurements to be made of the areas of different conservation value lost or put at risk by each Variant. The use which was made of these findings in the elimination of Variants is described later.

The work of the Terrestrial Ecology Working Group was of great value in establishing the ecological importance of the four alternative reservoir sites, especially since two of the sites had not been documented comprehensively before. It demonstrated that the effects of construction impacts on existing habitats would be most marked in the case of Haweswater and Borrowbeck, of limited significance at Hellifield. In Morecambe Bay the estuarial impacts were recognized as of paramount importance, but as a direct result of the study a new and previously unidentified SSSI has been declared by the Nature Conservancy Council.

The second area of ecological work on water storage sites concerned the review of potential bunded reservoir sites in Morecambe Bay. Here the study was able to draw upon the extensive research undertaken by NERC in relation to the previous Morecambe Bay Barrage Proposals.

No new survey work was undertaken for the present studies, although updated information on channel movements, fish landings, and wader population counts were obtained from a variety of sources.

Previous work in both Morecambe Bay and the Wash had revealed patterns in the behaviour and distribution of invertebrate species, and their predators, which are closely related to tidal heights, and the nature of the substrate. Natural phenomena such as storms, floods, and frosts had been shown to have marked short-term effects upon these patterns and distributions although their influences are usually reversible or cyclical unlike man-made changes which tend to be irreversible. For the purposes of the study, a number of ecological guidelines were developed on the basis of this observed behaviour of estuarial inter-tidal ecosystems.

Firstly, it was established that any engineering work which interfered with the areas of highest tidal velocity in the low-water channels of major rivers entering the Bay would be more likely to lead to serious and cumulative changes in both the areas and rates of accretion and sedimentation in much wider areas of Bay. A second, and related factor was that development of bunded storage reservoirs on the lower inter-tidal sandflats covered by all states of the tide would be more likely to have a lasting adverse impact on wader populations, through destruction of their main feeding grounds. This is due to the fact that the greatest concentrations of invertebrates such as *Corophium*, *Hydrobia*, *Macoma*, *Cardium*, *Arenicola* and *Nereis* are found in the area lying between low water and mean high water of spring tides 0 to 15 ft OD. The population density of each of these species varies within this broad band with *Corophium* and *Hydrobia* predominating in the upper areas above +9 ft OD, while the worms, *Arenicola* and *Nereis* are found more frequently below 9 ft OD.

The five main species of waders which winter in Morecambe Bay, namely the Oystercatcher, Dunlin, Redshank, Knot and Curlew, feed selectively on the invertebrate fauna. Consequently, the precise location of a bunded storage reservoir could have a significant effect upon both the numbers and species of birds displaced.

A third factor which was judged to be of particular importance in assessing the likely degree of ecological disturbance was the relationship of the reservoir sites to existing wader roosts although for most species of bird this was seen as less critical than the potential effects on feeding grounds.

Fourthly, a number of construction impacts were considered such as the effects of pumping sand from substantial 'borrow pits' dredged either inside or outside the proposed reservoir site; the effects of laying pipelines across the inter-tidal and salt marsh areas of the Bay; and the consequences of diverting river channels or altering their hydraulic gradients.

The fifth issue to be considered was the possible effect of increased siltation on the salt marshes fringing the shore and the spread of *Spartina* in sheltered bays, as a result of modifications in tidal flow brought about by the configuration of the bund walls themselves.

Finally, the opportunities to ameliorate or integrate adverse impacts by creating new ecological habitats, and the possible effects of other changes in land use such as increased recreational activity, had to be considered. Each of the different options for storage in Morecambe Bay was subjected to a detailed review under these headings as the study proceeded and the ecological analysis played a very important role in the elimination of variants.

A study of drawdown characteristics formed a third area of work on the water storage sites which was most important in making the final assessment of potential environmental impacts for each scheme although it did not contribute significantly to the elimination of variants.

This study was undertaken jointly by the Authority's Hydrology team and Land-Use Consultants and involved the production of computer-simulated reservoir water level movements for each of the fourteen short-listed variants from which the extent, frequency and duration of drawdown could be predicted.

A review of relevant literature was undertaken on the effects of water level fluctuations on plant growth and survival, and consideration was also given to the visual, and recreational consequences of water level changes.

(d) Adequate routing

The fourth area of ecological impact to the studies consisted of a thorough review of all known sites of conservation value, identified by NCC and the County Naturalists Trusts in areas through which pipelines and tunnel aqueducts might be routed. Critical zones identified in the course of the study, which had an important bearing on the elimination of Variants, included the limestone hills forming the divide between the Eden and Lune Catchments and the mosslands fringing Morecambe Bay, together with the limestone outliers of the Arnside and Silverdale AONB.

6.4.9 Use of environmental data

At the outset of the study the use of a simple checklist matrix of the type described in the Aberdeen manual was found to be the most convenient way of ensuring that all areas of potential environmental concern were covered at each site and of highlighting possible variations between sites which would require different methods of survey and analysis.

Each matrix consisted of a list of the development characteristics on the vertical axis, which were matched by the site characteristics along the horizontal axis. They were filled in by noting in the appropriate square the relative level of significance of any environmental impact which was likely to arise through the combination of a particular set of development and site characteristics. The following notation was used to record the significance of any interaction: 0 = no impact; ? = doubtful impact; 1 = some minor impact; $\frac{1}{2}$ = indeterminate level of impact; 2 = major impact.

Expanded matrices were developed in the form of Record Sheets to provide supporting evidence for the elimination of scheme Variants in the early stages of the study. Their introduction was an attempt to reduce the great bulk of information collected on each of the scheme variants to manageable proportions, and to present the Environmental Impact Panel with a comprehensive array of data on which to base its decisions. In practice, however, even the record sheet proved to be unwieldly as a means of communicating preliminary environmental conclusions, as opposed to basic factual information, and reliance was subsequently placed on the use of short written summaries of the findings of the Working Groups, and on reports on the merits and demerits of individual scheme Variants.

In the published study report a modified version of the summary matrix has been used to record the main categories of impact which would stem from development of any of the fourteen shortlisted Variants. Wherever possible, short verbal and graphic descriptions of potential impacts are given in the text of the report and an attempt has been made to make this as readable as possible.

At all times, a conscious effort was made to bear in mind the ultimate use to which the information would be put. The stated objectives of the study were to present both the engineering and environmental findings in an objective, credible, comprehensive and comprehensible form, so that they would be readily intelligible to the informed lay members of the Water Authority, whose ultimate responsibility it will be to select a preferred scheme for development. Under these circumstances, it was decided that no attempt should be made to develop a system for weighting the importance to be attached to the different forms of environmental impact, or to seek to compare explicitly the relative merits and demerits of the four schemes. Indeed, given the diverse environmental character of the four schemes the complexity of the supply and transmission systems, the variations in technical performonce and cost, and the timescale for the eventual decision, it is questionable whether a system of ranking and weighting could be devised, within the framework of an EIA, which would gain the level of public acceptance required to justify the attempt. In the final analysis, a largely subjective and political judgement is required in determining the degree of the importance which is attached to environmental gains and losses in the disparate areas of ecology, agriculture, recreation, landscape and local community welfare.

At the same time, it is clearly desirable that this element of judgement should be based upon the best available information and in this respect the authors feel that there is no substitute for systematic and thorough surveys of all the respective environmental issues by adequately trained and competent professional staff. By providing the essential framework for such investigations and a vehicle for expressing the results without bias and in an objective manner as possible, environmental impact analysis should become an increasingly valuable tool not only in project appraisal but in long-term planning and policy formulation for natural resource development.

6.5 ASSESSING THE IMPACTS ON PLANTS OF MAJOR HIGHWAY DEVELOPMENTS

D. M. Colwill and J. R. Thompson

6.5.1 Introduction

The trunk road network and particularly the motorway network has been developing rapidly in the UK over the past 25 years. These essentially interurban routes often carry heavy traffic flows – in some cases more than half a million vehicles per week – and they do so through a variety of terrain.

All aspects of their impact on the environment need to be considered very carefully. The report of the Advisory Committee on Trunk Road Assessment chaired by Sir George Leitch (1977) has proposed a framework to enable the highway engineer to take into account the environmental changes associated with new routes and this framework includes the assessment of the effect of the route on the road user, on people living nearby, on people working in the area (including farming) and on public open space. The intrinsic value of the area is also taken into account.

Among the aspects which need to be taken into account in considering the total impact of a highway are its effect on the environment of the area through which it passes and, indeed, of improved areas within the highway boundary itself. These are important for the landscape impact of the road as well as for conservation. The factors which contribute to the total environmental impact of the highway are numerous, for example, visual intrusion, noise, the emissions from the vehicle exhausts, both gaseous and particulate, and the pollutants associated with vehicle movement – dust, salt, spray, polluted run-off and air turbulence. There are also some problems encountered during the construction phase alone; for example, in chalk terrain road-side vegetation and trees can become covered with a thick layer of dust and be badly damaged.

In this paper, detailed discussion will be confined to three items: (a) lead pollution, (b) gaseous emission, and (c) salt used for de-icing roads. Some consideration of aspects other than pollution is also included.

6.5.2 Lead pollution

Lead in the form of tetraethyl and tetramethyl lead is added to petrol to improve the combustion of the fuel. It is emitted from the vehicle mainly in the form of fine particles. During the 1970s there was a policy of controlling the use of lead additives in petrol which led to a reduction in the maximum level from $0{\cdot}84\ g\ l^{-1}$ to $0{\cdot}45\ g\ l^{-1}$ with a commitment to $0{\cdot}40\ g\ l^{-1}$ in 1981. Future policy on lead in petrol is currently under review in the light of recommendations of a Department of Health and Social Security Working Paper on Lead in the Environment (DHSS, 1980) and lead additives will be reduced to $0{\cdot}15\ g\ l^{-1}$ by 1985. This report was about lead and health but its recommendations have implications for the soil and the atmosphere. The literature on the contamination of roadsides by lead from petrol is extensive but little was known until recently about the extent of contamination on major roads in the UK. Samples taken during our investigation into the levels of de-icing salts were also analysed for lead. Some examples of the lead contents of soils in central reserves are given in Table 6.1. Large differences were found between sites probably depending on the volume of traffic and the age of the motorway (Colwill *et al.*, 1978).

Table 6.1 Total lead in soils of central reserves

Location	*Lead content (ppm)**
M1 Rotherham	397 ± 31
M1 Barnsley	484 ± 43
M4 Hounslow	3341 ± 206
A423(M) Maidenhead	706 ± 49
M63 Eccles	879 ± 103

* Values with 95% confidence limits.

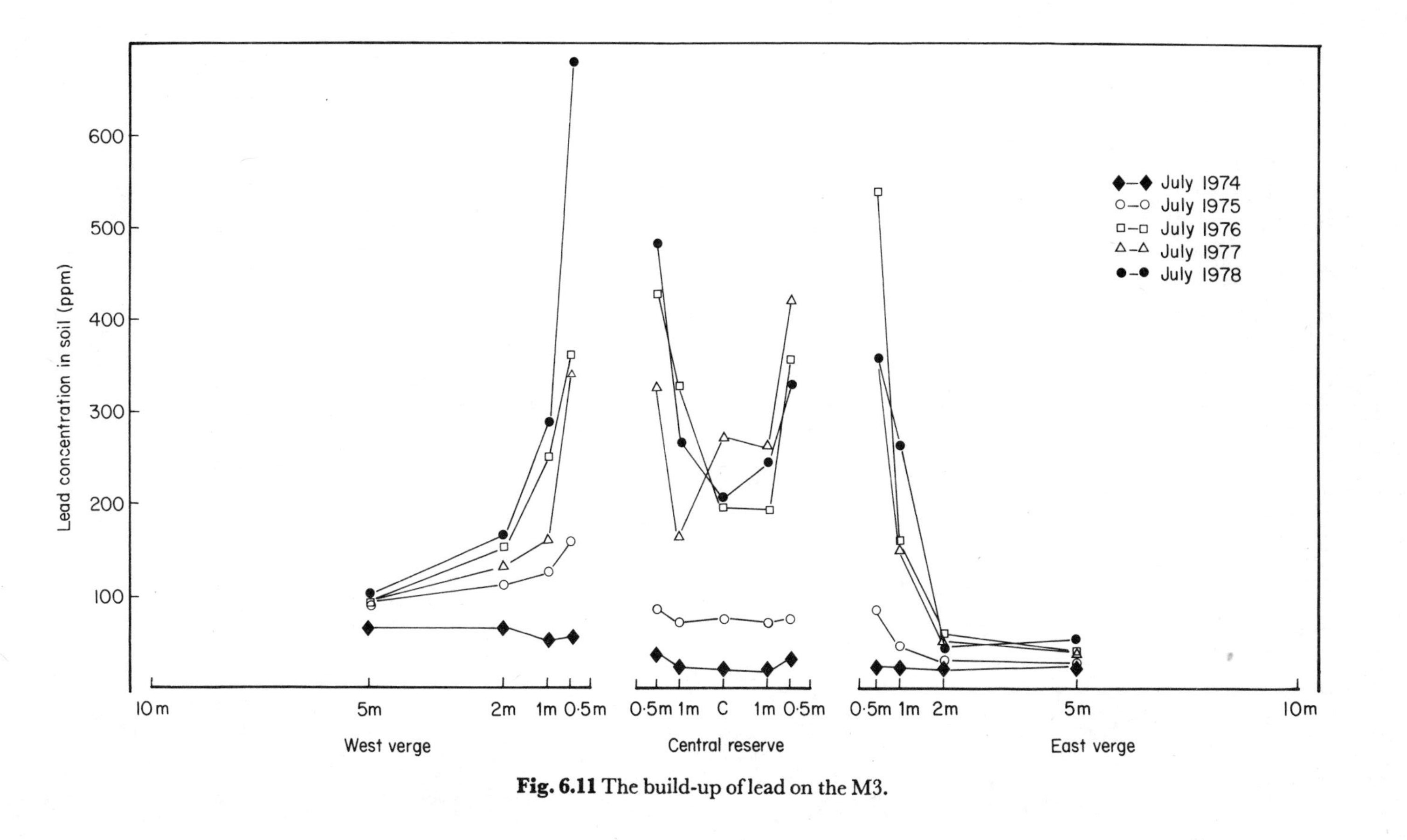

Fig. 6.11 The build-up of lead on the M3.

Transects across motorways have shown that the major contamination occurs within 2 metres of the hard shoulders.

The accumulation of lead in soils on the M3 motorway has been monitored at two locations every year since immediately before the road was open to traffic. The data from one site are presented in Fig. 6.11 and show a continuing accumulation of lead in the soil which increased most rapidly between 1975 and 1976, a period of severe drought. Analysis of soil profiles in 1977 and 1979 showed that the top 5 cm contained much more lead than other parts so it is concluded that although the transport of lead down the soil profile would appear to be assisted by rainfall, it is not a significant feature.

Although high concentrations of lead are found within the motorway boundary, recent work by Little and Wiffen (1978) has shown that only 10–40% of all lead emitted is deposited within 100 m of the highway, the remaining material being in the form of fine particles which are dispersed over much larger distances. These amounts contribute to the general background levels but are unlikely to produce any significant ecological effects.

There are no air quality standards for lead in the UK, but an EEC directive proposed that annual mean levels should not exceed 2 $\mu g\ m^{-3}$ and this was supported by the DHSS Working Party Group.

Lead has been used as an indicator of vehicle pollution and a method of predicting the 2 $\mu g\ m^{-3}$ contour around highways, put forward in the Jefferson Report (DOT, 1979), has been widely used. The method is applicable to rural motorways. In the urban situation with more complex road networks the pollution impact of the highway is more accurately represented by the more recently developed carbon monoxide model described later.

6.5.3 Gaseous pollution

Of the gaseous pollutants, the greatest interest has focused on carbon monoxide, hydrocarbons and the oxides of nitrogen. The emissions of these pollutants from petrol-engine vehicles is controlled, in addition the emission of visible smoke by diesel engines is prohibited. The three gaseous pollutants are known as the primary pollutants, they can react in the atmosphere together and with other materials to form so-called secondary pollutants, like ozone and peroxyacetyl nitrate (PAN), which are generally more damaging to plants than the primary pollutants. We shall confine our consideration to primary pollutants because we are concerned with the roadside situation where there has been insufficient time for the formation of large amounts of secondary pollutants.

Carbon monoxide results from the incomplete combustion of fuel and has well established effects on people but little effect on vegetation.

A wide range of hydrocarbons are emitted by vehicles and ambient air measurements often include the naturally occurring methane. Many of these materials contribute to the malodorous nature of exhaust gases and to oily deposits on surfaces. The hydrocarbon with the most pronounced effect on vegetation is ethylene.

Nitric oxide is the principal oxide of nitrogen in vehicle emissions, the other oxides, mainly nitrogen dioxide, are present in small amounts but are much more damaging to vegetation than nitric oxide (Mansfield, 1979).

Pollution monitoring data at four sites in Coventry (Hickman *et al.*, 1976) were used to show that carbon monoxide can be used as an indicator of the level of vehicle pollution and a method predicting carbon monoxide levels around highways has been developed (Hickman *et al.*, 1979). The inputs have been confined to the information generally available to the highway engineer. The programme is best considered in four sections:

(a) The road layout: A computer program has been written to handle a large number of straight road sections. A curved road may be represented as a connected series of sections and a multi-lane road as a set of parallel sections; in this way it is possible to consider the impact of complex road networks. There are at least ten points in each section which are considered as sources of pollution in the subsequent dispersion calculations.

(b) The pollutant emission rates: As the road system has been considered as a series of short sections the vehicle emissions can be considered in some detail. As vehicle speed has a marked influence on emission rates, the mean speeds are estimated for each section of road which, with the traffic flow data give the total emissions for each section.

(c) The pollutant dispersion: The standard Gaussian plume method is used to calculate the dispersion of the pollutant. Simplifying assumptions are made that the emissions are at ground level and that ground level predictions are required. The calculation requires an input of atmospheric stability and wind speed.

(d) The presentation of predicted levels: A method has been developed of presenting predicted pollutant levels in the form of contours around the road system. Levels are calculated for a grid of receptor locations from which the contours are generated. The road system may be drawn on the contour map and by selecting a suitable scale, the contour plot can be used as an overlay for the site plan.

The computer program has been checked against measurements at two sites, one urban and one rural, and some modifications have been made to the theoretical model. The revised prediction method is useful in the comparison of the environmental impact of alternative road schemes with the existing situation in order to identify the areas where pollution will be increased or decreased. The accuracy of the values predicted will be established during the course of further investigation.

6.5.4 De-icing salt

Winter maintenance practices may also have a considerable environmental impact. A detailed study of this problem has been made and as a result an assessment of its impact can now be made with greater certainty.

The current recommended rate of spreading de-icing salt is 10 gm^{-2} as a preventative measure when frost is forecast and 40 gm^{-2} when snow is falling or when conditions are severe. The previously recommended rates of 17 and 54 gm^{-2} may still apply as many areas have yet to acquire the necessary vehicles to be able to spread at such low rates. Variations in the spreading frequency and rate together contribute to a very variable pattern of salt application from year to year and from area to area within the same year (see Fig. 6.12 and Table 6.2).

Table 6.2 Comparison of rates of salt applied in three winters by three Authorities (tonne/lane km)

Winter	*Winter maintenance authority*		
	A	B	C
1969–70	12·26	4·60	25·87
1975–76	3·97	0·16	7·76
1976–77	11·53	1·87	12·93

In order to assess the effect of salting on roadside areas, a sampling network of 60 sites was set up to monitor levels at six-monthly intervals (Colwill *et al.*, 1978). The programme started in April 1974 and extended to October 1976. Analyses of the top 5 cm of the soil were made for sodium and for chloride and very large differences were found from area to area. Nine sites were so arranged that they formed an east/west transect across the Pennines and sampling at these sites was carried out every month for three years. The variations from month to month in the levels of sodium at the sites could be enormous. In general, at all the sites there was a tendency for levels to increase up to March and April and then reduce to a minimum in about October. Levels in the central reserves were very much higher than on the verges, where at distances greater than 2 m from the hard shoulder there was little effect of salt.

Professor Rutter at Imperial College has developed a model which predicts the levels of sodium and chloride in soil for a given rate of application of de-icing salt enabling the assessment of the effect of a particular winter to be carried out. The model envisages the salt being transferred to the central reserves mostly as spray, and then it is leached through the soil by rainfall. This suggests that the effects of a given application of salt are greater in the

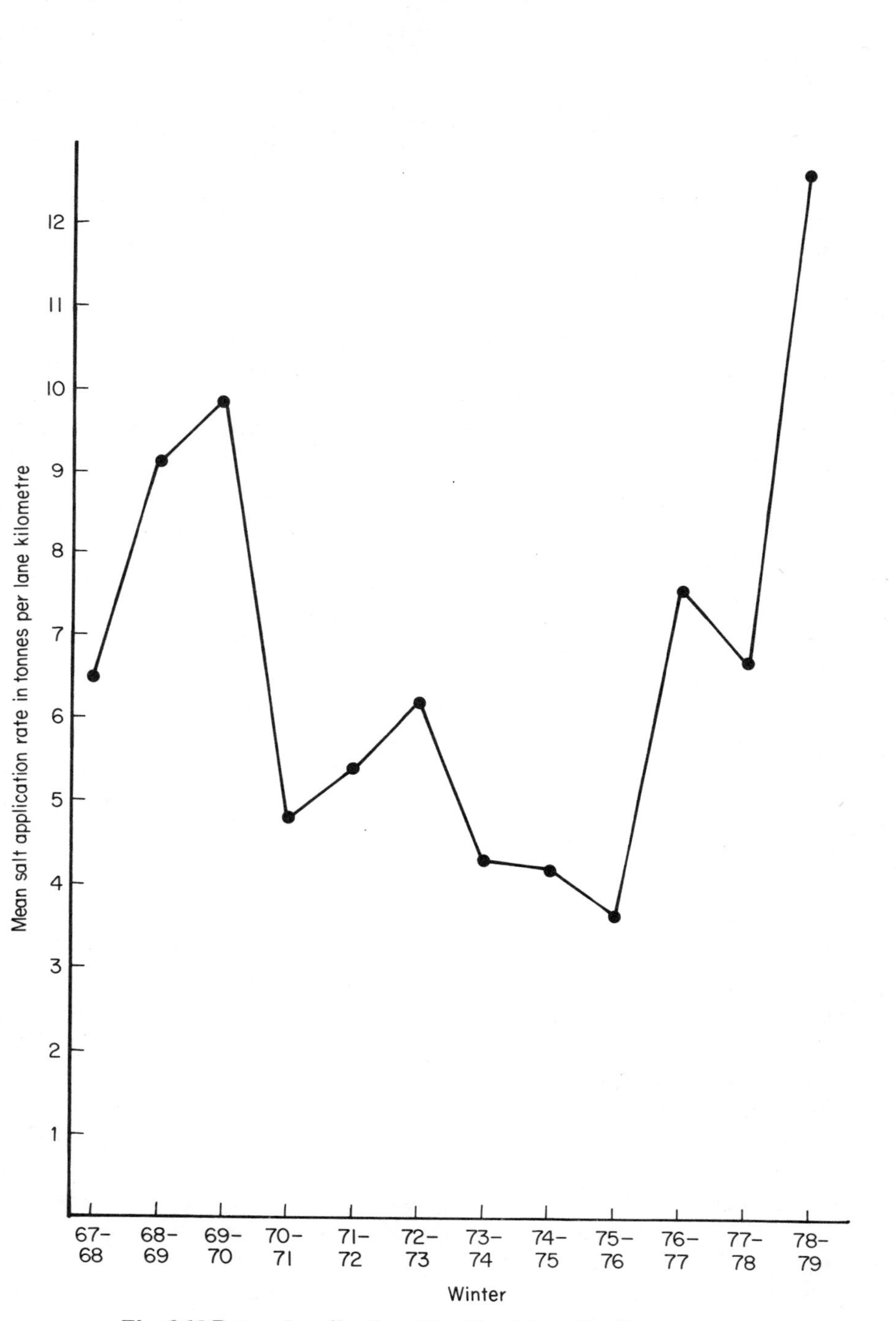

Fig. 6.12 Rates of application of de-icing salt on English motorways.

drier south and east than in the wetter north and west because there is less rainfall to wash the salt through the soil.

The monitoring work was followed up with some large-scale experiments with shrubs growing in large containers in order to assess whether the levels of salt in the soil or in spray would affect the shrubs commonly used in landscaping. The effect of salt in spray was found to be slight but salt in soil, at the highest concentrations, killed many species. The threshold levels for damage were about 1400 ppm sodium and 650 ppm chloride for adverse effects on the more sensitive species.

By taking a large number of transects it has been possible to relate the concentration at the centre of the central reserve to that at a given distance from the hard shoulder. For the critical concentration of 1400 ppm sodium to occur at 2 m from the hard shoulder a concentration of 4600 ppm sodium would be required in the middle of the central reserve and this level was not reached during the entire monitoring programme. Thus, damage to plants would be expected to occur infrequently very close to the hard shoulder and hardly at all at greater distances. However, recent circumstantial evidence conflicts with this finding because in March 1979 Sitka spruce 15 m from the hard shoulder of the M6 and scots pine in various locations about 5 m from the hard shoulder were observed to have substantial leaf damage. During this season there were the highest rates of application of salt ever recorded; as the damage occurred on one side only it was attributed to spray. Reports from Germany, where similar levels of salt are used, indicate widespread damage to spruce near roads. In the UK deciduous species dominate the planting on the verges and we have never observed salt damage to them. Thus, unless evergreen species are used close to highways the impact of de-icing salt will be minimal.

6.5.5 Other ecological impacts

While a substantial part of the impact of a major road on plants is connected with the degree of exposure to pollutants there are other aspects which need to be considered. These include the amount of land taken by the road and during construction, the conservation value of the land, the choice of species for roadside planting, the management of verges and the improvement and creation of habitats.

The amounts of land taken for a motorway network are large, in 1974 there were 1102 miles of motorway in use in England and Wales with approximately 14 000 areas of land as verges. Quantifying the ecological value of this land is complex and various approaches here have been suggested (e.g. Tubbs and Blackwood, 1971). Although there is disagreement about the procedure for assessment, there is general agreement that the conservation value of land decreases with increasing human interference whether it is agricultural, industrial or domestic. The beneficial or detrimental value of a road may be viewed in this context. A road passing through an intensive industrial or agricultural area may have little detrimental effect on that area's conservation value – indeed it may be beneficial. On the other hand a road passing through an area of natural vegetation is more likely to have a detrimental effect on its conservation value.

The choice of species used for planting the verges and central reserves also has a significant bearing on their subsequent conservation value. The use of alternative mixture of grass seed with and without top soil has been discussed by Bradshaw and Roberts (1979). A large number of native trees and shrubs have been used on verges with considerable success.

The management of roadside verges is extremely important. Since 1975 funds have been drastically reduced and some deterioration has occurred. The verges should not be seen as extensions of town parks and a policy of regular but frequent cutting can achieve the maximum diversity (Way, 1979). Opportunities to create or to improve habitats have been taken and much depends on the ability of the designer to perceive that the opportunity exists. These measures often involve little or no expenditure.

Many of the verges on minor roads are now recognized for their conservation value and some may even be Sites of Special Scientific Interest. Experience on motorways is rather limited because of the shorter time scale but there are some encouraging signs. Kestrels (*Falco tinnunculus*) are very commonly seen; primrose (*Primula vulgaris*) and cowslip (*Primula veris*) occur on the M1. In some areas the salt resistant *Puccinellia distans* has spread in a band next to the carriageways for many kilometres, *Senecio squalidus* and *Atriplex hastata* are also spreading. Populations of insects and particularly caterpillars have increased greatly in some areas and have resulted in considerable damage to host plants. Examples are the infestation of a beech hedge (*Fagus sylvaticus*) by caterpillars of *Phalera bucephela* and of hawthorn (*Crataegus monogyna*) by caterpillars of *Euproctis similis*. Reasons for these outbreaks have recently been investigated by Port and Thompson (unpublished) who demonstrated that there were elevated levels of nitrogen in vegetation close to the hard shoulder of a motorway. These were thought to have been derived from oxides of nitrogen emitted by automobiles.

6.5.6 Conclusions

It is clear from this brief review that assessing the impact of a major highway on vegetation is very complicated and difficult to quantify. Aspects such as the concentrations of pollutants and the levels of de-icing salt can be predicted with some certainty and related to threshold levels. Other aspects are not yet amenable to this type of treatment and in some cases it is not known whether they have a detrimental or beneficial effect. However, the detrimental impact of a highway on vegetation can be minimized by careful attention to detail during the planning and construction and subsequent maintenance stages of its development and use.

REFERENCES

Allen, D.S. (1979) *Seaducks in the Moray and Dornoch Firths, Scotland, Winter of 1978/79.* Unpublished report to NCC and RSPB.

Anderson, P. (1975) An ecological survey of the Sankey Valley, St Helens, Report to Merseyside County Council.

Anon (1975) *The Ecological Survey of Cumbria – Handbook of Field Methods,* Institute of Terrestrial Ecology, Merlewood.

Atkinson-Willes, G.L. (1963) *Wildfowl in Great Britain,* Nature Conservancy Monograph No. 3, HMSO, London.

Bradshaw, A.D. (1976) Pollution and Evolution in *Effects of Air Pollutants on Plants* (ed. T.A. Mansfield), SEB Seminar Series, Vol. 1, Cambridge University Press, Cambridge, pp. 135–160.

Bradshaw, A.D. (1979) *J. Roy. Town Planning Inst.,* **65,** 85–8.

Bradshaw, A.D. & Roberts, R.D. (1979) in *Department of the Environment TRRL Supplementary Report SR 513,* Transport and Road Research Laboratory, Crowthorne.

Bunce, R.G. & Shaw, M.W. (1975) *J. Ecol.,* **63,** 597–613.

Catlow, J. & Thirlwall, C.G. (1976) *Environmental Impact Analysis* DOE Research Report, No. 2, Department of the Environment, HMSO, London.

Clark, B.D., Chapman, K., Bisset, R. & Wathern, P. (1976) *Assessment of Major Industrial Applications – a Manual,* DOE Research Report, No. 13, Department of the Environment, HMSO, London.

Clarke, R.D.G., Gordon, P.H. & Smith, J. (1969) *The Crossing of the Three Firths,* Privately printed, Glasgow.

Coleman, A. (1975) *Landscape Analysis and Review of Development on Merseyside,* Report to Merseyside County Council.

Colwill, D.M., Thompson, J.R. & Ridout, P.S. (1978) *Department of the Environment TRRL Supplementary Report SR 217,* Transport and Road Research Laboratory, Crowthorne.

Currie, A. (1971) *A Programme for Conservation, with Special Regard to Wildfowl in the Cromarty Firth,* Unpublished report to NC.

Currie, A. (1974) *Marine Pollut. Bull.,* **5**(8), 118–9.

Department of Transport (1979) *The Assessment of the Environmental Impact of Highways,* HMSO, London.

DHSS (1980) *Lead and Health,* HMSO, London.

Gouge, R.L., Symes, G.L. and Buckley, A.D. (1977) *Paper B,* Institute of Water Engineers and Scientists, Summer Meeting 1977).

Handley, J.F. (1980) *Int. J. Remote Sensing,* **1** (2) 181–95.

Handley, J.F. (1982) in *The Resources of Merseyside* (eds Gould, W.T.S. & Hodgkiss), Liverpool University Press, pp. 83–110.

Harris, A.L. & Peacock, J.D. (1969) *Sand and Gravel Resources of the Inner Moray Firth,* NERC, Inst. Geol. Sci., Report No. 69/9, HMSO, London.

Hickman, A.J., Bevan, M.G. & Colwill, D.M. (1976) *Department of the Environment TRRL Report LR 695,* Transport and Road Research Laboratory, Crowthorne.

Hickman, A.J., Colwill, D.M. & Hughes, M.R. (1979) *Department of the Environment TRRL Supplementary Report SR 501,* Transport and Road Research Laboratory, Crowthorne.

Jack Holmes Planning Group (1968) *The Moray Firth* (The Holmes Report), Report to HIDB.

Jackson, H.C. (1979) *Biol. Conservation*, **16** (3), 177–93.

Knetch, J.P. & Freeman, P.H. (1979) *J. Env. Man.*, **9**, 237–46.

Leitch, G. (1977) *Report of the Advisory Committee on Trunk Road Assessment*, HMSO, London.

Little, P. & Wiffen, R.D. (1978) *Atmos. Environ.*, **12**, 1331–43.

Mansfield, T.A. (1979) in *Symposium on the Impact of Road Traffic on Plants* (eds D.M. Colwill, J.R. Thompson and A.J. Rutter), Transport and Road Research Laboratory, Crowthorne.

Merseyside County Council (1976*a*) Merseyside Structure Plan, Draft Report of Survey, Natural Resources, Merseyside.

Merseyside County Council (1976*b*) Merseyside Structure Plan, Draft Report of Survey, Environmental Pollution, Merseyside.

Merseyside County Council (1979) Merseyside Structure Plan, Draft Report of Survey, Land Resources, Merseyside.

Miller, C.E. (1979) The St. Helens Case Study, *Research Paper No 2*, University of Manchester, Pollution Research Unit.

Moore, N.W. (1962) *J. Ecol.*, **50**, 369–92.

Mudge, G.P. (1978) *Seaducks in the Moray and Dornoch Firths, Scotland, Winter of 1977/78*, Unpublished Report to NCC and RSPB.

Mudge, G.P. & Allen, D.S. (1980) *Wildfowl*, **31**, 123–30.

Nature Conservancy (1972) *A Prospectus for Nature Conservation within the Moray Firth*, NERC, London.

Nature Conservancy Council (1977) *Nature Conservation and Agriculture*, NCC, London.

Nature Conservancy Council (1978) *Nature Conservation within the Moray Firth Area. North West Scotland Region*, NCC, London.

O'Riordan, T. (1971) *Perspectives on Resource Management*, Psion, London.

Ormiston, J.H. (1973) *Moray Firth: An Agricultural Study*, HIDB, Special Report No. 9.

Porter, E. (1973) Pollution in Four Industrialised Estuaries, *Royal Commission on Environmental Pollution*, HMSO, London.

Prater, A.J. (1981) *Estuary Birds of Britain and Ireland*. T. & A.D. Poyser, Calton.

Purves, D. (1977) *Trace Element Contamination of the Environment*, Elsevier, Amsterdam.

Ratcliffe, D.A. (1977) *A Nature Conservation Review*, 2 Vols, Cambridge University Press, Cambridge.

Roberts, J. (1975) *Woodland on the Wirral Peninsula*, Report to Merseyside County Council.

Scottish Wildlife Trust (1969) *Wildfowl in the Moray Basin* (Plus Appendix), Typescript.

Shoard, M. (1980) *The Theft of the Countryside*. Temple Smith, London.

Smith, J.S. (1979) *An Assessment of the Effects of Development on Farming in the Moray Firth Sub-Region, 1972–78*, HIDB, Special Report No. 16.

Strategic Plan for the North West (1973) Department of the Environment, HMSO, London.

Tubbs, C.R. & Blackwood, J.W. (1971) *Biol. Conserv.*, **3**, 169–72.

University of Aberdeen, Department of Geography (1970–1981) Moray Firth Development, Ecological Survey, Interim Reports Nos. 1–11, University of Aberdeen.

Vick, C.M. (1975) Air-Pollution on Merseyside, PhD Thesis, University of Liverpool.

Vick, C.M. and Handley, J.F. (1975) *J. Inst. Parks and Recreation Admin.*, **40**, 39–48.

Vick, C.M. & Handley, J.F. (1977) Environmental Health, May 1 1977, 115–7.

Way, J.M. (1979) in *Department of Environment TRRL Supplementary Report SR 513*, Transport and Road Research Laboratory, Crowthorne.

CHAPTER SEVEN

Project Planning

7.1 OVERVIEW

This section deals with the role of planners and ecologists in project or development control. In contrast to earlier sections the processes here are concerned with assessment and evaluation of the consequences of specific projects in identified locations so that both environmental assessment, impact factor identification and impact analysis are, or should be quantifiable processes.

In countries with formalized EIA procedures ecological information may be required at three distinct levels in the project design and planning process:

(a) The initial identification of potential sites and comparison of site options.
(b) The evaluation of process design options.
(c) The detailed assessment of the consequences of the development at the selected site.

As discussed more fully by Effer (Section 7.2) the planning and ecological methodologies are consequent upon the requirements of these levels and the nature and cost of the proposal. In the Canadian system for example, level (a) may involve a desk-based consideration of the broader aspects of site conditions, levels (b) and (c) may involve in the case of complex power installations detailed site, laboratory and computer investigations whilst some smaller-scale developments with readily predictable impacts may require no formal analysis and be accepted under the 'Class Assessment' procedure (Effer, Section 7.2). Throughout EIA processes the key question to be answered is whether a predicted change, consequent upon a proposed development, is important and has everything been done to mitigate it. EIA systems therefore require a detailed analysis of the environmental consequences – usually social, physical and biological impacts – on the proposed development, alternatives to it and alternative sites and the presentation of the analysis as a clear statement as the basis for discussion. The aim of EIA procedures is not to make the decision but, rather, to ensure it is taken from a position of informed knowledge of the consequences of each option.

Traditionally, under the British planning system, the developer need only prepare a planning application with details of the developmental features with little, or no, analysis of the environmental consequences. The planning authority then considers the application in relation to its conformity with Structure and Local Plans and in the light of discussions with affected parties which may range from local residents to statutory bodies, depending upon the nature of the application and the practices of the particular authority. Consideration of environmental consequences is dependent upon the content of the relevant development plans and on issues raised during public participation. Thus considerations of a proposed petrochemical complex on Merseyside or Canvey Island, Essex would be based on very different criteria. The assessments of such factors as air quality, land pollution and water resource quality and examination of likely ecological consequences of future industrial developments should be considered. In contrast, the relevant development plans for Canvey Island do not consider such information, instead laying an emphasis on transport, social population and land-use criteria. In the absence of a detailed Subject Plan such as that carried out in the Mersey Marshes of Cheshire (Booth, Chapter 2) assessment of a proposal in Canvey Island with an already large petrochemical industry would be based on criteria for more rural areas such as requirements for parking facilities, together with major hazard risk assessments by the Health and Safety Inspectorate.

This is perhaps an extreme example. But it serves to show the disparities which exist within the British planning system. The main recourse for interested parties in the Canvey Island case would be to press for the application to be 'called-in' by the Secretary of State and subjected to a Public Inquiry. Environmental consequences could be considered at the Public Inquiry although the Developer is under no obligation to prepare an EIS unless specifically requested by the planning authority, nor to provide the relevant information to enable opponents to develop such an analysis.

The wide-ranging and prolonged arguments which have been a feature of recent public inquiries have forced some large industries to consider whether some orderly analysis of the overall impact of a planned development might not speed up the planning process. At the very least, such an analysis would identify areas of agreement by all parties and could limit the debate to the key areas of concern. In the UK, however, a number of large industries have developed procedures for incorporating impact analysis into the present planning process. Some such as BP produce internal evaluation at an early state in the planning of all new developments. Others, such as British Gas, produce an evaluation which incorporated the views of statutory and non-statutory agencies (e.g. National Trust, CPRE, etc.). Some developers favour the establishment of a representation committee as was set up for BP's Sullom Voe terminals, which not only oversees the preparation of the environmental evaluation but also follows through the environmental monitoring and auditing of the construction and operational phase (see Syrrat, Section 7.4).

Amongst the UK examples of impact analysis there appears to be a reasonable degree of agreement regarding the basic components of the process, namely: (a) a baseline survey of physical and biological features of the area; (b) an understanding of the inter-relationships between these factors; (c) an assessment of the imposed changes in the area in the absence of a development; (d) a description of the development; (e) an assessment of the ecological changes resulting from the development and (f) a description of techniques and design changes necessary to minimize the impacts. There is also agreement that the evaluation should be done as early in the design phase as possible so that environmental considerations can be accommodated where feasible.

The components of the evaluations have been undertaken in a variety of ways. Under the North American systems, the geographical and ecological features of the study area and the characteristics of the project would be described using overlay or other mapping procedures. The significant impacts of the development are identified using a matrix of development characters against ecological and other parameters. The projected change in any particular parameter is given a numerical weighting and the summation of these weightings allows the impacts of different sites or development options to be evaluated (see Effer, Section 7.2). Some of these techniques have been incorporated into the approach used by some UK developers but numerical comparisons have generally not found favour. Instead, the critical changes resulting from the development and ameliorative measures are generally identified on the basis of experience at other similar developments, expert opinion within the industry or concerns identified during consultations with outside agencies. This approach leans heavily in well-established industries such as the power industry on a well-reasoned understanding of the impacts of existing developments (see Howells and Gammon, Section 7.3). Producing an environmental evaluation for novel developments will clearly be more difficult particularly for those with short 'lead times.' In the Scottish experience with oil and gas terminal developments, a system of more formalized EIA was adopted.

Under the system that seems to be developing in the UK, the most difficult problem in environmental evaluation of assessing the significance of a projected impact need not be explicitly addressed. The impact is usually quantified as far as possible and the value judgement and weighting of the need for the development against the value of the impact factor in the national interest is made by Public Inquiry Inspectors and Central Government Ministers.

Ecological considerations also play an important part in the process of effluent control which, in the UK at least, is under the control of statutory authorities but is increasingly becoming part of environmental evaluation.

The paper by Hammerton (Section 7.6) details the control of discharges to inland and coastal waters whilst Diprose and Pumphrey (Section 7.7) outline the procedures and constraints in controlling discharges to the atmosphere. In both cases, the authors are in favour of control by environmental quality objectives coupled with the 'best practicable means' approach rather than fixed discharge limits as proposed by the EEC. In both cases the main difficulties in quantifying impacts lie in understanding chronic long-term effects and dealing with mixtures of pollutants (see Davison 7.8).

Under whatever guise, an orderly assessment of the overall impact of a planned development is increasingly being seen by industry as an important component of the preparation for planning applications. Those industries which carry out impact analysis cite benefits ranging from (a) better designed developments when viewed in the national interest, (b) improved relations with planning authorities and statutory agencies and (c) achieving consents within prescribed time-scales. Nevertheless there is considerable opposition to formalized EIA requirements as it is argued that these will be costly, prolong the time to planning consent, necessitate the publication of confidential commercial considerations and require consideration of a range of impact targets which would be irrelevant for a particular project. Many are in favour of the flexible system of impact analysis which appears to be developing under the present planning legislation in the UK.

The crucial question throughout this debate must be whether a proposed system produces an informed decision on the importance of a predicted change irrespective of the location or type of development. Advocates of a more formalized EIA system in the UK point out, that such a system would remove, or reduce, many of the disparities of the present system. They also point out that many EIAs carried out in the UK differ markedly from American examples by not producing quantitative assessments of the consequences, nor do they consider alternatives to the development or alternative sites and that, therefore, a more formalized system would allow more informed comparison of site options and alternatives within a site to the development.

7.2 ECOLOGICAL INFORMATION AND METHODOLOGIES REQUIRED FOR ENVIRONMENTAL ASSESSMENTS OF CANADIAN POWER GENERATION INSTALLATIONS

W. R. Effer

7.2.1 Introduction

For a large generating station the time between the definition of need for a new project and the time that project produces 'first steam' is about 14 years in Canada. Project development is not a single-step process but consists of several distinct phases and activities, each of which becomes operational on receipt of favourable responses to, and completion of, previous steps. Planning and development of power-generation installations, and, indeed any activity with a long lead-time, is therefore, essentially a process of successive steps. Environmental studies must be arranged and designed to meet the requirements of the various stages. For example, in the early phases where alternative sites are being compared, it would be time-consuming and inappropriate to develop a finely-tuned data acquisition system which is not going to be used or needed at that particular stage for making decisions. In some of the early US EISs long impressive lists of species were prepared, with virtually no assessment of how any of these species were to be affected by the project. Descriptive, base-line information has no place in an environmental assessment if the potential effects of construction and operation on each section of the environment is not subsequently described and analysed.

This paper examines this theme by reference to worked examples of environmental studies carried out during development of Ontario Hydro's power plants. Environmental studies which were carried out during the various stages of the development of Ontario Hydro's power generation installation include:

(a) Site Selection. Office studies use existing environmental data, becoming more detailed throughout the various phases of site selection, culminating in the recommendation to Ontario Hydro management of a site based on environmental factors. This recommendation may or may not be accepted for the overall site recommendation which goes into the environmental assessment process.
(b) Project Approval. At least one year's on-site studies at one or more previously acquired sites, results of which are included in the project environmental assessment document in support of the proposal for a specifically designed generating station.
(c) Pre-Operation Studies. Started at the site at least three years prior to start of plant operation and used to prepare a base line of information against which results of the operational studies will be compared.
(d) Operational Studies. These are carried out to assess the effects of the operation of the station based on the information obtained in the pre-operational phase.

Examples will be taken from only (a) and (b) above because only the results from these phases are actually used in the preparation of environmental assessments. Pre- and post-operational studies are verifications which may have little or no influence on the design and operation of a facility but may have implications for future developments. Examples from actual projects will be described to underline the general approach for obtaining necessary information and its treatment.

7.2.2 Site selection

Over a period of two years, in 1976–8, an office study identified the most environmentally attractive site on the North Channel of Lake Huron for an energy centre, consisting of up to 12 000 MW of electrical generation, and a heavy-water plant. The process followed a set of draft guidelines provided by the Ontario Ministry of the Environment under the *Environmental Assessment Act* and consisted of three phases involving progressive elimination of areas leading to identification of potential sites and then to selection of the final site. At the end of each phase environmental results are integrated with engineering, economic, social and transmission aspects of the study to agree on the scope and direction of the next phase. This phased progression illustrates the point that factors other than environmental have to be considered at each stage and it is essential not to get too far ahead with unnecessarily detailed studies. The process can be best described as a series of parallel ongoing studies with cooperative decisions on where and how to proceed, after cross-linking at the end of each phase.

(a) Phase I – zone identification

The study area (about 150 km of lake shoreline) was reduced to zones, based on the number of constraints identified from broad environmental and legal and quasi-legal considerations.

(i) Major environmental constraints
National parks and provincial recreation areas.
Cooling water intake and discharge.
Unique natural (geological or geomorphological) features.
Sensitive biological areas (marshes, wetlands, sensitive vegetation, etc.).

(ii) Legal constraints
Federal and provincial acts, regulations and guidelines relevant to environmental concern.

(iii) Quasi-legal constraints
Official plans, zoning restrictions and bylaws.
Indian lands. It is Ontario Hydro's policy to refrain from encroaching on Indian lands for the siting of generating facilities. Therefore, all Indian reserves are identified as major constraints.
Atomic Energy Control Board guidelines for the siting of heavy water plants.

Application of these constraints resulted in elimination of areas, which left a number of zones in which potential sites were to be identified.

(b) Phase II – potential site identification

Specific environmental constraints were used to identify areas most suitable for choice as potential sites. In the Northern Channel study, the application of these constraints resulted in the identification of five potential sites. In this phase, even though an environmental constraint may be identified in a particular area, it may be retained if discussion with engineering and other staff show that effects can be mitigated by specific design changes. An example of such cooperation between environmental and engineering staff was that at one site the offshore area was shallow and had a productive fishery. Environmental staff proposed that the only way this site could be retained as a candidate would be to place the cooling water discharge offshore. This was more costly than the reference shoreline discharge design but it was accepted by engineering staff as a feature of stations that may be built on that site. The increased costs were included in the economic comparisons of sites.

Possible site boundaries and plant layout are prepared in this phase. The environmental factors used to assess constraints include existing air quality, dispersion conditions, water quality, water body conditions, existing land-use patterns and potential effects of atmospheric and aquatic emissions as detailed below.

(i) Air Quality. The existing air quality, future changes due to development, main sources of atmospheric emissions and potential effects on existing air quality.
Atmospheric dispersion conditions. The potential influence on atmospheric plume behaviour, including wind characteristics, lake-breeze effects, terrain, specific topographical constraints and atmospheric stability.
Specific conflicts. The potential influence on localized areas of sensitive vegetation, agricultural crops and recreational areas, as well as proximity of dairy herds.

(ii) Water Quality. The existing water quality, main sources of discharge, potential effects on existing quality and future changes due to development.
Water-body conditions. The potential influence on thermal plume dispersion including offshore depths, availability of cooling water and physical shoreline characteristics (open or confined).
Specific conflicts. The potential influence on municipal or other water intakes and water resource activities such as spawning beds, nurseries, commercial and sports fisheries, waterfowl staging and nesting areas, marshes and wetlands.

(iii) Land. Agriculture and Forestry. The potential influence a change of land use would have on present land use and the capabilities for agriculture and forestry.

Recreation and wildlife. The intensity of present land use and capabilities for recreation and wildlife, including parks, park reserves, cottage areas, fishing and hunting camps, and important wildlife species.
Aesthetics. The potential influence on historical, cultural and natural features of the landscape, including historical hiking trails, unique geological features, Indian reserves and burial grounds.

(c) Phase III – site comparison

Site inventories based on available literature were used to identify the environmental concerns for atmospheric emissions, aquatic emissions and the facility effect on land use. Environmental factors investigated were essentially those used in Phase II but in considerably more detail.

(i) Atmospheric emissions

Dispersion. The capacity of the site atmospheric environment to disperse air emissions is assessed. Dispersion potential is considered in terms of atmospheric stability, winds, temperature inversions, mixing conditions and heights. Climatology and the influence of topography on atmospheric dispersion potential is assessed.

Air quality: degradation. The influence of atmospheric emissions on existing air quality is assessed. The existing air quality is considered, as well as potential changes due to existing sources of emissions and other planned or possible future development in the area.

Air quality: assimilative capacity. The capacity of the atmospheric environment to assimilate emissions is assessed. This factor considers existing air quality, as well as potential changes due to existing emission sources and other planned or possible future development in the area.

Population. The potential influence of atmospheric emissions on population within about 50 km of the site is assessed. This factor considers distribution and densities of population at various distances from the site, including consideration of more sensitive groups such as hospitals and sanatoriums, as well as other populations affected by maximum ground-level concentrations (8 to 16 km distance).

Agriculture and vegetation. The potential influence of atmospheric emissions on agriculture and vegetation within about 50 km of the site is assessed. The distribution and densities of susceptible vegetation at various distances is considered, including farming, commercial gardening, greenhouses, forests, tree plantations and woodlots, particularly any affected by maximum ground-level concentrations (8 to 16 km distance).

Property and industry. The potential influence of atmospheric emissions on property and industry within approximately 50 km of the site is assessed. The occurrence of sensitive industries and property at various distances is considered, particularly any affected by maximum ground-level concentrations (8 to 16 km distance).

Recreation and wildlife. The potential influence of atmospheric emissions on recreation and wildlife within about 50 km of the site is assesssd. This factor considers the influence on temporary and permanent populations occupying recreational areas, particularly any areas affected by maximum ground-level concentrations (8 to 16 km distance), including consideration of waterfowl, wildlife and water quality with respect to existing uses.

(ii) Aquatic effluents

Dispersion conditions. The capacity of the site to disperse thermal, chemical and radioactive effluents discharged to the aquatic environment is assessed. Winds, currents, shoreline characteristics, islands and offshore topography are considered.

Water-body conditions. The capability of the site to provide cool water which will facilitate meeting regulatory requirements with respect to maximum thermal discharge temperature and ambient water-body temperature (surface and at depth) is considered. This involves a consideration of offshore topography and potential plume overlapping and recirculation.

Water quality: degradation. The influence of aquatic emissions on existing water quality is assessed. The existing physical, chemical, microbiological and radiological characteristics are considered, including main sources of existing effluents and future changes due to other developments.

Water quality: assimilative capacity. The capability of the aquatic environment to assimilate aquatic emissions is assumed. The existing physical, chemical, microbiological and radiological characteristics, main sources of existing effluents and future changes due to other developments are considered.

Fishing and spawning. The influence of a once-through cooling system, and thermal and miscellaneous effluents on fisheries is assessed. Spawning areas, nurseries, migratory routes, commercial fishing and sports fishing are considered. It is assumed that elevated temperatures are undesirable, particularly to cold-water species, but deep intakes will reduce the difference between discharge and ambient surface temperatures. Concern is also assumed for entrainment potential, duration of thermal exposure and impingement with respect to fisheries.

Other aquatic life. The influence of once-through cooling systems and thermal and miscellaneous effluents on benthos, plankton and aquatic macrophytes is assessed. This factor considers population, biomass, diversity and dominant organisms of benthos and plankton. The potential for entrainment, and the growth and fragmentation of periphyton and aquatic macrophytes are also considered.

Recreation and wildlife. The influence of once-through cooling systems, and thermal and miscellaneous effluents on recreation and wildlife is assessed. The loss of ice cover with respect to increased water fowl activities and decreased winter sports such as snowmobiling is considered.

Local intakes. The compatibility of thermal and miscellaneous effluents with present and future water supplies is assessed. The proximity of present and future residential, municipal and industrial water supplies and intakes is considered.

Local discharges. The compatibility of thermal and miscellaneous effluents with present and future liquid discharges from other local sources is assessed. The proximity of present and future residential, municipal and industrial discharges is considered.

(iii) Facility effects

Site land use. The influence of the facility on the existing site environment is assessed in terms of present and planned land uses. This factor considers the site area in terms of existing land-use types (number of homes, farms, cottages, commercial, etc.), recreational, historical and other unique values and agricultural values.

Site: topographical modifications. The influence of facility construction on existing site topography is assessed. This factor considers offshore reclamation, water courses, marshes and wetlands, and physical changes associated with site preparation (clearings, rough grading, drainage, excavation, filling, etc.) for powerhouse or plant, switchyard area, fuel storage and ash disposal areas, and cooling water intake and discharge structures.

Site biology. The influence of the facility on the existing natural environment of the site and immediate vicinity is assessed. This factor considers wildlife and waterfowl activities and habitat, vegetation, sensitive species, and any biological uniqueness in the area.

Impact area: land use. The potential of the facility to attract other development, resulting in increased pressure to change existing land uses in an area, is assessed. This factor considers existing land use and plans for development or maintenance of agriculture, forestry and recreation, including consideration of present and future industrial and residential development plans.

Impact area: environmental stress. The potential of the facility to attract other industrial and residential development, thus further stressing the physical and natural environment, is assessed. This factor considers present and future development plans in terms of industry and population.

Impact area: aesthetics and historical. The aesthetic compatibility of the facility with the natural surroundings is assessed. The visual and noise influences, as well as historical significance of areas or trails affected by the facility, are considered. This includes a consideration of the visibility of the stack and plume from recreational and historical areas.

Sites were compared on a numerical basis using a number of environmental parameters each of which is assigned a number or weight. Each factor is assigned an environmental factor of concern; 0–minimal, 1–slight, 2–moderate, 3–significant, 4–large, 5–very large, which, when multiplied by the factor weight gives a number for that factor and potential site. Summation of values for each site gives a number which allows the site to be rated in order of preference. Tables 7.1 and 7.2 illustrate the method.

Table 7.1 Parameter study weight distribution for the development options

	Development options		
Parameter	*Fossil generation*	*Nuclear generation*	*Heavy water plant*
Atmospheric emissions	35	20	30
Aquatic emissions	35	55	30
Facility effects	30	25	40
Total study weight	100	100	100

In Tables 7.3, 7.4 and 7.5 are shown the Phase III site comparisons for a fossil station, a nuclear station and a heavy-water plant respectively.

From this study, Dobie Plant and Burton Island were proposed as the two most environmentally preferred sites for an energy centre. The proposal included a sensitivity analysis of the changes in values which would have to be assumed in order to affect changes in the ratings.

In criticisms of this study, ecologists and others have objected to the use of numbers in comparing environmental factors. We have requested alternative suggestions and to date have had none that could be used in a practical sense to compare and eliminate alternatives. Because of the large number of individual and separate environmental factors used to assess the total environment there is little need to go into greater detail than is required to compare sites. Only substantial error in the assessment of one factor or an accumulation of several smaller errors can affect the composite site rating. Neither of these is very likely to happen if good general information is available. Good information is usually obtained by discussion with local naturalists, fishermen, foresters and representatives of local authorities. Assignment of weights to the various factors can be a contentious exercise and must be carried out by a multi-disciplinary team. Levels of concern should

Table 7.2 Distribution of relative factor weights

	Development options		
Individual site factors	*Fossil generation*	*Nuclear generation*	*Heavy water plant*
1. Atmospheric emissions			
(a) Dispersion: atmospheric conditions	7	4	6
(b) Dispersion: topography	7	3	5
(c) Air quality: degradation	4	1	3
(d) Air quality: assimilative capacity	3	1	2
(e) Population	5	4	6
(f) Agriculture and vegetation	5	5	3
(g) Property and industry	1	1	1
(h) Recreation and wildlife	3	1	4
Sub-total	35	20	30
2. Aquatic effluents			
(a) Dispersion condition	6	9	5
(b) Waterbody conditions	5	8	5
(c) Water quality: degradation	4	5	4
(d) Water quality: assimilative capacity	3	4	2
(e) Fishing and spawning	7	10	6
(f) Other aquatic life	5	7	4
(g) Recreation and wildlife	2	3	1
(h) Local intakes	2	6	2
(i) Local discharges	1	3	1
Sub-total	35	55	30
3. Facility effects			
(a) Site: land use	7	7	9
(b) Site: topographical modifications	5	3	5
(c) Site: biology	4	3	4
(d) Impact area: land use	5	5	11
(e) Impact area: environmental stress	5	5	5
(f) Aesthetics and historical	4	2	6
Sub-total	30	25	40
Study weights — Total	100	100	100

be agreed upon by at least three individuals from the same appropriate discipline.

This North Channel site selection environmental assessment document has been completed but has not yet been formally presented under the *Environmental Assessment Act*. It will be one of a series of documents which will form the project proposal.

Table 7.3 Environmental site comparison, fossil generating station alternatives

		Level of concern					*Values*				
Individual site factors	*Weight*	*Larry Island*	*Dobie Point*	*Burton Islands*	*DeRoberval Point*	*Hunt Point*	*Larry Island*	*Dobie Point*	*Burton Islands*	*DeRoberval Point*	*Hunt Point*
Atmospheric emissions											
(a) Dispersion – atmospheric conditions	7	2	2	2	2	2	14	14	14	14	14
(b) Dispersion – topography	7	1	1	1	1	3	7	7	7	7	21
(c) Air quality – degradation	4	1	3	3	3	1	4	12	12	12	4
(d) Air quality – assimilative capacity	3	2	1	1	1	5	6	3	3	3	15
(e) Population	5	4	1	1	2	2	20	5	5	10	10
(f) Agriculture and vegetation	5	2	1	1	1	4	10	5	5	5	20
(g) Property and industry	1	2	1	1	1	1	2	1	1	1	1
(h) Recreation and wildlife	3	1	2	2	2	4	3	6	6	6	12
Sub-total	35						66	53	53	58	97
Aquatic effluents											
(a) Dispersion conditions	6	3	2	2	2	4	18	12	12	12	24
(b) Water body conditions	5	2	2	2	3	3	10	10	10	15	15

(c) Water quality – degradation	4	1	2	1	1	1	4	8	4	4	4
(d) Water quality – assimilative capacity	3	1	1	1	1	1	3	3	3	3	3
(e) Fishing and spawning	7	3	4	4	4	3	21	28	28	28	21
(f) Other aquatic life	5	3	2	2	2	3	15	10	10	10	15
(g) Recreation and wildlife	2	2	1	2	3	3	4	2	4	6	6
(h) Local intakes	2	2	0	1	0	3	4	0	2	0	6
(i) Local discharges	1	1	0	0	0	1	1	0	0	0	1
Sub-total	35						80	73	73	78	95
Facility effects											
(a) Site: land use	7	2	1	1	3	5	14	7	7	21	35
(b) Site: topographical modifications	5	5	3	4	4	5	25	15	20	20	25
(c) Site: biology	4	1	2	2	2	3	4	8	8	8	12
(d) Impact area: land use	5	1	3	3	3	5	5	15	15	15	25
(e) Impact area: environmental stress	5	2	3	3	3	3	10	15	15	15	15
(f) Aesthetics and historical	4	3	2	2	3	4	12	8	8	12	16
Sub-total	30						70	68	73	91	128
Total	100						216	194	199	227	320

Table 7.4 Environmental site comparison, nuclear generating station alternatives

		Level of concern					*Values*				
Individual site factors	*Weight*	*Larry Island*	*Dobie Point*	*Burton Islands*	*DeRoberval Point*	*Hunt Point*	*Larry Island*	*Dobie Point*	*Burton Islands*	*DeRoberval Point*	*Hunt Point*
Atmospheric emissions											
(a) Dispersion – atmospheric conditions	4	2	2	2	2	2	8	8	8	8	8
(b) Dispersion – topography	3	1	1	1	1	2	3	3	3	3	6
(c) Air quality – degradation	1	1	3	3	3	1	1	3	3	3	1
(d) Air quality – assimilative capacity	1	1	1	1	1	1	1	1	1	1	1
(e) Population	4	3	1	1	2	2	12	4	4	8	8
(f) Agriculture and vegetation	5	1	2	1	0	0	5	10	5	0	0
(g) Property and industry	1	1	1	1	1	1	1	1	1	1	1
(h) Recreation and wildlife	1	1	1	1	1	1	1	1	1	1	1
Sub-total	20						32	31	26	25	26
Aquatic effluents											
(a) Dispersion conditions	9	3	2	2	2	4	27	18	18	18	36
(b) Water body conditions	8	2	2	2	3	3	16	16	16	24	24

(c) Water quality – degradation	5	1	2	1	1	1	5	10	5	5	5
(d) Water quality – assimilative capacity	4	1	1	1	1	1	4	4	4	4	4
(e) Fishing and spawning	10	3	4	4	4	3	30	40	40	40	30
(f) Other aquatic life	7	3	2	2	2	3	21	14	14	14	21
(g) Recreation and wildlife	3	2	1	2	3	3	6	3	6	9	9
(h) Local intakes	6	2	0	1	0	3	12	0	6	0	18
(i) Local discharges	3	1	0	0	0	1	3	0	0	0	3
Sub-total	55						124	105	109	114	150
Facility effects											
(a) Site: land use	7	2	1	1	3	5	14	7	7	21	35
(b) Site: topographical modifications	3	2	1	1	1	3	6	3	3	3	9
(c) Site: biology	3	1	1	1	1	2	3	3	3	3	6
(d) Impact area: land use	5	1	3	3	3	5	5	15	15	15	25
(e) Impact area: environmental stress	5	2	3	3	3	3	10	15	15	15	15
(f) Aesthetics and historical	2	1	1	1	1	2	2	2	2	2	4
Sub-total	25						40	45	45	59	94
Total	100						196	181	180	198	270

Table 7.5 Environmental site comparison, heavy water plant alternatives

		Level of concern					Values				
Individual site factors	*Weight*	*Larry Island*	*Dobie Point*	*Burton Islands*	*DeRoberval Point*	*Hunt Point*	*Larry Island*	*Dobie Point*	*Burton Islands*	*DeRoberval Point*	*Hunt Point*
Atmospheric emissions											
(a) Dispersion – atmospheric conditions	6	2	2	2	2	2	12	12	12	12	12
(b) Dispersion – topography	5	1	1	1	1	3	5	5	5	5	15
(c) Air quality – degradation	3	1	3	3	3	1	3	9	9	9	3
(d) Air quality – assimilative capacity	2	1	1	1	1	3	2	2	2	2	6
(e) Population	6	4	1	1	1	2	24	6	6	6	12
(f) Agriculture and vegetation	3	2	1	1	1	4	6	3	3	3	12
(g) Property and industry	1	2	1	1	1	1	2	1	1	1	1
(h) Recreation and wildlife	4	1	2	2	2	3	4	8	8	8	12
Sub-total	30						58	46	46	46	73
Aquatic effluents											
(a) Dispersion conditions	5	3	2	2	2	4	15	10	10	10	20
(b) Water body conditions	5	2	2	2	3	3	10	10	10	15	15

(c) Water Quality – degradation	4	1	3	1	1	1	4	12	4	4	4
(d) Water quality – assimilative capacity	2	1	1	1	1	1	2	2	2	2	2
(e) Fishing and spawning	6	3	4	4	4	3	18	24	24	24	18
(f) Other aquatic life	4	3	2	2	2	3	12	8	8	8	12
(g) Recreation and wildlife	1	2	1	2	3	3	2	1	2	3	3
(h) Local intakes	2	2	0	1	0	3	4	0	2	0	6
(i) Local discharges	1	1	0	0	0	1	1	0	0	0	1
Sub-total	30						68	67	62	66	81
Facility effects											
(a) Site: land use	9	2	1	1	3	5	18	9	9	27	45
(b) Site: topographical modifications	5	3	5	5	2	2	15	25	25	10	10
(c) Site: biology	4	2	3	3	2	3	8	12	12	8	12
(d) Impact area: land use	11	1	2	2	2	5	11	22	22	22	55
(e) Impact area: environmental stress	5	1	2	2	2	1	5	10	10	10	5
(f) Aesthetics and historical	6	3	2	2	3	4	18	12	12	18	24
Sub-total	40						75	90	90	95	151
Total	100						201	203	198	207	305

7.2.3 Project environmental assessments

To illustrate the approach used in the environmental assessments of specific projects, two examples will be taken from a study based at Chats Falls, a site proposed for the development of a 4 × 600 MW nuclear station. At this site there is an existing hydraulic station and it is proposed to locate the nuclear station on the headpond created by the hydraulic station. The two examples described are (a) the collection of field data on fish in the Ottawa River with a calculation of the potential thermal effect on the indigenous fish species downstream of the station, and (b) the use of field data to assess the impact of station site layout on the terrestrial ecology of the site.

(a) Thermal limitations on the aquatic system at Chats Falls

The Ontario provincial water quality criteria state that heated discharges to inland waters are not permitted unless the effluent enhances the usefulness of the water resource without endangering the production and optimum maintenance of aquatic life. The Ottawa River downstream of Chats Falls is a productive river with 33 species found in a recent survey (Anon, 1979). The heat from the generating station would be discharged to the headpond above the hydraulic turbine inlets so it has been assumed for this study that full-mixing would occur immediately downstream of the hydraulic station.

In 1977 a study (Knox, 1978) arrived at an acceptable temperature rise on the river. This study involved the summer period only, and a rise of 2·1 °C in the fully mixed river downstream of the hydraulic station was calculated to be one which could be supported and used as a basis for preliminary engineering and environmental analysis. This temperature rise was arrived at by the following steps.

(i) Selection of important or representative organisms in the river community, which included fish and macroinvertebrates. Important or representative organisms are those dominating the community either by number or by mass or whichever may be available to the commercial or sports fisheries. Endangered or economically valuable species are included. The list of important species and reasons for their selection are given in Table 7.6.

(ii) Development of a thermal effluent guideline. Although optimum and lethal temperature data for early life stages is very limited, five representative fish species were used and a literature review provided a good estimate of the upper temperature limit for 24 hour exposure of the juvenile/adult stages. As growth rate is considered to be the most temperature-sensitive physiological response, this was used for development of an acceptable temperature rise which was calculated to be

Table 7.6 Important or representative organisms in the Ottawa River (from Knox, 1978)

Species	*Reasons for selection and references*
Brown bullhead	Total numbers, 38% of sport fish yield
Yellow perch	25% of sport fish yield
Black crappie	14% of sport fish yield
Yellow pickerel/Sauger	Spawning of Chats Falls, 12% of sport fish yield
Lake sturgeon	Spawning of Chats Falls, endangered
Emerald shiner	Total numbers and biomass
Northern pike	Spawning at Chats Falls, 6% of sport fish yield
Gammarus sp.	Numbers and biomass

one third of the difference between the upper lethal temperature and the optimum temperature for growth (IJC, 1976). Both the summer six-monthly mean (17·1 °C) and the maximum daily temperature (26·0 °C) were used as assumed acclimation temperatures. The differences between the various acclimation temperatures and associated lethal temperatures obtained from the literature were plotted against

Table 7.7 Differences between upper lethal and acclimation temperatures at two river temperatures (from Knox, 1978)

Species	*Difference upper lethal and acclim. temp. of* 17·1 °C	*One-third diff.* °C	*Difference upper lethal and acclim. temp. of* 26·0 °C	*One-third diff.* °C
Brown bullhead (Channel catfish)	13·5	4·5	8·2	2·7
Emerald shiner	11·8	3·9	6·3	2·1
Northern pike	10·4	3·4	6·9	2·3
Walleye (Sauger)	11·4	3·8	5·0	1·7
Yellow perch	10·5	3·5	5·6	1·9
Mean		3·8		2·1

acclimation temperatures for each species. The differences between the upper lethal and the two temperatures of 17·1 °C and 26·0 °C can be read from the curve and are shown in Table 7.7. An acceptable temperature rise was considered to be the mean value for the five fish species. Based on the more conservative approach of using maximum river temperatures the upper temperature allowable would be (26·0 + 2.1) or 28·1 °C.

Table 7.8 Calculated MWAT criteria at Chats Falls hydraulic station outfall

Week	*MWAT*	*Limiting factor*
1–2	4·9	Cold sensitivity
3–10	4·7	Cold sensitivity
11–13	4·9	Cold sensitivity
14	5·4	Cold sensitivity
15	6·9	Cold sensitivity
16	7·5	Spawning – Northern Pike
17	9·0	Spawning – Sauger
18	11·0	Spawning – Redhorse
19	12·0	Spawning – Muskellunge
20	14·0	Egg incubation – Muskellunge
21	14·5	Spawning – Iowa Darter
22	16·0	Spawning – Golden Shiner
23	18·0	Spawning – Rock Bass
24	20·0	Spawning – Largemouth Bass
25	20·5	Egg Incubation – Rock Bass
26	22·0	Spawning – Branded Killifish
27	23·5	Spawning – Brown Bullhead
28–42	24·0	Growth – Yellow Perch
43	22·3	Cold sensitivity
44	19·2	Cold sensitivity
45	17·0	Cold sensitivity
46	14·1	Cold sensitivity
47	11·1	Cold sensitivity
48	9·1	Cold sensitivity
49	6·3	Cold sensitivity
50	5·6	Cold sensitivity
51	5·2	Cold sensitivity
52	5·0	Cold sensitivity

Based on this study, it was recommended that 2·1 °C be used for further analysis. Another environmental constraint was the percentage of river flow through the generating station at the lowest river flows for minimizing downstream effects of impingement and entrainment. Following a United States Siting Committee recommendation (NAE, 1972) we decided that at the lowest river flows a maximum of 15% should be considered environmentally acceptable. Based on the two environmental constraints of temperature rise and entrainment percentage an engineering and economic study showed that in order to have the station available all year at full load, two of the four generating station units would need to be put on closed cycle cooling.

A further more detailed environmental study (Christie, 1979) considered all the fish species found to occur in the river and determined the maximum

Table 7.9 Comparison of mean weekly temperature, maximum weekly temperature, MWAT and MDAT (°C)

Week	*Mean*	*Maximum*	*MWAT*	*MDAT*	*Week*	*Mean*	*Maximum*	*MWAT*	*MDAT*
1	0·6	0·6	5·0	13	27	20·8	23·6	23·5	25
2	0·6	0·8	5·0	13	28	21·4	23·3	24·0	26
3	0·5	0·6	5·0	13	29	22·1	23·6	24·0	28
4	0·5	0·7	5·0	13	30	22·1	23·3	24·0	28
5	0·5	0·6	5·0	13	31	22·2	23·6	24·0	28
6	0·5	0·6	5·0	13	32	22·3	23·6	24·0	28
7	0·5	0·6	5·0	13	33	22·2	24·0	24·0	28
8	0·5	0·6	5·0	13	34	21·6	22·8	24·0	28
9	0·5	0·6	5·0	13	35	21·0	23·8	24·0	28
10	0·5	0·7	5·0	13	36	20·0	23·5	24·0	28
11	0·6	0·6	5·0	13	37	18·9	21·1	24·0	28
12	0·6	0·7	5·0	13	38	18·0	19·8	24·0	28
13	0·6	1·0	5·0	13	39	16·3	19·4	24·0	28
14	0·9	1·5	5·0	13	40	14·6	17·9	24·0	28
15	1·7	4·1	6·5	13	41	12·9	14·7	24·0	28
16	3·7	7·8	8·0	14	42	11·3	14·4	24·0	28
17	5·0	8·9	9·5	15	43	10·1	13·3	22·0	28
18	6·8	8·8	11·0	16	44	8·4	12·1	19·5	26
19	8·1	9·9	12·0	17	45	7·2	10·1	16·0	24
20	10·2	13·4	13·5	18	46	5·6	7·8	14·5	22
21	12·7	16·9	15·0	19	47	4·0	5·1	11·5	20
22	14·5	18·4	16·5	20	48	2·9	3·9	9·0	18
23	16·4	17·6	18·0	21	49	1·4	3·3	6·5	16
24	17·6	19·0	19·0	22	50	1·0	2·9	6·0	14
25	18·5	20·4	20·5	23	51	0·8	2·8	5·0	13
26	19·5	22·9	22·0	24	52	0·7	2·4	5·0	13

weekly average temperature (MWAT) and the maximum daily average temperature (MDAT) allowable, for each week of the year in order to protect the most sensitive fish species during their most sensitive period of either growth, reproduction or cold exposure. The rationale and methodology for this approach are based on work by the US Environmental Protection Agency (IJC, 1976) and a recommendation to the International Joint Commission by the Water Quality Advisory Board.

In Table 7.8 is given the calculated MWAT for each week of the year together with the limiting factor deciding the MWAT. Ten species were found to be limiting the MWAT in different weeks during the year due to spawning sensitivity, two species exerted a weekly limit due to sensitivity during egg incubation and one species due to growth limitations. Table 7.9 shows that the maximum weekly temperature in the river (taken from 11 years of data) exceeds the MWAT during several weeks of the year. We can conclude from this that some of the native fish species may be under stress on those occasions when unusually high temperatures have naturally occurred in the river.

Results show that more severe constraints exist than the 2·1 °C temperature rise arrived at by the first study. In the next study we will determine in what periods of the year under the various river flow rates there may be an actual temperature constraint on station operation if the MWAT data are used. The river flow has a marked seasonal and yearly variation so the MWAT constraints will only be of concern at the lowest river flow conditions for each time of the year. Then we will find out where the sensitive fish species are during their life cycle so that it may be possible to adopt the less conservative assumption that all activities of all fish species do not occur in the warmest areas immediately downstream of the hydraulic plant.

(b) Site layout and the terrestrial ecology at Chats Falls

In an environment assessment for a project on a previously acquired site, we must assume that in the previous process of site selection sufficient environmental data were obtained to establish that construction is not going to be held up by the finding of the truly rare or endangered species as has occurred several times in the United States.

An inventory of species and their seasonal abundance on and near the site is a necessary first step for the assessment. As stated before, inventories are of little or no value if there is no assessment of the impact of the project on the species found.

A description of the ecology of the area should follow the inventory and should include:

(i) The significance of the site in relation to the region. For example, how rich is the site in species diversity and how typical is the site and its ecosystems of other areas in the region?

(ii) The importance of species found on the site. For example, what is their regional significance, how close are they to the edge of their normal distribution; what is their relative rarity: are they endangered?

(iii) The use of the site by transient or migratory species in relation to other sites in the area.

(iv) The relative value of site features in relation to each other. For example, is a stand of trees which is a nesting area for a relatively rare bird species more important than the marsh which is not so well used as others in the region outside the site?

(v) The past and present uses of the site by man and how the value of its terrestrial ecosystem may have been affected.

At this point, the ecologist should be in a position to discuss this information with the engineers involved in the site-layout and conceptual engineering of the project. If any of the terrestrial studies reveal that some of the species or areas on the site have regional importance, these facts should be clearly stated in a report. The ecologist should also have made up his mind which sections of the site he will be prepared to sacrifice, if necessary, in order to preserve others. The report should be prepared in sufficient time for engineering to be still receptive to discussion on mitigative procedure such as alternative site layouts, and adjustments of construction schedules to meet seasonal needs of some species. It has been our experience that if such concerns are raised sufficiently early in the preparation of the environmental assessment, the site layout engineer is usually receptive to suggestions. It is also surprising how often the alternatives to conventional ways of doing things turn out to be no more complicated, inconvenient or expensive.

The terrestrial ecologist should divide his concerns into those which are due to construction and those which are due to operational effects. The impacts of construction activities are acute in that they are often very abrupt, occur over a relatively short period and are often irreversible, whereas operating impacts are chronic and last for the operating life of the project but may be reversible. Each phase has to be considered separately in an environmental assessment but the total impact should not be ignored. It is of little value, for example, to protect the marsh area during construction by preserving drainage patterns and by putting up fences to keep out construction staff if the noise of the plant operation makes the marsh uninhabitable. On the Chats Falls site the terrestrial survey (Anon, 1979) showed that much of the area had previously been used for agriculture and that rapid reforestation was taking place in areas where farms had been abandoned. Also, the marsh area on the site was formed by the previous construction of the hydraulic station in 1932. This information established the type of previous

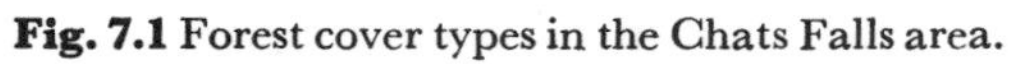

Fig. 7.1 Forest cover types in the Chats Falls area.

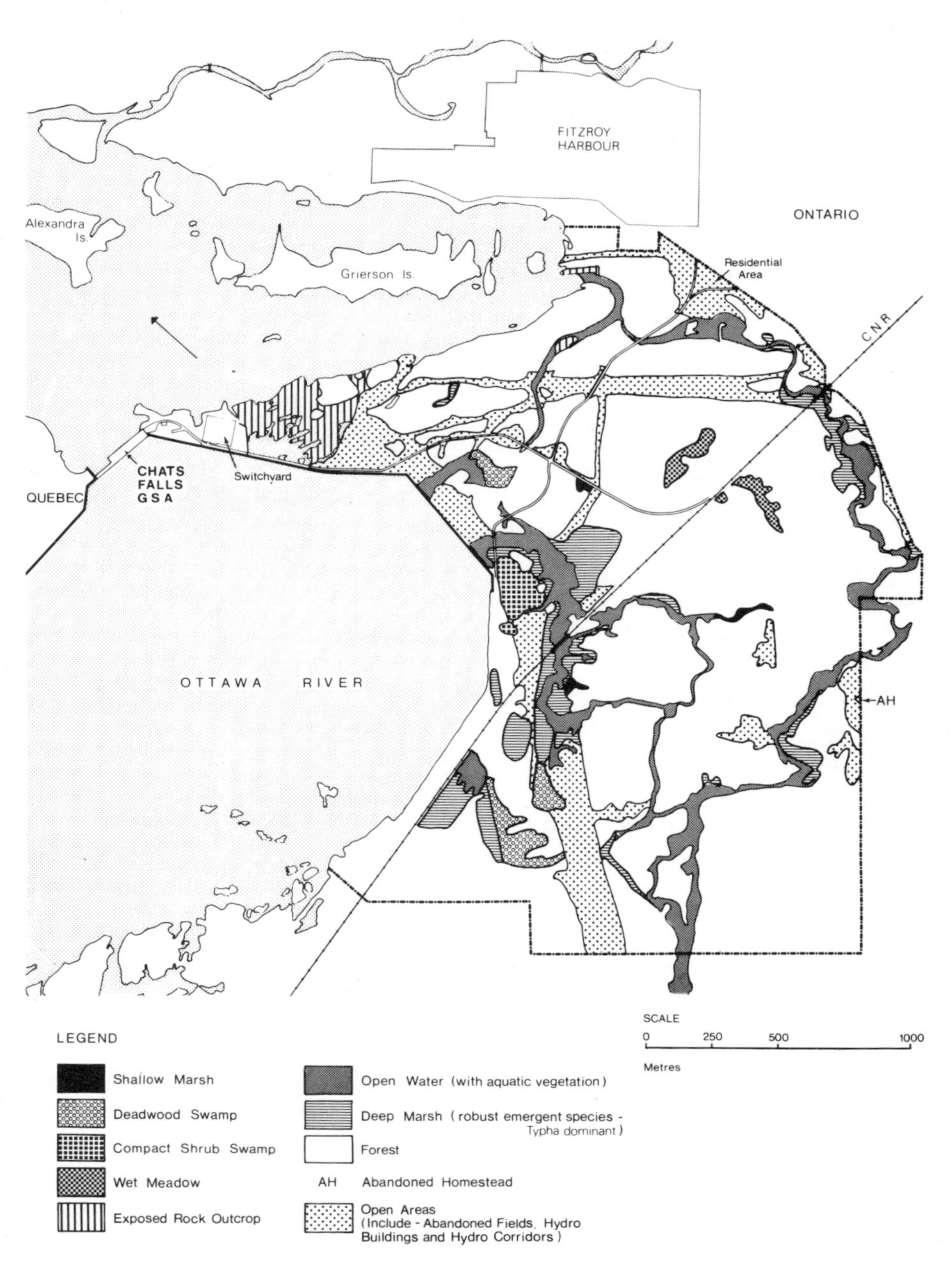

Fig. 7.2 Vegetation communities (excluding forests) in the Chats Falls area.

impact of Man's activities and helps to compare the value of the site with other more established areas in the region. Two-thirds of the site is forest (Fig. 7.1) with the remainder consisting of wetlands, abandoned fields, barren rock outcroppings and a small residential area (Fig. 7.2). A river channel, railway line, access road and transmission lines cross the site. The average age of the forest coverage is 50 years, and hardwoods are beginning to replace the conifers. Of the 221 plant species found on the site, 47 are considered to be regionally rare, sparse or uncommon and are spread fairly evenly across the site.

Wildlife found on the site is typical of the region. Of the 175 species of birds, two-thirds are non-resident transients, and 21 species are classified as regionally significant, 17 of them presently blue-listed in the United States. The use of the headpond of the hydraulic station by waterfowl and shore birds is low compared with other nearby areas. The man-made marsh is poorly utilized. Very small numbers of species of mammals, amphibians and reptiles (24, 13 and 10 respectively) were found. None is considered endangered or threatened in Ontario, but six species are considered to be regionally significant. Two species were found to be in an extension of their previously established range.

After a mapping of the site based on these observations, we concluded that because 45% of the site would have to be cleared of vegetation during construction, certain site impacts would occur irrespective of location of the project on the site. These were:

(i) Forest cover in the site will be reduced by about 37% irrespective of a wide range of possible alternative site layouts. Of the 47 regionally significant plant species 40 are associated with the forest cover and the majority of this population will not be disturbed. Nineteen of the 47 species are associated with the mixed wood forest (Fig. 7.1) which will be reduced to about half its original area on the site. The conclusion from these data is that the majority of the regional plant species will be retained in their natural state on the site.
Due to the large percentage of cover removed from the site, construction activity will initially significantly reduce wildlife populations. Noise during construction will temporarily displace sensitive species such as the redheaded woodpecker, red-shouldered hawk and the white-tailed deer. During the construction period, therefore, the carrying capacity of surrounding areas may be exceeded for some species. No detailed analysis of the effect of this population shift has been made or whether it will be temporary or permanent for each species. The majority of regionally significant bird species will be forced to avoid the site due to construction but mortality will be low because they are mostly transients and there are other large water areas nearby.

(ii) About 80% of the site wetlands will be affected by construction activities and this is largely independent of a reasonable range of site layouts. We conclude that the loss of this large percentage of wetland cannot be avoided. However, because they were created by the previous hydraulic station construction and they are poorly utilized, their loss is considered to be small in relation to larger more actively used areas off the site in the region.
Reduction of wetland area will largely eliminate reptiles and amphibians from the site, but because of the low species numbers and populations and because there are more productive wetlands off-site, this loss is not considered important to the region. The regionally significant reptile and amphibian species which were found to be in an extension of their previously established range will most probably be displaced from the site by elimination of the wetland.

A large berm running parallel to the railway line will be erected during the construction phase to protect the generating station and this has the additional benefit of separating much of the property from active construction and operation (Fig. 7.3). This physical isolation of about a quarter of the site will provide a relatively undisturbed area which will allow species sensitive to human activities such as deer, mink, coyote, black bear and long-tailed weasel to re-establish themselves. We have requested that this berm be constructed as early as possible in the site construction phase in order to isolate the southern portion of the site. On other sites we have observed recolonization by sensitive wildlife such as deer after construction is finished.

The active nest of a goshawk, a regional rare raptor, was discovered during surveys in the southern corner of the site. Part of the site was acquired at the time of the bird's discovery with a condition that the former owner be allowed to log the area. It was therefore recommended that the area immediately adjacent to and including the nesting site, (comprising not only the active nest but other inactive nests), not be touched by the logging operation and that the logging operations be completed prior to the nesting season. Also, a proposal to relocate the existing transmission line around the proposed station included an alternative alignment which would cross the nesting area. It was recommended that another alternative which relocated the line across the forebay be considered, because of the low use of the forebay by waterfowl and the sensitivity of the southern section of the site. These recommendations were accepted.

After the construction period, the reduced noise and other site activities will encourage several species to move back into landscaped areas of the developed site north of berm but some, such as nesting waterfowl and mink will most likely be eliminated by the operating noise in otherwise acceptable areas. These areas will mostly be near the cooling towers where noise level

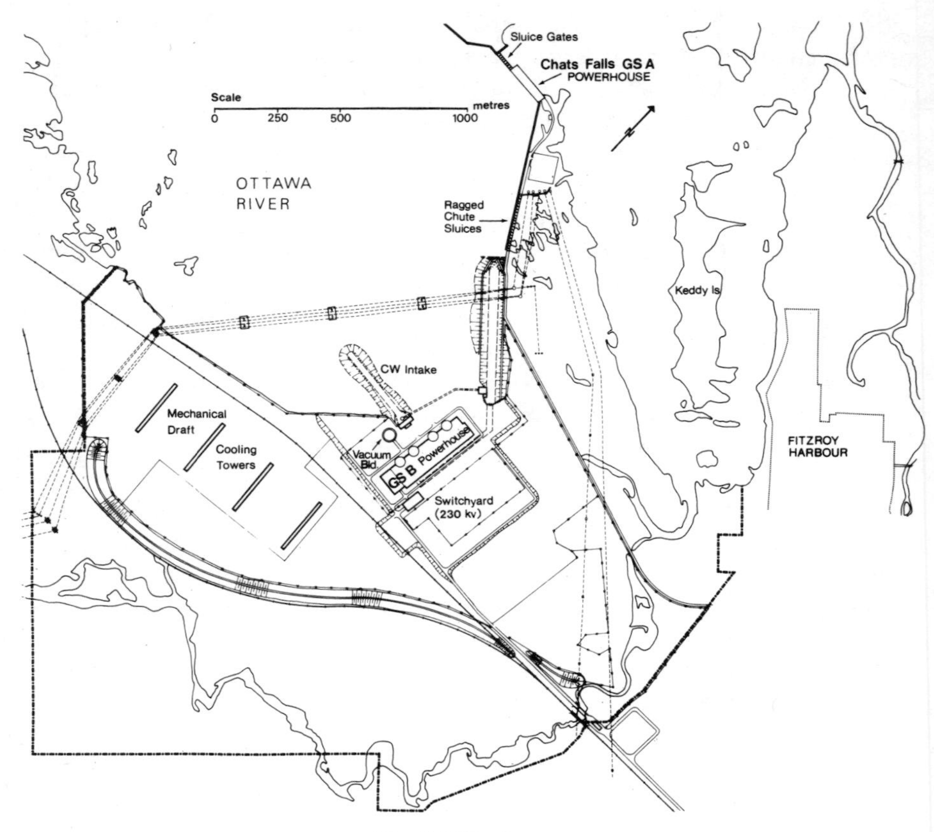

Fig. 7.3 Site layout.

should be acceptable outside about a distance of 100 m. Based on experience at other sites on migration routes, loss of birds due to collision with buildings is expected to be very slight due to the low profile of all the structures.

7.2.4 Conclusions

This very general description of two studies at the Chats Falls site highlights some principles which are important in the preparation of an environmental assessment:

(a) Studies, their results and recommendations have to be carried out in sufficient time to influence design in the site layout and conceptual design phase, that is, when the environmental assessment for seeking approval of the project is being prepared.
(b) The environmental and engineering studies should go along in parallel

during this phase, with each influencing the other as studies progress and become more detailed and committed.

(c) Environmental solutions may require some compromise with engineering and economic considerations.

(d) Assumptions used as a basis for prediction may have to be made more rigorous or may even be allowed to be relaxed depending on the course the study takes.

(e) Mitigation measures already agreed to during the EA preparation should be described in detail. Also, those impacts should be described which are considered to be significant but where mitigative measures are not proposed. Reasons for such actions may be cost or lack of adequate knowledge of how to engineer around the problem.

(f) A clear explanation of how each of the claimed environmental impact or lack of impacts has been established. For example, were assessments made solely by reference to previous literature, by personal experience with other similar situations, by modelling, by professional feeling for the situation or by what?

(g) What is not capable of prediction should be stated, with reasons.

Assessments of the kind included in the Chats Falls Environmental Assessment may still leave some concerns. The regulatory agency may then request further studies to be undertaken as a condition of project approval. The results of such studies are required for more specific design approvals and permits. In Ontario, for example, the *Environmental Protection Act* and the *Ontario Water Resources Act* require that the proponent obtain permits to take and discharge water and to emit contaminants to the air. Except for being laid down as a condition of approval under the *Environmental Assessment Act,* approvals based on these such studies really fall outside the environmental assessment process.

In countries contemplating a form of environmental assessment, the value and effectiveness of the process may strongly depend on the level of detail which the regulatory agency requires before it gives approval to proceed with the project. If the project is large, complex and being developed over a long period of time, it is essential to have a staged, progressively more detailed procedure for review and approval as now exists in Ontario. Public and regulatory debate on the need for the project may confuse the issue and downgrade the value of the EA process itself.

Many biologists who have a regulatory role find difficulty in reviewing an environmental assessment if some environmental damage or loss is stated to be unavoidable. This difficulty arises because the biologist is trained to have a responsibility for maintaining the environment in a natural state. He is therefore reluctant to trade off a part of the environment against other claimed benefits. However, it is much better for him to learn how to make a reasonable compromise than to take an uncompromising position and then later to be forced into accepting a situation which may be worse than what he could have originally negotiated. In an environmental assessment process we may have ecologists working for the proponent and reviewing regulatory agency. In this sense, the two are in adversary position, but they should have similar professional views. If the ecologist acting for the proponent finds the best set of solutions when all other factors such as cost and engineering are considered, he should be able to tell a story which the ecologist of the reviewing regulatory agency should have little difficulty in agreeing with.

7.3 ROLE OF RESEARCH IN MEETING ENVIRONMENTAL ASSESSMENT NEEDS FOR POWER STATION SITING

G. D. Howells and K. M. Gammon

7.3.1 Introduction

The electrical industry recognizes the responsibility for the environmental consequences and, through its statute, is obliged to give proper consideration to safeguarding environmental quality. It would be unreasonable to expect that modern power developments have no environmental impact – these are major schemes in every sense – during construction, commissioning, operating and finally decommissioning phases – and a large modern 2000 MWe station will have substantial impact on its environment during each one of these phases. This paper sets out the way in which environmental constraints have been met, and legal obligations discharged, taking as examples some of the ecological studies at old and new stations, and tracing over the past 30 years the evolution of present policy towards environmental issues and its dependence on research which can be directed towards identifying, clarifying and at times overcoming environmental problems.

While older industrial policies have limited environmental activities to purely defensive strategies, such as monitoring for compliance with consents, and to detect and assess damage, the CEGB has also long supported a programme of basic research directed to achieving greater understanding of phenomena and processes applicable to a wide range of site, operating and climatic conditions. This, supplementing monitoring emissions and discharges of existing stations, gives a predictive capability so that the probable consequences of operating at any particular site can be considered at the outset, and improved plans or designs proposed, and alternative options considered. A sound basis for environmental assessment is provided by the Board's longstanding experience of the effects of power station operation on the UK environment, and the knowledge gained at existing sites can be used to estimate, with some precision, the effects of further development at these sites, or at similar new sites. Environmental research within the Board (including both physical and ecological aspects) has continued over a period of more than 25 years, advised by specialist Panels and working closely to the needs of the Board's Planning and Operating Divisions.

7.3.2 Organization of electricity supply industry in England and Wales and its legislative control

The basic organization of the Electricity Supply Industry within England and Wales is shown in Fig. 7.4. The Central Electricity Generating Board

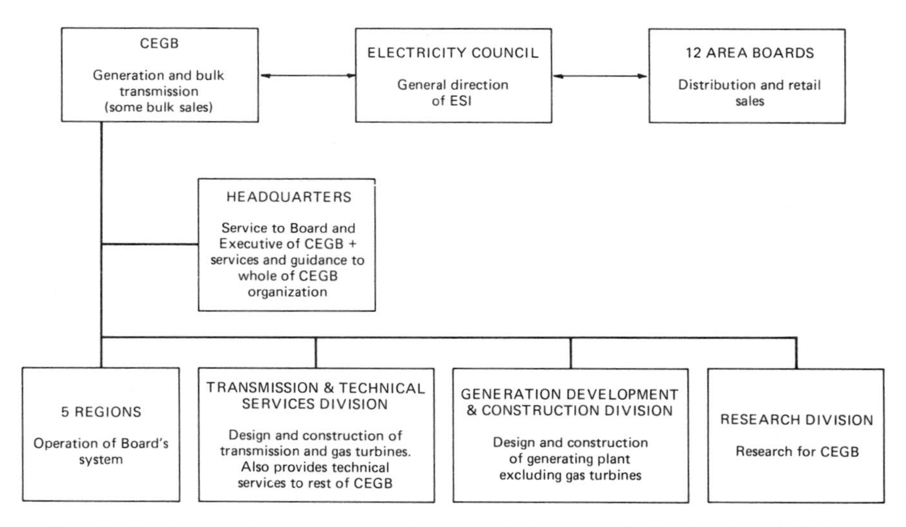

Fig. 7.4 Basic organization of electricity supply industry in England and Wales.

(CEGB) own and operate the power stations and main Grid Transmission Lines. Electricity is supplied in bulk to Area Boards who are responsible for its distribution and sale to consumers. This is one of the largest integrated electricity systems in the world and the board provides staff for construction and operation, as well as departments for internal planning and research.

Within CEGB Headquarters, Planning Department (internal organization has now (1982) changed, although the function remains) provides specifications and plans for the development of the supply system in England and Wales as a whole. These include identifying and evaluating technical, economic and environmental aspects of potential sites and proposing new power stations and transmission lines. One of our tasks is to recognize the gaps in knowledge or understanding needed for this task and with Research Division to set up the work required. Planning Department remain as executive controllers of each particular scheme up to release for construction.

Before dealing with specific environmental matters it is helpful to outline the legislation under which CEGB develops new plant as well as the basic planning procedures and constraints in siting.

Table 7.10 Main legislation covering development of power stations

Electric Lighting Act 1909	Consent of Secretary of State for Energy to develop a site with a specific power station
Town and Country Planning Act 1971	Planning direction of Secretary of State for Energy covering details of Buildings, land use, etc.
Nuclear Installations Act 1965	Nuclear Site Licence from Health and Safety Executive to permit use of site for nuclear reactor
Radioactive Substances Act 1960	Approval of Secretary of State for the Environment and Ministry of Agriculture, Fisheries and Food to store and move radioactive waste
Dumping at Sea Act 1974	Licence from Ministry of Agriculture, Fisheries and Food to dump substances in UK waters below high water mark
Coast Protection Act 1949	Consent of Secretary of State for Trade to works below high water mark
Water Resources Act 1963	Licence from relevant Water Authority to abstract cooling water
Rivers Prevention of Pollution Act 1951	Consent of relevant Water Authority to discharge cooling water and trade effluent
Land Drainage (Amendment) Act 1976	Consent of relevant Water Authority for changes to land drainage
Control of Pollution Act 1974	Licence from County or District Council for waste disposal. Consent from relevant local Authority relating to control of noise. Other pollution matters

The consents needed to build and operate a power station are shown in Table 7.10; a large number deal specifically with environmental matters and indicate the number of pollution control authorities, each with detailed knowledge and experience. The United Kingdom has already a comprehensive and effective system to protect the environment. The introduction of an additional environmental impact assessment procedure would not necessarily improve the attention given to environmental questions by responsible developers and authorities. Experience in other countries shows that delay and waste of resources can result.

Two of the statutory obligations under which CEGB operates are that it must maintain an efficient, co-ordinated, safe and economical system of generation and transmission while having regard for the environment. The latter obligation has equal merit with the former and over the last 30 years the Board has evolved ways of taking environmental matters into account throughout their work.

Each project is considered on its own merits so the degree of attention varies to meet needs and to make the best use of resources. We would deprecate any proposals that laid down rigid environmental standards on a national basis without regard to particular characteristics of a site as this might divert resources without compensating benefit (for example, if a fixed temperature limit on cooling water discharges led to installation of cooling towers at estuarine or coastal sites.) This could have been the result of some EEC Directives as drafted; their amendment resulted from many years of sound scientific research and monitoring of the biological effects of warmed discharges.

7.3.3 Siting considerations

(a) Engineering and economic constraints

New major power stations are expected to have 1000–4000 MWe output and will use coal or nuclear fuel. Their engineering and economic siting constraints are outlined in Table 7.11 (Gammon, 1979). The major requirements can be summarized as a flat area with suitable foundations, close to an unfailing water supply, economic fuel supply, and with reasonable access to the 400 kV transmission grid. Nuclear stations can be sited economically in more remote areas as fuel delivery and waste removal costs are less (one tonne of uranium used in a thermal reactor produces the same electrical output as about 14 000 tonnes of coal).

To maintain an efficient supply system, balancing output and demand, some energy storage system is essential. The only proven plant with major capacity is hydroelectric pumped storage as at Ffestiniog or Dinorwic in North Wales. Pumped storage does not burn fuel but, as the name implies, is a means of storing energy using the water held in two reservoirs at different heights. At times of low system demand, power is used to pump water from the lower reservoir into the upper where it is stored until high demand periods when it is released into the lower reservoir, turning turbines and generating electricity as it falls. Although pumped storage is only 75% efficient, this still enables savings of cost and fuel to be made. It also provides an efficient and very fast-reacting standby plant. Dinorwic power station

Table 7.11 Technical and economic siting constraints for power stations

Parameter	*Type of power station*		
	Coal (2000 MW)	*Nuclear* (2000 MW)	*Pumped storage* (1800 MW)
Fuel delivery			
(a) Proximity to source	Close	Distant	No fuel necessary
(b) Transport	Rail or Sea	Rail and road	However must have initial supply of water and a supply for make-up
(c) Amount and period	5×10^6 tonnes/year (about 20 trains/day)	Low volume and infrequent	
Cooling water			
(a) Direct cooled	60–100 cumecs (50–80 million gallons/hour)		Insignificant
(b) Tower cooled	To 4 cumecs (80 million gallons/day) **Supply must be unfailing**		
Land			
(a) Terrain	Reasonably level site above flood level but not so high as to incur large cooling water pump costs		Needs 2 areas immediately adjacent but differing in height by at least 300 m.
(b) Site area	150 ha	80 ha	Power station itself underground, some offices above ground, main area requirement is for reservoirs (85 ha at Dinorwic)

(c) Geology	Foundations must be capable of supporting very heavy loads	Must be virtually no differential settlement. Reactors can weigh up to 100 000 tonnes	Rock must be suitable for tunnelling and be able to support dam abutments and small buildings
Access			
(a) Construction materials	1–2 million tonnes over 6–8 years		1–2 million tonnes over 6–8 years
(b) Abnormal loads	About 70, 6 of 250 t to 500 t. Occasional during operation	About 70, up to 18 of 250 t and up to 6 greater than 500 t	About 100, 10 over 150 t Occasional during operation
Aircraft safeguards	Height, restrictions on chimneys, cooling towers, buildings within defined areas around airfields		
Nuclear siting		Sufficiently remote from large areas of population to comply with licensing authorities' requirements	
Waste disposal	Up to 1 million tonnes/year	Low level of controlled radio-active effluents and omissions	
Transmission connection		Preferably close to transmission system	
	Normally 2 400 kV lines to nearest suitable 400 kV national grid point		Normally 1 400 kV line to 400 kV grid point

under construction in North Wales, for example, will be able to run from no-load to 1800 MWe in 10 s.

Siting constraints for pumped storage are entirely different from those of other stations. The basic requirements are for two reservoirs immediately adjacent (within about 2 km) at different heights (at least 300 m between them). This means pumped storage must be sited in areas of high land.

Other new power stations are expected to include the use of smaller coal-fired units (less than 1000 MWe), perhaps with combined heat and power output, especially for re-establishing old urban sites. Developments in other energy storage systems and alternative generation methods, such as wind, wave and tidal power, are kept under review. None of these are expected to be ready to make a major contribution to supply for many years.

(b) Environmental considerations

Many aspects of constructing and operating power stations, and of their associated transmission connections, result in interactions with the environment. Each case must be studied and assessed to determine what the overall effect might be. Many developments may involve ecological constraints and some representative case studies will be described in more detail in this paper. Other environmental constraints which are taken into consideration include visual intrusion of large buildings and structures, noise during construction and operation, traffic disturbance, and socio-economic effects, and are not dealt with here.

(i) *Land-use*

Significant land-use constraints imposed by other national needs limit potential sites (Fig. 7.5). More than 40% of the total land area of England and Wales is protected under Statute by National Parks, Green Belts, Nature Reserves, and Areas of Outstanding Natural Beauty, while existing urban development occupies a further 10%. Along the coastline protected land and urban development take up over 60% and 25% respectively. Clearly this limits potential sites and development in these protected areas calls for special measures to minimize environmental impact. The topography and infra-structure of a site is also important in determining how major components, for example, road/rail access, pipe runs, coal stores, ash lagoons, will be located. It may be noted that 4 of 40 stations developed between 1961 and 1975 were sited in areas of high amenity (Ffestiniog, Trawsfynydd, Ratcliffe and Dinorwic).

(ii) *Atmospheric dispersion*

In a fossil-fuelled station, effective dispersion of stack gases calls for a sufficiently open site to ensure that emissions are quickly diluted to acceptable levels of concentration and then dispersed by air currents into the whole

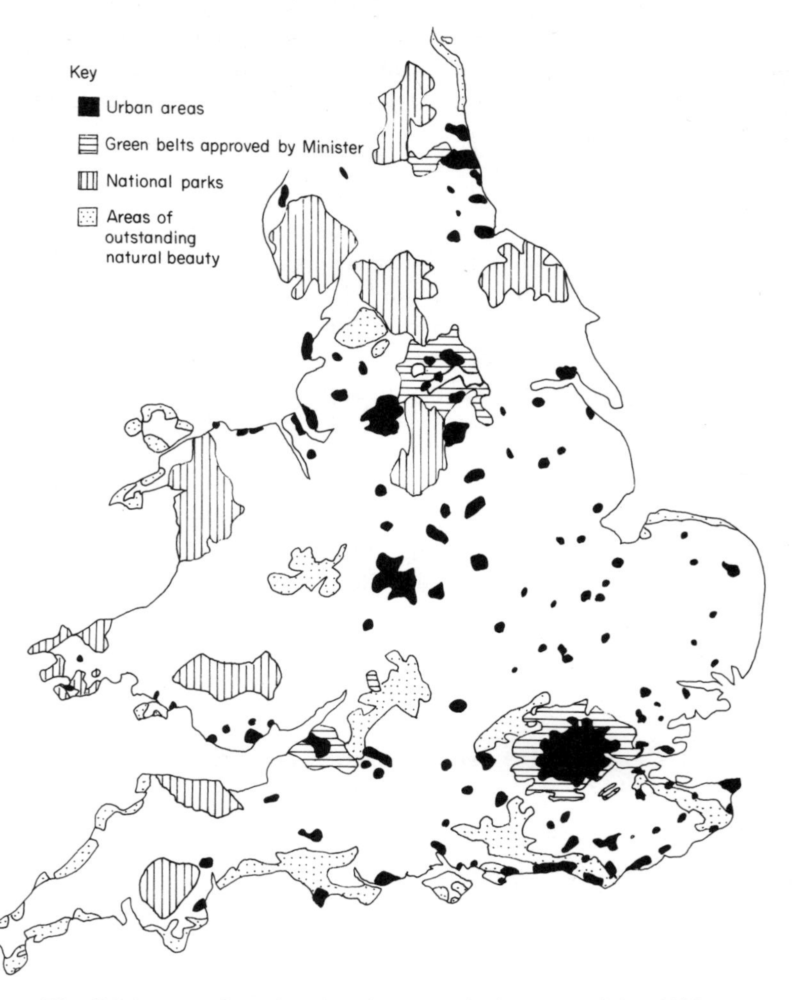

Fig. 7.5 Protected land and major conurbations as at May 1978.

volume of troposphere downwind. Dilution factors of $>10^6$ between stack discharge and the point of maximum impingement is the usual design objective (Clarke *et al.*, 1971). Similar considerations apply to the dispersion of the cooler, lower, cooling tower plumes, which although innocuous, are an occasional cause of nuisance. Emissions must be designed to meet agreed regulatory consents, and must achieve these in operation. A substantial experience of research and monitoring has been accumulated; dispersion, dilution and subsequent fate of emissions have been studied in great detail (Scriven and Howells, 1977), and there are now programmes of investigation into the long-term effects of low-level pollutant concentrations on human health, and on flora and fauna. Prior to operation, baseline monitoring of air pollutants at the site establishes 'before' ambient air concentrations to be set against 'after' ambient levels of stack emission constituents such as SO_2 and dusts.

(iii) *Cooling water*
For the most part, cooling in turbine condensers is achieved by water circuits – direct, once-through systems at coastal or estuarine sites, and by recirculation through towers at inland sites where water supply is limited – a single 500 MWe generating unit would require more than the dry weather flow of many of our small UK rivers. The volume of this cooling water is very large for direct-cooled stations (up to 100 $m^3 s^{-1}$ for a 2000 MWe station) and still quite large (up to 4 $m^3 s^{-1}$ for 2000 MWe) for tower cooled stations (Macqueen and Howells, 1978). This volume is also used to dilute and discharge other station wastes at concentrations acceptable to the appropriate regulatory authority. When a new station is proposed, hydrological studies are made to predict cooling efficiency, as well as to anticipate any effects, either physical or ecological, that may be caused by cooling water flows. These studies are initially investigative, but may lead into regular monitoring, if this is seen to be needed, and may stimulate long-term generic research programmes, as, for instance, at Trawsfynydd and Bradwell.

7.3.4 Research for ecological assessment – past, present and future cases

The procedures for developing power stations are shown in Table 7.12. A similar procedure is adopted for transmission lines, although technical constraints are less stringent and the amenity constraints can be many and varied over a route. It is our view that these procedures, which have been developed over many years, provide the necessary safeguards for amenity and other interests in a satisfactory and economic manner. With knowledge accumulated over many years, and for a range of site conditions, the steps in Board procedures are being modified – for instance, the 'Area of Search' has become a continuing review of known sites, and is merging with the 'Detailed Investigation' stage as more information is available and collated.

Some examples from past, present, and future sites are given to illustrate how environmental questions have been identified and answered by research investigations during the course of station development and operation.

(a) Coal-fired stations on the river Trent

The River Trent represents the largest UK concentrations of inland power stations, with a maximum generating capacity of 10 GWe. The stations are

Table 7.12 Planning procedures for power station developments

1. Area of search	Area of country, where need foreseen, existing studies and known resources reviewed Area's potential as well as technical and environmental constraints determined Views sought from Government Departments, Countryside Commission, Nature Conservancy Council and local Planning Authorities All locations examined to produce short list of sites worthy of investigation
2. Detailed investigation	Public announcement that particular sites to be investigated Owners and occupiers approached and detailed investigations started Trial bores; topographic, hydrographic and environmental surveys Studies of impact and amenity Detailed discussions with Government Departments, local authorities and other statutory authorities. Their views invited on merits and impacts of development and factors to be investigated. Reactions of individuals, societies, local authorities, etc. received With results of detailed investigations relative merits of each site assessed Balance has to be drawn between technical/economic and amenity factors. Any long-term use purchased by agreement and notified to public and local authorities
3. Selection of site and application for consent	When specific type and capacity of station required in a region the most suitable site is selected and application for consent under *Electric Lighting Act* 1909 made Form B submitted to local planning authority to obtain formal views Notice of application given to public Public inquiry may be required before Secretary of State makes decision to go ahead with project
4. Design and construction	Once consent given final design arrangements made and site preparatory work starts Main contracts for the 6–8 year construction period issued once financial sanction given by Secretary of State for Energy

old and new, some direct cooled, but most with cooling towers, and the river is used for many other industrial purposes. This major industrial river has changed substantially in the last 20 years, from a poor quality river (Class 4 in DOE classification) to one of mostly Class 2 quality, coincident with an improvement in sewage and industrial discharges, and replacement of old direct cooled stations by new cooling tower stations. The Trent and its valley has been studied by many agencies including the CEGB, and many investigations of the effects of chimney emissions and of warmed discharges in terrestrial and aquatic ecosystems have been reported. The results of these investigations are invaluable in determining the effects of present stations, and in predicting the impact of new stations at similar sites elsewhere.

(i) *Fishery studies* (Sadler, 1980; Brown, 1973; Langford, 1978)
Following short-term experiments on fish avoidance and reduced survival at high temperatures, studies of fish distribution were made upstream and downstream of heated effluents in the Trent and other East Midland rivers. Thermal discharges to the Trent raise overall temperature to about 6 °C above ambient, and during dry, hot summers, temperatures may reach 26 °C for short periods. Field studies have demonstrated that native 'coarse' fish in the Trent and other lowland rivers are apparently unaffected by discharges – growth, reproduction and abundance are unchanged. The only effect observed has been a possible delay in seasonal upstream migration of some species which seem to prefer warm effluents as river temperature drops in the autumn (Fig. 7.6). Studies of sonic-tagged fish show that, contrary to the belief that these fish species are rather static, they indulge in quite far-ranging up and downstream excursions probably related to changing flow conditions, and are free to move in and out of thermal discharges at will.

(ii) *Aquatic ecology*
A secondary effect of raised temperature is to reduce the quantity of dissolved oxygen, already low in a river like the Trent which receives organic waste discharges. The introduction of cooling tower stations to the Trent brought substantial improvements deriving from their capacity to reaerate water during its passage through the towers – about 15 tonnes oxygen day^{-1} per 1000 MWe are provided by single pass of cooling water in conventional, natural draft towers. The effect of the tower has been to reverse the 'sag curve' reported for oxygen levels downstream of a thermal discharge (Whitehouse and Foster, 1974).

The limited capacity of the River Trent, and the need to meet agreed temperature constraints, has led to the development of operational control (Anon, 1971). River temperature data are fed to a regional grid centre where the mix of generation between direct cooled and cooling tower stations can be selected, with load reductions as needed to meet discharge limits.

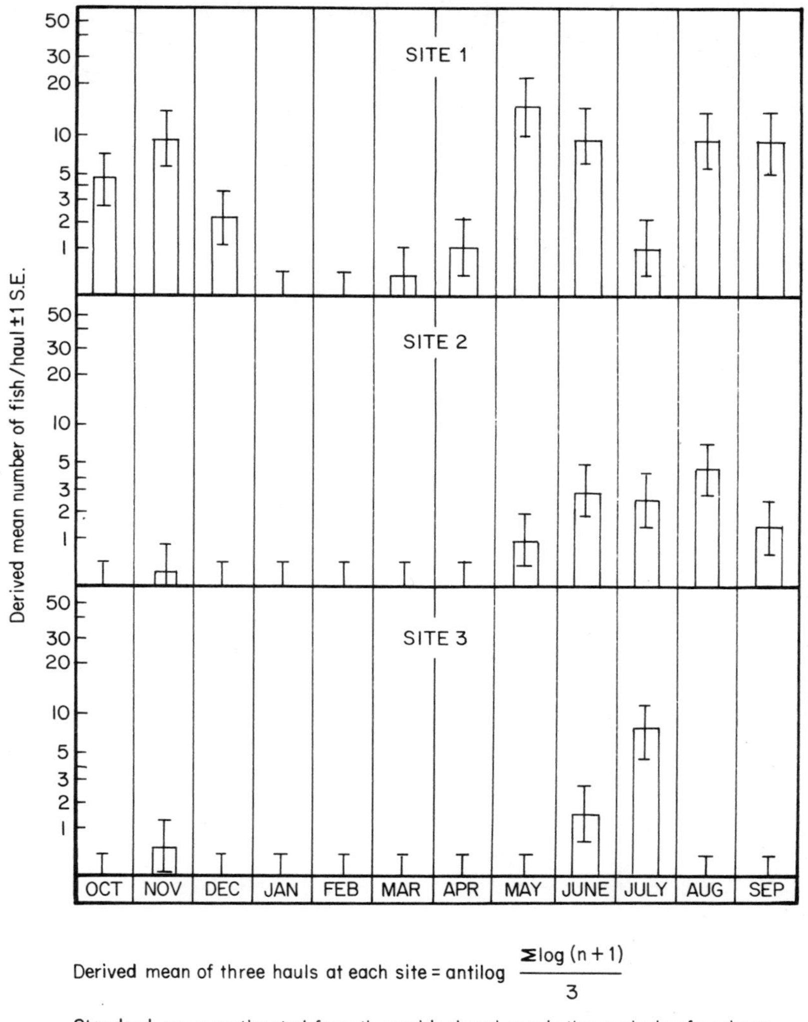

Derived mean of three hauls at each site = antilog $\frac{\Sigma \log(n+1)}{3}$

Standard errors estimated from the residual variance in the analysis of variance calculations

Site 1 = downstream, site 2 and site 3 = upstream

Fig. 7.6 Abundance of roach at Castle Donington. October 1977–September 1978.

The complex interactions between temperature of discharge and other water quality conditions in the Trent have led to investigation of biological indicator species (Aston, 1973). Intensive autecological studies of characteristic fauna – *Asellus aquaticus*, tubificids, and the leech (*Erpobdella octoculata*) have shown that population density and length and weight of individuals are largest at locations with highest organic loadings. Some reproductive functions – emergence of young from eggs, and life cycle – were shortened in warmed reaches (Fig. 7.7).

The quality of the Trent has improved, in parallel with recent industrial exploitation of its flow for cooling and discharge. The subtle biological

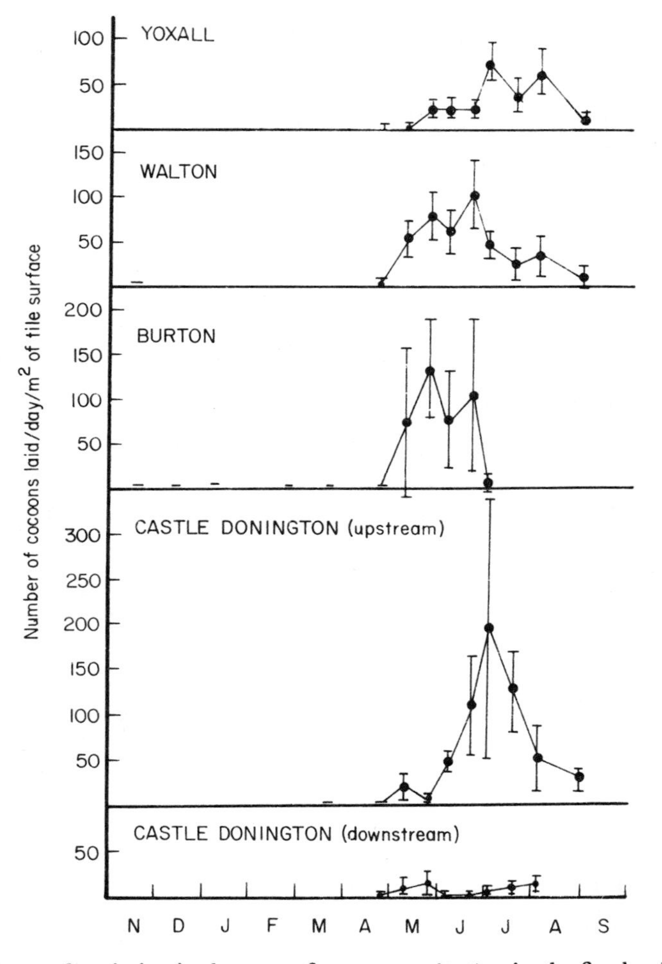

Fig. 7.7 Seasonal variation in the rates of cocoon production in the five leech populations showing means and total ranges (Aston & Brown, 1975).

changes observed are consistent with this overall improvement. As our understanding of the river is built up, our capacity to predict the effect of similar operations grows, so that the consequences of new station operation at inland sites can be forseen.

(iii) *Atmospheric studies*
As part of a routine programme, ambient air downwind of Trent Valley power stations is sampled for a period beginning two years prior to operation, and continuing more or less indefinitely during operation, to establish baseline and 'operational' SO_2 ambient air concentrations. Monitoring of this kind was started as long ago as 1950, and has shown that around large concentrations of generation as in the country as a whole, there is an annual trend of decreasing ground level SO_2 concentrations (Martin and Barber, 1966, 1967; Barber and Martin, 1976). Associated research investigations have established the localized dispersion of SO_2 in 1963 and 1970, the dispersion of water droplets from cooling tower plumes, and the wash-out of air pollutants in rain (Martin and Barber, 1974; Martin, 1979; Maul *et al.*, 1980). None of these investigations has suggested that acute ecological effects occur, although the question of subtle effects is under study.

(b) Nuclear generation at Bradwell

This first commercial nuclear station (Magnox) aroused great public interest and major environmental effects were envisaged by objectors at the Public Inquiry held early in 1956. It was argued that use of the estuary for cooling would cause major ecological changes (Table 7.13) (Gurney, 1956).

At this time (1956) environmental assessment was scarcely recognized. The claims of disaster put forward were based on limited experimental and field data, not always pertinent to the real conditions. A substantial programme of work was initiated, in response to the arguments raised, to satisfy objectors (and the general public) that the fears for ecological disaster would not be realised.

Hydrographic work showed that temperature increases were in line with those predicted – a local increase close to the outfall was confined to the surface and was of strictly limited extent (Talbot, 1966). Biologically, studies confirmed that the oysters were important – their history in the Blackwater is characterized by large fluctuations in commercial landings, largely as a result of unpredictable recruitment. Massive losses occurred during the cold winter 1962/63, following which the industry has had limited success, presumably unrelated to station operation since the outfall remains an active site for settlement and growth (Coughlan, 1966). Studies of the general ecology of benthos and littoral fauna show that such changes as can be quantified are restricted to areas in the immediate vicinity of the outfall near the Barrier Wall (Hawes, 1968; Hawes *et al.*, 1974; Bamber and Henderson, 1981). The changes seem to have been too slight to justify more intensive or longer-term investigations. Certainly the grosser predictions raised at the Inquiry can be dismissed and the spectre of 'thermal pollution' laid to rest,

Table 7.13 Areas of concern identified at Bradwell Public Inquiry, studies undertaken, and results

Expected problem	*How effect would operate*	*Studies undertaken*	*Results, consequences, comments*
1. Isotopes – discharge to estuary	Accumulation in sediments and filter feeders. Transfer food items (e.g. oysters)	MAFF/CEGB statutory study – data reported routinely by MAFF, levels below National and International standards for ^{65}Zn	No effect on oyster industry
2. Use of water for cooling			
(a) Chlorination	Direct toxicity of Cl and production of persistent residuals (halo-organics)	Oyster larval distribution to determine relation to P.S. discharge (1960)	Larvae abundant on N. side of estuary. CEGB discharge affects 25% of all larvae
(b) Heating	Rapid heating in cooling water (now called *entrainment*)	Entrainment experiments on oyster larvae, jointly with MAFF/CEGB (1974 and 1975)	Larval distribution confirmed. Entrainment affects some larvae, but large proportion viable
3. Heated discharge to estuary (effects on general ecology)	Elimination of benthic species, physiological stress on estuary fauna, limit to geographical distribution of northern	Benthic studies (5 year intervals). Shore fauna (with MAFF/CEGB 1956 and 1972). Hydrography (MAFF) 1966). Temperature studies	'Rain of dead plankton' *not* confirmed. Estuary bed not anaerobic. Littoral fauna close to station changed somewhat (but coincident changes to

	species, synergism with other toxic pollutants	(1964, 1965, 1966, 1974–75). Zooplankton (1960–1964)	whole shoreline). Study of benthos inadequate to quantify long-term subtle trends, in relation to other factors (climate, other industry) and natural biological fluctuations
	Greater *Teredo* infestation	Studies of *Teredo* infestation in this and other estuaries	Extent related to waste wood, not water temperature
4. Commercial fisheries			
(a) Oysters	(a) Cumulation of poisons	Oyster landings reported 1957–1972 (MAFF)	Massive fluctuation unconnected with P.S. operation
	(b) Market instability	Commercial fish landings reported annually (MAFF)	
(b) General fishery	(c) Failure to reproduce	Larval herring study (1978–1980) (CEGB) – identify spawning grounds and time, hatching and dispersal of larvae, larval growth and mortality population dynamics. Entrainment of larvae	Detailed and specific study in relation to possible 'B' station, using existing station as 'test bed'. Studies of generic interest for coastal sites with estuary or off-shore intake
(c) Herring fishery	(d) Larval kill		

although the legal requirement that the estuary be kept under surveillance, jointly with MAFF, still stands.

This continuing responsibility, and the possibility that the site can be developed further, has led to a reappraisal of what is ecologically important in the Blackwater at Bradwell. Since station construction the estuary has improved in quality and has become a satisfactory herring nursery, with a small but successful, commercial fishery. The decline of herring stocks elsewhere has given this stock greater importance than would have been so a few years ago. The Thames herring is a sub-group of the North Sea herring – it has a spawning ground just off the mouth of the Blackwater (the Eagle Bank) as well as other grounds elsewhere in the Thames. The major present concern is whether larvae hatching from the Eagle Bank enter the Blackwater and become entrained in the cooling circuit of the present power station. Further, if greater volumes of water were used in the future would this result in an unacceptable loss of larvae? Models based on the estuarine hydrography and an assumed uniform distribution of larvae predicted a power plant caused mortality of 80–40 % (depending on water volume extracted). To clarify this question, a study was made in 1978, 1979 and 1980 to establish the extent and size of spawning; growth, mortality and distribution of larvae; relation of Blackwater herring to the Thames stock; and entrainment by the power station (Henderson and Whitehouse, 1980).

The study has shown that spawning occurs over a three week period in spring, partly determined by water temperature. Hatching occurs within 6–20 days and the larvae are then distributed by tidal movements. Only half the larvae hatching from the Eagle Bank enter the Blackwater (others being distributed to neighbouring estuaries such as the Wallet). Distribution within the Blackwater and the known water requirements of the station suggest that 5–10 % of larvae would be entrained with cooling water during the expected three weeks of larval development. In the event, sampling in the screen forebays failed to find larvae in the expected quantity, suggesting that this was an overestimate – the true value obtained was less than 1 %. The reasons why larvae are entrained in fewer numbers than expected is of particular significance (with implications for intake design) and are being investigated this year in greater detail.

(c) Pumped storage – Tintwistle

The possible environmental impact of a pumped storage scheme in and above the Longdendale Valley at Tintwistle (near Manchester) is being considered prior to final site selection. Environmental constraints on the general engineering proposals outlined are being identified and the major significant questions pertinent to environmental impact formulated. Many aspects are to be studied and work has started already on finding how the fragile high

Table 7.14 Ecological studies at Tintwistle

Area of concern	*Questions*	*Source and nature of available information*	*Actions*
Ecological status	What communities present?	Aerial photographs, local naturalists, reports etc.	Commission field surveys of vegetation, small mammals, birds
	What is ecological value of the area?	Comparison with other sites (SSSIs, etc.)	Discussions with Planning Authority, NCC, etc.
Peat erosion	What is current rate of erosion?	Comparison of historical aerial surveys and peat trap loadings	Include research on: (a) rates of peat loss in streams (b) effect of burning, grazing, climate, etc.
	What are causes of erosion?	Published literature	
	What is preferred method of peat disposal from upper reservoir site (40 ha)?	Experience of other sites (e.g. Sullom Voe Oil terminal) and commercial peat companies	Review alternatives (disposal on-site, off site, sale to commercial company, burning, etc.
Revegetation	What techniques available for stabilizing peat and dam faces with commercial grasses?	Experience from Dinorwic, Sullom Voe	Commission review of experience at other sites
	What techniques available for revegetation with native species?	Limited information available. Mainly from Pipeline revegetation	Initiate 3–4 year research programme on revegetation of moorland with native species
	What techniques available for revegetating reservoir margins?	Experience from reservoirs relevant but draw-down not as frequent. Experience at Ffestiniog	Investigate rock terracing with some revegetation for margins of lower reservoir
Water quality	What is present water quality?	Some information available from Water Authority	Commission chemical and limnological survey to provide base data. Predictions of effects can then be made
	Will water quality be affected during construction? Any long-term water quality effects?	Experience from other reservoir schemes including pumped storage	

altitude moor ecology may be preserved. Some areas of concern so far recognized are set out in Table 7.14, together with the questions being considered and the current status of investigation. Other effects on the environment such as visual intrusion, noise, social infra-structure are expected to need investigation.

Some major areas of ecological concern, and the research studies being initiated, are described below in more detail.

(i) *Ecological status*

An assessment of the upper reservoir site (500 m) has shown it to be a seriously eroded blanket mire with up to 2·5 m of peat. A relict oak woodland covers part of the southward facing Longdendale valley slope. Field surveys are under way to establish the ecological characteristics of the area. Sightings of birds and mammals are made during these surveys, but local records can be used to provide more annual and seasonal detail. 'Conservation value' will be difficult to categorize objectively and the present aim is simply to gather sufficient information for a judgement to be based soundly.

(ii) *Peat handling, erosion and revegetation*

The upper reservoir site (490–540 m) lies in an exposed position and consists of eroded peat overlaying Millstone Grit. Between 50 and 70 % of the moor has been denuded of vegetation and deep eroding gullies dissect the area – the erosion is variously attributed to changes in climate, ground water level, air pollution, burning, sheep grazing. Studies have been initiated to measure the present rate of erosion and to understand its cause. Peat stabilization and revegetation techniques will be explored in the field and laboratory, and conditions compared with other revegetated sites in the UK. The extreme climate and erosion conditions at Tintwistle pose particular difficulties which will have to be overcome if revegetation is to be successful.

(iii) *Water quality*

Longdendale provides water for domestic supply, industrial use, river compensation and recreation. Existing information on water quality is somewhat limited and data are being gathered by sampling at 18 locations throughout the catchment to establish a baseline and to provide the basis for predicting the water quality and biological changes expected if the scheme were to go ahead.

7.3.5 General conclusions

The examples given show an evolving policy. Early studies, as at Bradwell, undertaken in response to a generalized concern for the environment, were not well focused to answer particular questions. The subsequent programme of research was broad and legally justified 'to keep the estuary under surveillance'. The major investigations have followed station construction and operation. The Trent studies provide an example of continuing investigation as old stations are replaced by new, and where the effects of several emissions and discharges are combined.

For the future, new investigations in the Blackwater Estuary are underway before the nature and size of any new development are firm. A specific question has been posed, focusing resources on what we believe to be the most important issue. At Tintwistle, our investigations are exploratory, prior even to final site identification, although we hope that the results of our studies will prove to be valuable at this or a similar suitable site for pumped storage.

This general account of CEGB procedures, and the case-studies described show how research needs can be identified at any stage in the planning, siting, construction or operating phases, and how research can be directed with economy, to resolve particular identified issues. It is clear that for this industry, early claims of ecological disaster, for example at Bradwell, were unfounded, but the general and unspecific investigations set up in response to this claim could only detect major changes, not the more subtle effects that could be important if, for instance, a single species, possibly rare or not present all of the time, were considered. The lesson we have learnt is to identify potentially important questions more precisely, to take account of changing conditions and changing values, and to begin the task of identifying important ecological problems early in the planning process so that the needed research can be effective at every stage through to operation.

The history of CEGB studies of environmental questions over the past 25 years has shown the value of continuing effort – on basic, fundamental studies of generic value; to accumulate an experience of the wide range of problems set by a diversity of sites; and to monitor the success (or otherwise) of particular options taken during construction for environmental protection. This experience also allows the costs of environmental measures, including the costs of ecological surveillance, to be estimated and taken into consideration as an integral part of development costs.

It may be valuable in the present debate to look at the Board's procedures for environmental assessment and to speculate how these might be affected by a more formalized system. We believe that experience over many years leads to clear identification of the significant environmental impacts, so that skilled resources can be concentrated on the most important issues rather than diverted to more general, but less thorough, investigations, extended over a longer time-scale. How much public participation should be invited, as distinct from negotiation between professionally involved agencies concerned with pollution control, is a matter for debate. The Board believe conscientiously that they provide sufficient public information but if more is justified, this can be provided within present procedures. Experience does not lead the Board to expect those seeking to frustrate new projects would readily accept the limitations on debate imposed by rigid planning timescales. The advantages and disadvantages of a formal EIA procedure, setting time limits at the same time as inviting wider discussion, needs to be weighed carefully. The paramount objective must be the optimal protection of the environment compatible with needed developments, without undue delay or diversion of resources.

7.4 ASSESSING THE IMPACT OF MAJOR ON-SHORE OIL INSTALLATIONS: THE EXAMPLE OF SULLOM VOE

W. J. Syratt

7.4.1 Introduction

Since the 1960s a great deal of offshore oil and gas exploration has been taking place around the coast of Britain, particularly in the North Sea. Early oil finds were east of Scotland, in both the British and Norwegian sectors, but during the 1970s most of the larger finds have been concentrated north-east of Shetland in an area known as the East Shetland Basin.

It was soon realized that with fields the size of the Brent and Ninian in this area, the most economical method of producing and transporting the oil would be by pipelines to the nearest land mass. There a terminal could be built for the stabilization and onward shipment of the oil.

The land mass is Shetland and the terminal is nearing completion at Sullom Voe. This paper describes the landward developments and the measures which have been and are being taken to protect the Shetland environment.

In particular the advantages of adopting a policy of consultation between industry, conservation interests and local and central government are discussed. This has lead to a realistic appraisal of potential problems and provided practical solutions to them. This approach is reflected right through from the early design stages into the operating phase.

7.4.2 The oil terminal at Sullom Voe

The oil fields of the East Shetland Basin are approximately 200 km north-east of Shetland (Fig. 7.8). They are connected to Shetland by two 91 cm diameter pipelines which terminate at Calback Ness, a small headland which projects into Sullom Voe, where the terminal is built (Fig. 7.9).

The Sullom Voe Oil Terminal is currently one of the largest engineering projects in the world. The project is overall 99 % complete (December, 1981).

The first oil arrived through the Brent Pipeline from the Dunlin Field on 25 November 1978 and is now flowing through both pipelines from Brent, South Cormorant, Murchieson, Thistle and Dunlin (Brent line) and Ninian and Heather (Ninian line) at an average rate in excess of 800 000 barrels per day.

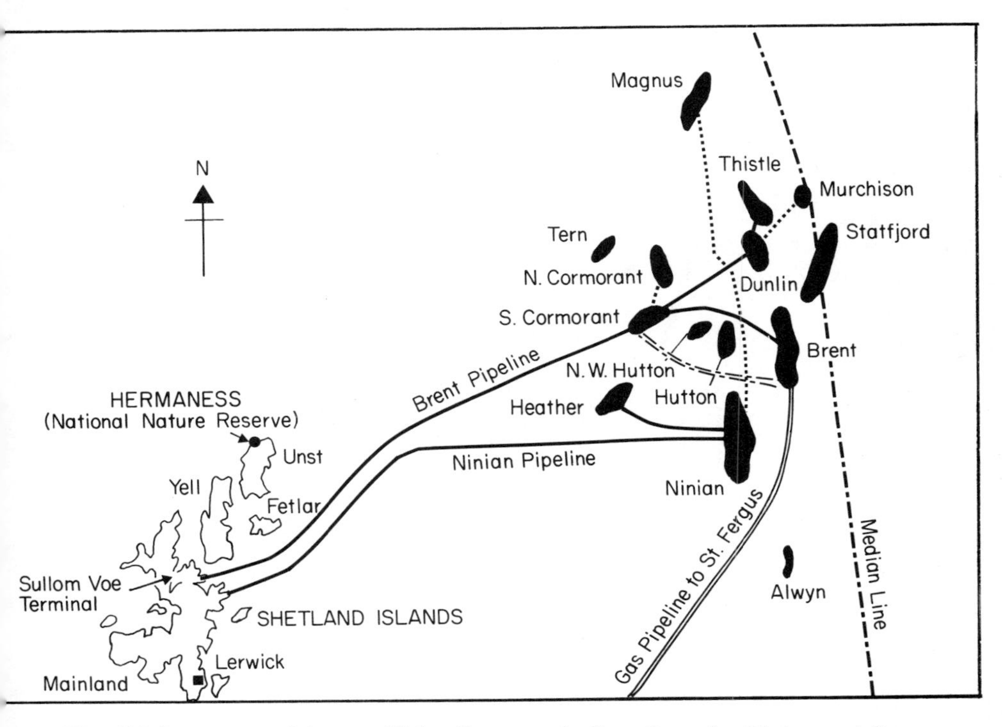

Fig. 7.8 Sea routes of the two 36 in. diameter pipelines from the Ninian and Brent Group Oilfields to the Shetland Islands.

The terminal performs a number of functions:

(a) It receives live (i.e. with gas) crude oil and a certain amount of production water. The existing pipelines are capable of delivering a nominal two million barrels of oil per day to the terminal. There is room at Sullom Voe to accommodate a third pipeline which would bring the potential of Sullom Voe up to a nominal three million barrels per day. However, at the present time it is envisaged that Sullom Voe will handle a throughput of 1·41 million barrels per day in 1982. At present there are no plans to increase this figure.

(b) Production water is extracted through two dehydration trains (trains consist of compartments strung sequentially together so that oil flows from one to the next) and the oil is de-gassed through five stabilization

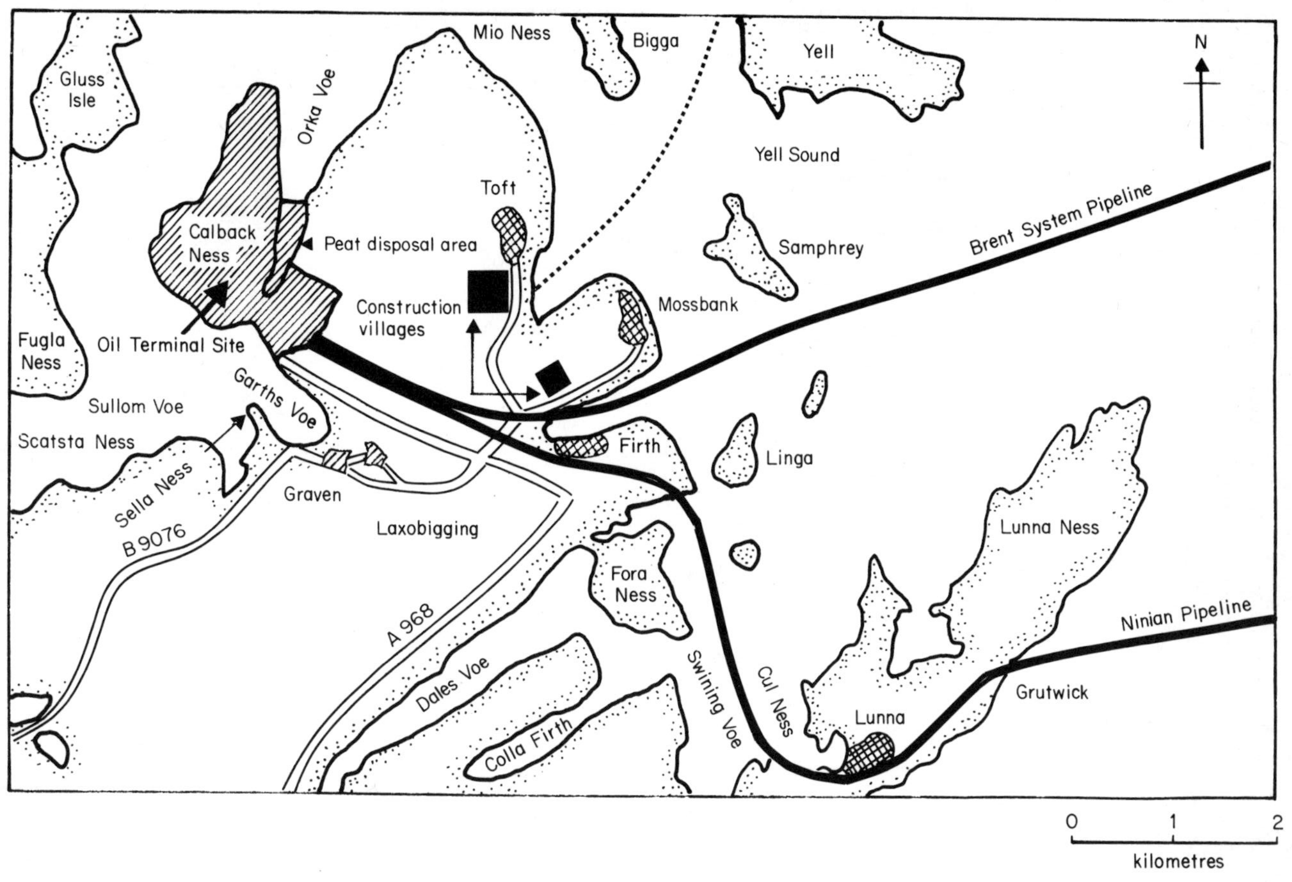

Fig. 7.9 Oil Terminal at Sullom Voe and pipeline routes.

trains (3 Brent, 2 Ninian). The stabilized oil is stored in 16 floating roof tanks each with a net capacity of 90 000 m³ (i.e. there is storage capacity for one week's production).

The gases taken off are compressed and fractionated into four components. Methane and ethane is fed to the power station gas turbines (see later). Propane and butane are recompressed and piped to a chill down plant where they are liquified for storage in five thermally insulated tanks each with a capacity of 30 000 m³. There is a surge train and tank for each pipeline. Excess gas is flared through two 120 m high common flares, with minor flares at the surge tanks and LPG storage area.

(c) It exports the stabilized crude oil and LPG. There are four jetties at Sullom Voe. No. 1 jetty is a combined crude oil and gas tanker jetty capable of taking vessels up to 120 000 tonnes, jetties 2–4 are for crude oil only and accept VLCC's up to 300 000 tonnes. Crude-oil loading pumps enable 30 000 tonnes per hour to be delivered on all four jetties with a maximum of 20 000 tonnes per hour at any one.

(d) It provides ballast water handling facilities of up to 40% of cargo throughput. Four fixed roof tanks of 66 000 m³ receive the ballast water from the tankers. After an initial settling period of 12 hours the content of each tank is drawn down through anti-vortex nozzles and pumped to the effluent treatment facilities. Oil in the reception tanks is skimmed off. The treatment plant consists of two API interceptors, three emulsion break tanks, five sand filters (four in use at any one time, the fifth on automatic backwash) and a final holding lagoon with 24–48 hours residence time. The system is capable of handling 5000 m³ per hour of ballast water. Site oily drainage water and water from the dehydration trains goes through the same plant. Recovered oil is passed to slop tanks for re-injection into the crude oil storage. The entire plant was commissioned in November 1979. Final discharge of the treated ballast water (10–15 ppm average oil content – 25 ppm maximum) is through a seabed diffuser located 470 m off the end of Calback Ness in approximately 25 m of water. Design dilution is 30-fold within 25 m of the diffuser. The receiving waters are flushed by an offshoot of the very strong Yell Sound tidal stream. Early results showed oil contents to the effluent to be in the region of 5 ppm, but have risen to the design figure as the amount of ballast water handled has increased.

(e) Power is provided by five, 24 megawatt gas turbines each with a waste heat boiler capable of producing 75 tonnes per hour of steam at 10·3 bar and 195 °C. The power output of the Sullom Voe Power Station is approximately four times that of the power station at Lerwick that provides the needs of Shetland. The turbines were initially fuelled by diesel and that continued until the processing plant delivered gas in 1981.

(f) In addition there is a considerable infrastructure needed to tie all the major functions together.

7.4.3 The environment of Shetland

Even by British standards the Shetland Islands are small, and rarely shown in their correct position on maps of Britain. Yet Shetland's position at the northern extremity of the British Isles, together with its distictive climate, complex geology, flora and fauna, and history, makes its wildlife of particular interest (Goodier, 1974). This has been recognized in the conservation policies of the Nature Conservancy Council and Shetland contains an above average number of conservation sites (Nature Conservancy Council, 1976). Forty Sites of Special Scientific Interest (SSSIs) – many of them grade one sites – have been notified to the Local Authority and there are four National Nature Reserves (NNRs). Three – Hermaness, Haaf Gruney and Noss – have been established for their outstanding coastal habitats with large seabird colonies and the fourth, the Keen of Hamar, for its arctic-alpine vegetation developed on serpentine debris.

The landscape of Shetland has distinctive features, especially in the open treeless countryside, with its scattered crofts, green farmland, rolling moors and magnificent coastal scenery, all under swiftly changing skies and influenced by the many moods of the sea. With a land area of 1440 km^2 and a coastline of 1450 km, no part of Sheltand is more than 5 km from the sea. There are few places in the British Isles where this kind of oceanic, pastoral and upland landscape exists on such a scale. The Shetland environment is described fully in Berry and Johnston (1980).

(a) The Sullom Voe Environmental Advisory Group

In 1972 Shetland Islands Council (formerly Zetland County Council) appointed consultants to identify sites that would be suitable for a major oil terminal. Eventually, in 1973, Calback Ness, a small headland within Sullom Voe was accepted as the site for development of the oil terminal (Livesey and Henderson, 1973; Zetland County Council, 1974) and the oil industry was directed to that area.

In 1974, before construction of the terminal began, the Sullom Voe Environmental Advisory Group (SVEAG) was formed as the result of a joint initiative of the then Zetland County Council and the oil industry. SVEAG included not only Council and oil industry representatives, but also members from the Nature Conservancy Council, the Natural Environmental Research Council and the Countryside Commission for Scotland, as well as two professors from Scottish universities both with special interests in Shetland. The group was essentially independent but was, of course, linked through its membership to the Shetland Islands Council and the oil industry who between them form the Sullom Voe Association Limited (a non-profit making organization comprising Shetland Islands Council on one hand and the Brent and Ninian Pipeline Groups, representing 32 oil companies, on the other) and Central Government.

It was not the group's responsibility to advise on the siting of the terminal, but how best to achieve minimum environmental disruption in an area chosen primarily on engineering and operating grounds, but with some environmental considerations. The group's terms of reference reflected this:

> 'The SVEAG will advise on environmental aspects of the developments associated with the oil terminal at Sullom Voe, Shetland, including the onshore sealine (out to the SIC limit), landfalls, terrestrial pipeline corridors, storage and other related installations and the tanker jetties and ship-handling facilities. The aim is to ensure that environmental considerations are taken into account in the planning, development and operation stages of the project. Responsibility for environmental aspects offshore (exploration, production and under-sea pipelines) will remain with the individual operators. The SVEAG will liaise with those operators on environmental aspects of the pipelines carrying oil from offshore to the terminal site at Sullom Voe.
>
> The role of the group will be advisory. It may set up sub-groups for specific tasks as required and call upon or co-opt expert assistance as deemed necessary.'

One of the group's main aims was to assess the environmental impact of the terminal development, not on a once-and-for-all basis before construction began, but while it was in progress. During the period the impact assessment was being drawn up, the group was fulfilling its other main aim, that of giving advice to the Council and constructors.

The formation of SVEAG and the way it operated, therefore, represented a unique step in conservation and environmental consideration for such a major development.

In June 1976 the group published its 133 page report *Oil Terminal at Sullom Voe: Environmental Impact Assessment* (SVEAG, 1976). This was presented to an international audience at a public seminar held in Firth Construction Village, Shetland. Briefly SVEAG identified two phases of potential environmental effects. The first would be during construction of the terminal, when the biggest impact would be felt on the land.

The first consideration was the extent of direct losses or major disturbances during the construction phase in the Inner Zone. The terminal site has been ocated on an area of derelict wartime buildings which avoided all the

Table 7.15 Environmental aspects of crude oil storage at Sullom Voe (1·2 million barrels per day throughput)

	Underground	*Above ground*
Visual impact	4×90 gantries above skyline. Visible from most aspects	Tankwalls but most landscaped to skyline and painted. Visible from Northmavine direction. Regular painting will be necessary in salt-laden air
Surplus material	Rock disposal 1·5 million m^3 of rock, peat and unsuitable material	Similar total volume of peat, rock and unsuitable material (present estimate 1·38 million m^3). Disposal of peat to Orka Voe less easy than for rock
	Land usage less but unlikely that land above caverns could be usefully employed for anything other than oil industry construction	Without crude oil caverns approx. 20 tanks of various types are already planned. Distances between tanks (as required by statutory bodies) may mean a large land area usage and impose limitations on where additional tanks could be placed
Construction	Danger from traffic involved in rock disposal etc., emerging from caverns. Dust problems (SiO_2) and poor rock structure (requiring concrete reinforcement) could be a potential hazard to workforce. Explosives handling. Rock conditions believed worse than at other sites	Similar traffic over a wider area of the site

Pollution	Oil may enter rock fissures in event of water pump failure. More water will require treatment before disposal. Hydrocarbon emissions. Water injection into bore holes may be required to maintain water-table	Possibility of leakage will be retained with impermeable bunds. Lower level of hydrocarbon emissions. Bunding is designed to accommodate the contents of one full tank. Major accidents involving more than one tank could lead to oil spilling outside bunds
Maintenance	Unacceptably hazardous to enter once sealed and filled with oil	Conventional maintenance on established procedures
Hazards	Explosive mixtures require controlling by inert gas blanketing, discharge of inert gas during filling	Possible intertank fire spread
Restitution and rehabilitation	Filled with seawater and sealed. Residual oil seepage into the sea through fissures may occur at very low rates over long periods	Tanks and pads can be removed and area rehabilitated, but terraces would remain
Major disaster (fire, explosion)	Possibly permanently out of commission given present technology. New facilities would be required following major accident	Repairable or replaceable on site

important fishery, seabird, seal colonies and statutory conservation areas. The environmental implications of storing crude oil in above-ground conventional steel tanks or underground caverns were considered in detail (Table 7.15) and the former selected although it was recognized that the various considerations (volume of rock and peat produced, emission of hydrocarbons, visual impact water and oil seepage, sabotage, health and safety etc.) could be weighted differently by different parties.

The Middle Zone, affect by pipeline corridors, landfalls, village and road construction, contained two SSSIs which would not be affected by construction. Sites of archaeological interest were surveyed and those in the Inner and Middle Zones were protected where possible. An emergency survey team was also organized in case any new sites were revealed by excavation operations.

There was also the disposal of large quantities of peat and sub-standard rock and moraine to be considered. The group recognized the difficulties in handling this material – the peat in particular – and finally recommended that it should be put in Orka Voe behind an armoured bund (Fig. 7.10). There it could neither escape to the sea nor could it flow and endanger livestock. A total of 3 million m^3 of peat and 7 million m^3 of rock debris have been placed behind the bund. This area and the wayleaves of the two pipelines have been rehabilitated with an improved pasture mixture which more than offsets the grazing land lost as a result of construction work. The area generally has been 'landscaped' and the buildings, tanks and units painted in such a way as to blend into the background against which they are set. Again SVEAG (1976) recommended detailed procedures for revegetation and rehabilitation of peat-disposal areas.

The second phase comes with the operation of the Terminal. Here the greatest impact would be felt in the marine environment. Recognizing the potential for oil spills from the handling of oil at the terminal and from tanker traffic in Yell Sound, SVEAG formed a working group to make recommendations on the type and extent of oil spill clean-up equipment which should be used in Sullom Voe (see below).

Tankers would arrive at Sullom Voe in ballast which may be 25–40% of the ship's carrying capacity. This is pumped ashore and the group made a recommendation that final oil levels in the treated ballast water should average 10–15 ppm, should not exceed 25 ppm and should be discharged through a diffuser to ensure a 100-fold dilution within 50 m of the discharge point. By commissioning hydrographic studies of the area the group was able to identify a suitable area for a diffuser that would ensure mixing with the surrounding water body. That area lies between Calbeck Ness and the island of Little Roe, where there is water movement at all states of the tide. The currents in this area are an offshoot of the main Yell Sound tidal stream. The local currents give a dilution capability in excess of 2000 : 1 – a figure

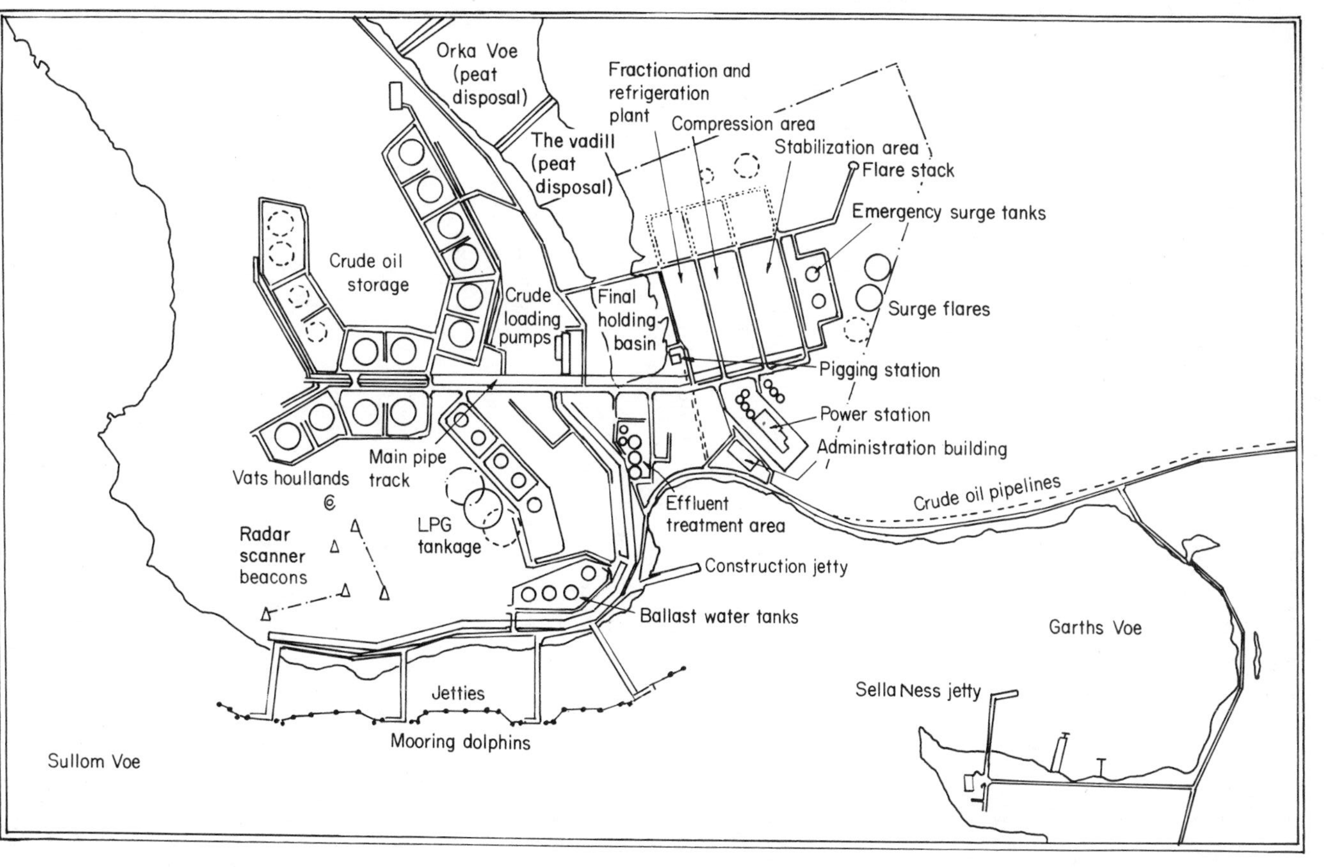

Fig. 7.10 Layout of the Terminal.

which assumes full throughput of the two pipelines and a 40 % ballast loading factor – thus ensuring adequate dispersion. The treated ballast water is at all times taken away by the tide from Sullom Voe into Yell Sound, where the dilution capacity is far greater.

7.4.4 Biological monitoring

SVEAG also set up ornithological and monitoring working groups which recommended that biological monitoring should take place in the area around the terminal. Seabirds, of course, receive wider attention because of their greater mobility and the potential danger of a tanker accident in the approaches to Shetland.

Baseline studies were recognized as an essential part of the monitoring programme. These studies have incorporated as wide a selection as possible of the habitat types of the area, and certain chemical parameters. In 1975 a start was made on monitoring the breeding success of the guillemot and gannet colonies on the National Nature Reserve at Hermaness, one of the most important seabird colonies in north European waters and, because of its proximity to oil-related activities, one which could be at greatest risk.

Winter 1975/1976 saw the start of the monitoring of over-wintering seaduck. The terminal area and the sea immediately east of Yell account for some 60% (6000 plus birds) of Shetland's over-wintering eider population. The same area is important for a number of other species. In 1976 baseline studies were commenced on rocky shores, soft shores, salt marshes, and benthos and lichens, the last being used as a technique for following changes in air quality. These studies are still taking place and will continue for some time into the future.

(a) The Shetland oil terminal environmental advisory group

Having advised on the environmental implications of the construction of the terminal, published its impact assessment, made recommendations on the types of oil-spill equipment that should be used at Sullom Voe, and having seen the start of the biological monitoring programme, SVEAG stood down. The group itself recognized that changes would have to be made in its constitution in order that it could operate successfully during the operation phase of the terminal. Since 1977, therefore, on-going environmental advice has come from a second group, the Shetland Oil Terminal Environmental Advisory Group (SOTEAG) which has one sub-group to deal with monitoring. The new group has an independent chairman. That he is one of the independent professors who sat on SVEAG has ensured continuity. SOTEAG has a wider range of environmental scientists, a great Shetland representation and local conservation bodies have a significant input to it.

The group's main function is to provide advice during the operating phase of the terminal. Through its monitoring sub-group SOTEAG will be responsible for the interpretation of data obtained from the ongoing biological and chemical monitoring programme.

(b) Planning for oil-spill clean-up

SOTEAG does not deal formally with oil pollution contingency planning, but has examined the ecological advice which has gone into the plan and upon which an oil spill clean-up strategy has been drawn up by the Sullom Voe Oil Spill Advisory Committee (SVOSAC). SVEAG originally envisaged the whole of Sullom Voe, Yell Sound and associated waters as forming a natural geographical area on which to base a contingency plan, and this view was followed by SVOSAC. In hindsight, following the events of the Esso Bernicia spill on New Years Eve, 1978, it has been shown to be a wise decision. An ecological survey (Syratt and Richardson, 1981) of the area was carried out to identify:

(i) Areas of water with biological interests, including fisheries, bird life, marine mammals, areas of potentially anoxic waters, shallow lagoons, etc.
(ii) Coastal types and habitats, including amenity beaches and harbours, salt marshes, soft shores, sand shores, rocky shores and shingle and sand ayres, spits, tombolos and bars.

The whole area divides naturally into four zones (Fig. 7.11). The first, Yell Sound North, is an area of relatively open water, flushed by a major tidal current and frequently experiencing heavy seas. Fisheries interests here are generally in deep water and for oil on water the use of dispersants is considered the most practical clean-up method, though it is recognized that there may be periods when containment and recovery could be carried out. Zone 2, Yell Sound South, contains many small islands and experiences one of the strongest tidal races in Shetland, reaching as much as 7–8 knots. Speed of reaction is essential in this zone and because of the very strong currents and numerous navigational hazards, dispersant spraying is the only practical technique for this zone. Zone 3 (Sullom Voe) and Zone 4 (Dales Voe complex) are more enclosed bodies of water, rarely receiving heavy seas. Both have a restricted water exchange and the extreme heads of the Voes have deep, still areas, often with a salt marsh complex. The head waters of Sullom Voe are 42 m deep and input of peat from run-off water is such that the sea bed and bottom layers of water undergo intermittent anoxic periods. Within these zones containment and recovery is recommended. In certain areas including the heads of the Voes, houbs and other shallow areas, it is recommended that dispersants must never be used. Around the jetties,

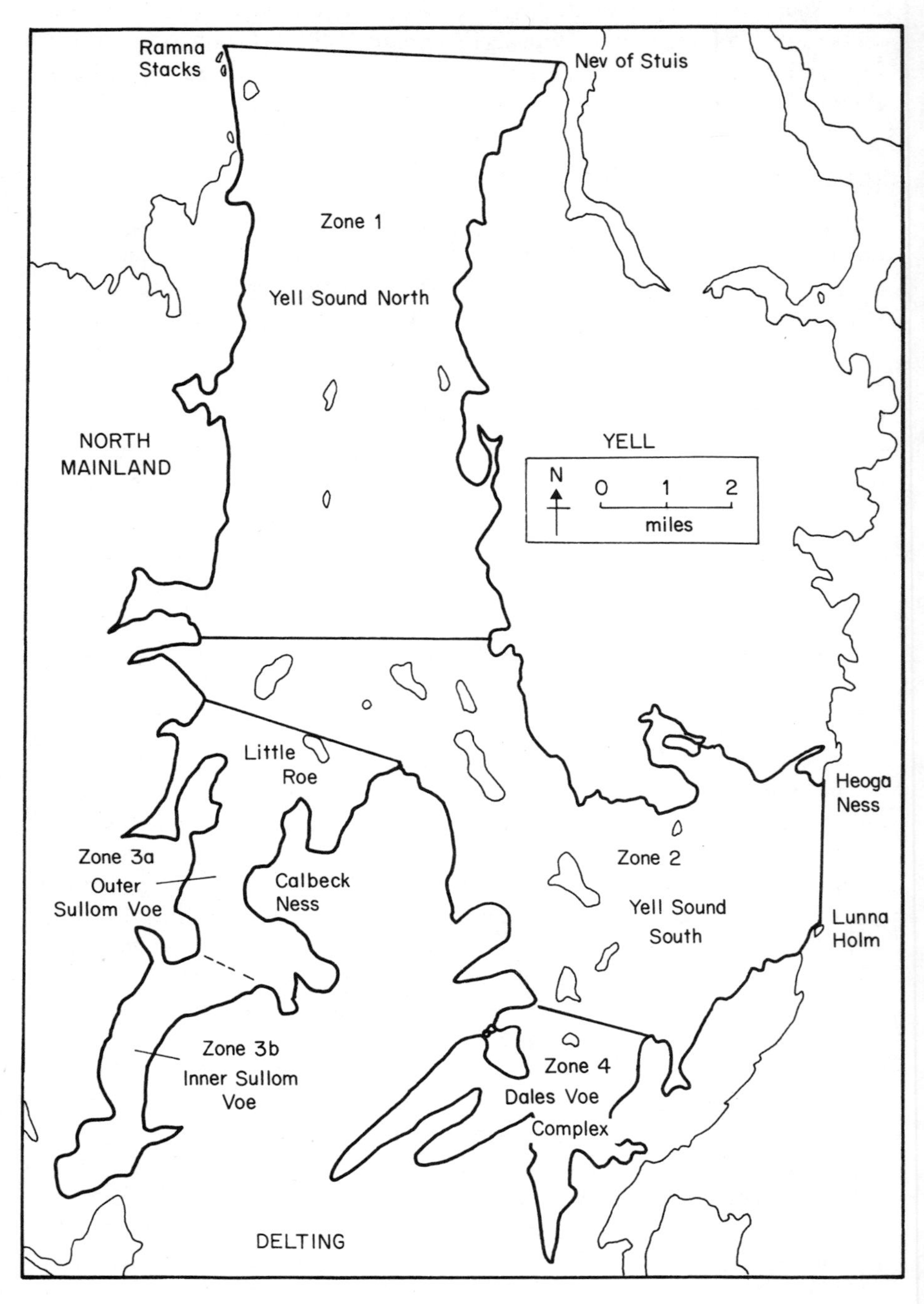

Fig. 7.11 Oil spill control area.

however, small handling spills will be treated with dispersant whilst on open water.

There is very little coastline in the oil spill control area that is classified for public amenity use and, therefore, very little that requires intensive clean-up techniques or the use of beach-sprayed dispersants. There are a number of salt marshes and species-rich soft shore associated with the houbs where the use of dispersants or heavy mechanical clean-up must be avoided, and a number of shingle and sand spits, ayres, tombolos and bars, so characteristic of Shetland's drowning landscape, where the use of heavy clean-up equipment would be similarly damaging.

Generally speaking, the strategy will be that if oil comes ashore it will be left to degrade naturally. Much of the coastline of the oil spills control area is rocky. About 50% of it is inaccessible from land, being more than a half mile from a road or below cliffs more than 10 m high. For Yell Sound North the inaccessible coastline is greater than 70%. The construction of a number of defensive booms to protect some of the more sensitive coastal areas in Sullom Voe is currently nearing completion.

It should be stressed that the oil spill plan catered specifically for crude oil spills. Sullom Voe is not a bunkering port and the risk of a fuel oil spill was considered extremely low. Nevertheless the first spill at Sullom Voe was just that, 1174 tonnes of Bunker C. A number of important points were raised by this spill not least of which is 'can oil spilled in the vicinity of the jetties be contained within Sullom Voe?' A system of spur booms appeared to offer the most practical solution and has been built.

Contingency plans for the land lines, which only total some eight miles have also been prepared. These follow standard clean-up practices. There are no areas of biological interest on or adjacent to the way-leaves, nor are there any major courses or standing bodies of water.

(c) Industry's own input – the environmental officer to the project in Shetland

Shell UK Exploration and Production were the original constructors of the terminal, but BP Petroleum Development Limited assumed that responsibility in November, 1975 and are the present terminal operators.

In 1976 BP posted a resident ecologist to Shetland as part of the Sullom Voe construction team. This was a natural and logical extension of the activities of BP's Environmental Control Centre (the body that is responsible for the BP Group environmental policy, recommendations, advice and assistance).

The role of the resident ecologist, since his posting, has been to:

(i) Co-ordinate the biological monitoring programme in Shetland and to make sure Industry's requirements are being met and that monitoring groups are properly serviced in the field.

(ii) Advise and assist the project on environmental matters, action to be taken, etc.

(iii) Make sure construction proceeds with due regard to the environment.

(iv) Act as focal point for environmental matters e.g. enquiries by the Local Authority, local conservation bodies, the public, etc., and representation on advisory committees.

His role has assumed a growing importance. Ecological considerations for oil spill plans were drawn up in consultation with local conservation interests, a suitable position for the ballast water discharge diffuser was located and accepted and rehabilitation of the pipeline routes was supervised. In addition, the biological monitoring baseline studies have been progressing. Now that the terminal is operating, his role may have changed, but not diminished. It is now his role to ensure statutory obligations are being met, as well as ensuring those aspects previously mentioned.

7.4.5 Conclusions

By any standards Sullom Voe is a large industrial development of world significance. Different countries have moved in different ways towards developing a method of analysis of the impact of industrial developments on the environment and the measures that may be taken to control it. The co-operation effort involved first with SVEAG and later with SOTEAG and SVOSAC has a number of advantages:

(a) It leads to a pooling of knowledge, expertise and resources.

(b) It draws together existing data at a very early stage and identifies areas where data are lacking.

(c) It allows collaborative discussions of ideas and procedures.

(d) It allows the understanding of problems from all angles, including commercial vis-à-vis environmental considerations.

(e) It leads to a unity of purpose and a breakdown of suspicions.

(f) It enables the project to be followed on a continuing rather than a one-off basis.

Within this context it is particularly significant to note that various bodies have their own 'environmental officers' in Shetland: the Nature Conservancy Council has an assistant regional officer; the RSPB has a representative, there is a local fisheries officer of the Department of Agriculture and Fisheries for Scotland and an agricultural adviser. The presence of these persons has meant that the Industry ecologist has not had to work in isolation and has, as a result, been able to operate more effectively: from the point of view of the Advisory Groups, it has meant that a great deal of first hand information has been available.

The experience gained from the activities of SVEAG and SOTEAG has shown that the collaboration of representatives of industry, local government, central government bodies, statutory conservation agencies and independent scientists is a practical way of approaching the problem of environmental impact analysis.

7.5 THE ECOLOGY OF OIL DEVELOPMENT IN SCAPA FLOW, ORKNEY

A. M. Jones

7.5.1 Introduction

The discovery of North Sea Oil in the early 1970s led to the search for suitable sites for the development of onshore facilities required for the exploration of this resource. Following the discovery and proving of the Piper Field, Occidental of Britain requested planning permission to develop part of the island of Flotta in the natural harbour of Scapa Flow, Orkney, as an oil storage and shipment terminal. Orkney Islands Council, therefore, asked the University of Dundee to prepare an environmental impact assessment on their behalf. The aims of this EIA were:

(a) To identify the nature and sources of potential pollutants and to put these in the context of any pollutants already present.
(b) To identify sensitive habitats/biota/ecosystems within the area of potential pollution.
(c) To review critically the locations and standards of emission of aquatic pollution.
(d) To recommend appropriate clean-up methods and to emphasize any problems.
(e) To recommend any further studies which might be necessary.

The proposal was a unique development in terms of scale and potential impact within the Orkney area, and the novel nature of the proposal restricted the use of traditional approaches to EIA. These limitations were added to by the short time scale available if EIA studies were not to cause undue delay in the planning process. This paper describes the role of the environmental impact assessment in this development and pays particular attention to the techniques adopted for environmental evaluation in response to the uniqueness of the development and the limited time available.

7.5.2 The nature of the development

The proposed development comprised a crude oil reception facility (degassing, storage tanks and ancilliary facilities) on the Golta Peninsula of Flotta (Fig. 7.12) and a loading facility initially comprising two Single Point Mooring towers (SPMs) located in Scapa Flow (Fig. 7.12). The SPMs were to be supplied by two sea-bed pipelines from the terminal. The terminal

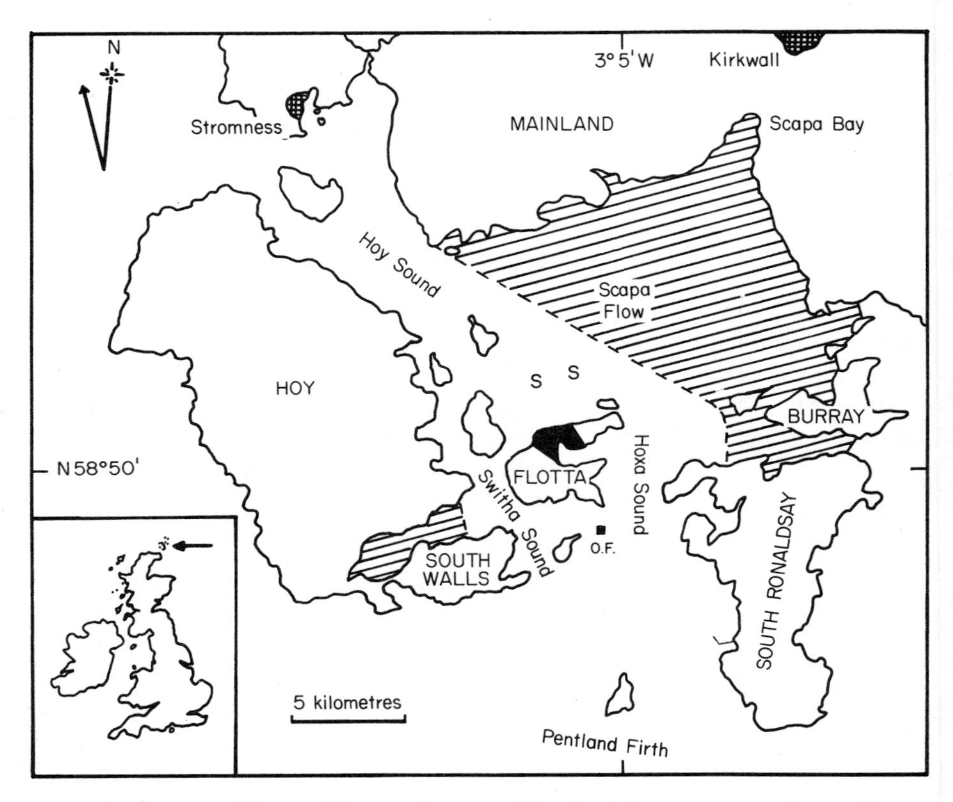

Fig. 7.12 Map of Scapa Flow study area. The shaded areas are regions with a restricted tidal circulation (S = single point mooring tower; O.F. sea outfall from treated ballast-water plant).

received crude oil from the Piper and subsequently, the Claymore Fields by a seabed pipeline from the production platform some 130 km east of Wick; the entire system is represented diagramatically in Fig. 7.13.

The Flotta development had several clearly definable potential sources of pollution namely: (a) spillage of oil due to major tanker accidents etc.; (b) spillage of oil due to accidents during loading operations at the SPMs;

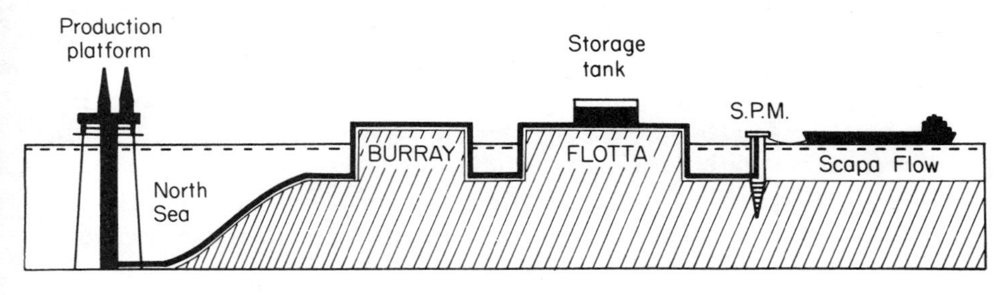

Fig. 7.13 Schematic representation of the oil reception, storage and trans-shipment operation of the Flotta terminal.

(c) release of treated ballast-water from the outfall (O.F. Fig. 7.12); (d) release of segregated ballast-water from Liquid Petroleum Gas Tankers.

7.5.3 Environmental assessment

The area of potential impact comprises some 300 km² of water and has a coastline of some 220 km (Fig. 7.12). The flora and fauna of the area were virtually undescribed except for the bird and sea mammal populations; the bird populations of Scapa Flow are recognized as being of national and international importance. The physical characteristics of the area were poorly known also, tidal information being from pre-second World War studies before the construction of the impermeable barriers between the eastern islands; these barriers prevent tidal exchange from the east, a feature which modified tidal circulation in Scapa Flow. The need for a rapid assessment demanded that field surveys had to be relatively brief and tidal and biological investigations were carried out in parallel (Jones and Stewart, 1974; Buller *et al.*, 1974). Tidal studies concentrated upon the nature of the circulation and upon the movement of potential oil slicks originating from the SPMs while biological studies were aimed at assessing the principal biological components and their broad distribution patterns within Scapa Flow and its approaches.

The principal conclusion from the tidal study (Buller *et al.*, 1974) was that the north-eastern section of Scapa Flow has a restricted circulation with a long residence time: such an area is clearly susceptible to pollutant accumulation resulting from chronic pollution. This feature (Fig. 7.12), combined with the dominance of winds tending to move spilt oil into this north-eastern region, poses one of the major environmental hazards resulting from this development.

The biological study (Jones and Stewart, 1974) showed the dominance of diverse littoral communities flourishing in the sheltered waters of this inland sea. The bottom fauna comprised mainly mixed tunicate, echinoderm and mollusc communities typical of bottom substrates composed of coarse stones and shell debris. The shellfish component, once important, has recently become less important due to the fishing-out of this region but it is still used as a foul-weather fishery and the fishing element was still relevant. The bird populations were already well known and were clearly the faunal component most directly at risk from surface oil slicks.

The EIA was able to ensure that the treated ballast-water effluent was released into the Pentland Firth tidal system where dilution and dispersion renders hydrocarbon concentrations negligible and where it is safely removed from the Scapa Flow system.

The segregated ballast-water release was made subject to specific harbour regulations ensuring that this is fully mixed water from adjacent sea areas, thus nullifying dangers due to the potential release of polluted estuarine waters or the transporting of alien species to Scapa Flow.

Table 7.16 Some of the factors influencing the fate and effect of spilt oil in the environment

Quantity of oil spilt
Type of oil
Chemical composition of oil
Nature of the oil's introduction to the environment
Air and water temperature
Water movements and tidal circulation
Depth of water
Climatic conditions (wind, precipitation, sunlight, etc.)
Duration of exposure to the atmosphere or hydrosphere

The chronic spillage of oil during terminal operations remained a potential threat, as did the potential catastrophic spillage from a major tanker accident. The principal threat of pollution, therefore, was spillage of North Sea crude oil on to the surface of the water where it posed its principal threats to the bird populations and to the littoral flora and fauna. The fate of spilt oil depends upon an array of complex variable factors (Table 7.16) and prediction of the impacts of oil spillages can only be made in the most general context. The fate of hydrocarbons and their potential environmental effects have been well reviewed by Blumer (1972) and Malins (1977) and it was clear that a monitoring programme should be instigated to observe environmental effects and to make appropriate recommendations for their mitigation.

The final role of the EIA was to evaluate the various clean-up techniques applicable to this area and to identify sites needing special protective measures in the event of an oil spillage. The restricted circulation in north-eastern Scapa Flow renders the use of chemical dispersal inadvisable since this does not remove oil from the environment; in fact, it makes it more readily available to the food web. However, the threat to major bird populations makes the use of the rapid chemical dispersal desirable under some conditions. The conflict between the protection of birds and amenity areas by the use of chemical dispersal techniques and the potential environmental dangers of this process to the food web and to fisheries in relatively still and shallow waters is evident and requires that each spillage be evaluated separately: no simple formula can be applied in this area and the success of anti-pollution schemes depends upon co-operation between scientific advisers and the authority responsible for clean-up activities.

7.5.4 Conclusions

The success of the environmental impact assessment in Orkney (Jones and Stewart, 1974) was founded on several important aspects of its application. Probably the most important was the liaison between the local authority and the University of Dundee, a relationship established at the outline planning permission stage so that all proposals could be considered at an early stage and modified as necessary. The second feature was the close relationship maintained with the developer so that problems and queries could be considered and solved by discussions between appropriate staff and their advisers. These features generally facilitated the planning stages while ensuring that each proposal had been considered in terms of its environmental impact.

The preparation of an EIA can be approached from two viewpoints. The use of a matrix analysis is frequently favoured but tends to produce complex analyses with little attempt at integration and rationalization of the conclusions. A comprehensive knowledge of the environment and of the pollutants' environmental pathways is also vital if the detailed matrix approach is to be considered valid; a matrix can only be as good as its information quality. The rapidly developing Orkney situation with only a single significant pollutant and a very rudimentary level of environmental detail available, favoured the use of a critical-path approach; here emphasis of the study is placed on those components clearly most directly at risk as determined from previous studies of the impact of oil pollution. Perhaps the most important feature, however, was the early recognition that many of the variables, not least the frequency and amount of oil spilt, cannot be predicted and that an essential component was to establish a monitoring programme (Jones, 1980) to evaluate the EIA recommendations and to observe continuously the development of any potentially adverse environmental trends. This programme began in Orkney in June, 1974, thus providing a three year baseline period, a feature made possible only by the request for an EIA at an early stage of the development.

The pressure for such developments in relatively remote parts of northern Britain has highlighted the deficiency in understanding of vital features such as the persistence, degradation rate and toxicity of North Sea crude oils under the climatic regimes of these regions. It is only through the activities of monitoring programmes that such observations may be made and thus fed back into future development and operational plans. An EIA can only provide predictive capability based on the state of the art; it is essential that this be continuously improved if future EIAs are to become more accurate and realistic. The EIA and its second phase, the monitoring programme, in Orkney has provided an improved background knowledge and understanding of this area against which any future developments can be more precisely calculated.

Acknowledgements

I wish to thank Orkney Council for their support in the studies reported here.

7.6. ASSESSING THE IMPACT OF MAJOR DEVELOPMENTS ON WATER RESOURCES

D. Hammerton

7.6.1 Introduction

It is perhaps not generally realized that one of the first and most painstaking studies of environmental impact in the United Kingdom was carried out by the Royal Commission on Sewage Disposal which was appointed in 1898 to report on methods for the treatment and disposal of sewage and industrial wastes. When this commission completed its work in 1915 it had produced no fewer than nine comprehensive technical reports which laid the foundation for modern water pollution control practice not only in this country but also in many countries throughout the world.

The 8th Report of the Commission, published in 1912, was of fundamental importance because it defined for the first time a set of scientific standards and tests to be applied to sewage and sewage effluents discharging to watercourses. In particular the now almost universally applied 'Royal Commission Standard' for sewage works effluents was recommended in this report. It gained wide acceptance not only because it bore the stamp of a Royal Commission but also because it was based on sound scientific principles.

It is instructive here to re-examine the findings of the Royal Commission because there can be little doubt that our present day concept of environmental quality objectives can be traced almost directly back to their 8th Report.

Table 7.17 Classification of rivers in accordance with their visible degree of cleanness, based on riverside observations under normal summer conditions. (From 8th Report of Royal Commission on Sewage Disposal, 1912)

Observed condition of river water	*Very clean*	*Clean*	*Fairly clean*	*Doubtful*	*Bad*
Suspended matter	Clear	Clear	Fairly clear	Slightly turbid	Turbid
Opalescence	Bright	Bright	Slightly opalescent	Opalescent	Opalescent
Smell on being shaken in bottle	Odourless	Faint earthy smell	Pronounced earthy smell	Strong earthy wormy smell	Soapy, faecal or putrid smell
Appearance in bulk	Limpid	—	Slightly brown and opalescent	Black looking	Brown or black and soapy looking
Delicate fish	May be plentiful	Scarce	Probably absent	Absent	Absent
Coarse fish	—	Plentiful	Plentiful	Scarce	Absent
Stones in shallows	Clean and bare	Clean	Lightly coated with brown fluffy deposit	Coated with brown fluffy deposit	Coated with grey growth and deposit
Stones in pools	Clean and bare	Covered with fine light brown deposit	Lightly coated with brown fluffy deposit	Coated with brown fluffy deposit	Coated with brown or black mud
Water weeds	Scarce	Plentiful Fronds clean except in late autumn	Plentiful Fronds brown coloured in places	Plentiful and covered with fluffy deposit	Scarce
Green algae	Scarce	Moderate quantities in shallows	Plentiful in shallows	Abundant	Abundant in protected pools
Grey algae	—	—	—	Present	Plentiful
Insects, larvae, etc.	—	—	—	Plentiful in green algae	Abundant in green algae

Table 7.18 Average analytical figures for river waters above and below sewage outfall. (From Royal Commission on Sewage Disposal, 8th Report, 1912)

Observed condition of stream	*5-day BOD 18·3 °C**	*4 hours OA from n/80 $KMnO_4$ at 80 °F*	*Ammoniacal N*	*Album-inoid N*	*Nitric N*	*Suspended solids*	*Chloride as Cl*	*Oxygen in solution when analysed*	
	ppm	*ppm*	*ppm*	*ppm*	*ppm*	*ppm*	*ppm*	*ml/l*	*ppm*
Very clean	1	2	0·04	0·1	0·5	4	10	7·8	11
Clean	2	2·5	0·24	0·25	2	15	25	6·5	9·3
Fairly clean	3	3	0·67	0·35	2·2	15	30	6·2	8·6
Doubtful	5	5	2·5	0·6	5	21	50	4·6	6·6
Bad	10	7	6·7	1·0	4	35	>50	Low	Low

* Usually referred to as Royal Commission river classification

7.6.2 River classification and the 'Royal Commission Standard'

In a comprehensive series of investigations the Royal Commission on Sewage Disposal first carried out a detailed study of water quality in a number of polluted rivers and used the results to propose a classification of river waters 'in accordance with their visible degree of cleanness based on riverside observations under normal summer conditions' (Table 7.17).

The next step was to correlate the physical and biological qualities with average analytical values for a range of chemical parameters as shown in Table 7.18.

The Royal Commission took the view that the five day Biochemical Oxygen Demand (BOD) test was the most reliable chemical index of river water quality and the BOD figures in Table 7.18 became known as the 'Royal Commission river classification'.

Table 7.19 Royal Commission on Sewage Disposal; 'limiting figures' for river waters

Test	*'Limiting figure', which, if exceeded, would probably result in nuisance ppm*
5-day BOD (18·3 °C)	4
4 hours N/80 permanganate value	4
Albuminoid nitrogen	0·45
Suspended solids	15
Dissolved oxygen when drawn	4 ml/l (5·7 ppm)*

* About 60% of saturation at maximum summer temperature (13·8°C)

Stemming from this classification the Royal Commission then calculated a set of values – termed 'limiting figures' – which if exceeded might give rise to nuisance (e.g. bad smells) and evident pollution as shown in Table 7.19. These limiting figures, it will be noted, fall between the values for the 'fairly clean' and 'doubtful' classes in the preceding table.

Klein (1957), to whom readers are referred for a more detailed consideration of this topic, points out that the BOD value of 4 ppm in this Table is very important because from this value the Royal Commission derived the 'normal' BOD standard which was recommended for the discharge of sewage effluents to inland rivers (Table 7.20).

Thus, if one assumes that a river water has a BOD value of 2 ppm immediately upstream of the sewage works effluent it can be calculated that, if the dry weather flow (DWF) of the stream is eight times the flow of the effluent then the effluent must not exceed 20 ppm BOD if the final mixture of stream effluent is not to exceed the limiting value of 4 ppm at 65 °F.

A 20 : 30 standard has now for many decades been the accepted standard for sewage works effluent in this country even when the dilution greatly exceeds 1 : 150 during dry weather. More stringent standards are normally imposed when the dilution is less than eight times the volume of effluent. The enormous progress which has been made in the last three decades in improving the quality of our inland rivers is of course very largely the achievement of the River Boards and their successors in England and Wales and of the River Purification Boards in Scotland which made full use of the legal powers in the *Rivers (Prevention of Pollution) Acts* of 1951 and subsequent legislation. At the same time there can be no question that – as far as sewage pollution is concerned – the authoritativeness of the Royal Commission

Table 7.20 Royal Commission on Sewage Disposal Standards recommended for sewage effluents discharging to inland streams

		Standards ppm	
Type of standard	*Volumes of clean diluting river water available*	*Maximum permitted BOD** (5 *days*, 65°*F*)	*Maximum permitted suspended solids*
Normal	8–150	20	30
Relaxed	150–300	—	60
	300–500	—	150
	Over 500	—	—
Strict	Less than 8	Standards not defined but to be decided on basis of local circumstances	

* Now modified to 20°C

standard was an extremely important factor and it is a great tribute to the thoroughness and rightness of the Commission's work that the standard, in its modified form as used today, still commands very wide acceptance.

Today, of course, standards for sewage works effluents especially in industrial areas cover many more parameters than BOD and suspended solids. In particular the consent conditions, especially where the volume of effluent is substantial, may include conditions covering ammonia, phenols, toxic metals and cyanide while consents for industrial discharges will include a much wider range of parameters.

In every case, however, the aim is to achieve satisfactory stream conditions which take account both of the self-purifying capacity of the receiving waters and of the use or uses to which it is being put. This requires that each discharge be treated on its merits according to the dilution it receives during low flow conditions, according to the pressure of other discharges in the vicinity and according to the subsequent uses of the river.

7.6.3 Modern practice in water pollution control

Let us now consider how the impact of complex major discharges are assessed and controlled by reference to actual cases in the West of Scotland – one in the freshwater environment and one in the marine environment.

The Clyde River Purification Board controls an area of 14 000 km^2 containing half the population of Scotland and approximately two thirds of its heavy industry. Within this area all discharges to inland waters and most discharges to the coastal waters must be licensed by a consent procedure under the *Rivers* (*Prevention of Pollution*) (*Scotland*) *Acts* of 1951 and 1965.

When the board was first set up 25 years ago it was severely handicapped in carrying out its functions because of an almost total lack of data concerning water quantity and quality: very few river gauging stations existed and no chemical analytical data was available, while biological monitoring had not even been thought of.

One of the first priorities of the new Board – in common with River Boards throughout the United Kingdom – was to set up a comprehensive monitoring network. Today the Board has 55 automatic river flow gauging stations, about 500 water quality sampling stations and 200 biological sampling stations. In marine waters the Board has about 100 routine sampling stations where water, sediment, biological and microbiological samples are collected. During 1979 over 12 000 samples were collected by the Board's staff for the purpose of recording environmental quality. In addition many samples are taken of sewage and industrial effluent discharges to ensure that these comply with the consent conditions.

Thanks to its comprehensive monitoring network, the availability of reasonable long and detailed records plus access to a vast research literature and above all the experience of its staff the board today is in an excellent position to assess the likely impact of new developments and to put realistic consent conditions on new discharges. Two examples will now be considered firstly in the marine environment off the Ayrshire coast and secondly in the lower reaches of the river Clyde.

(a) Example 1. A new marine outfall for ICI at Stevenston, Ayr

For many years the ICI Organics Division and the Nobel Explosives Company have discharged industrial wastes amounting to 15 000 m^3/day via 10 outfalls in the Garnock Estuary and 2300 m^3/day to Irvine Bay (Fig. 7.14). Because of the lack of dilution in the estuary and the long residence period these discharges caused serious pollution which, although not stopping the

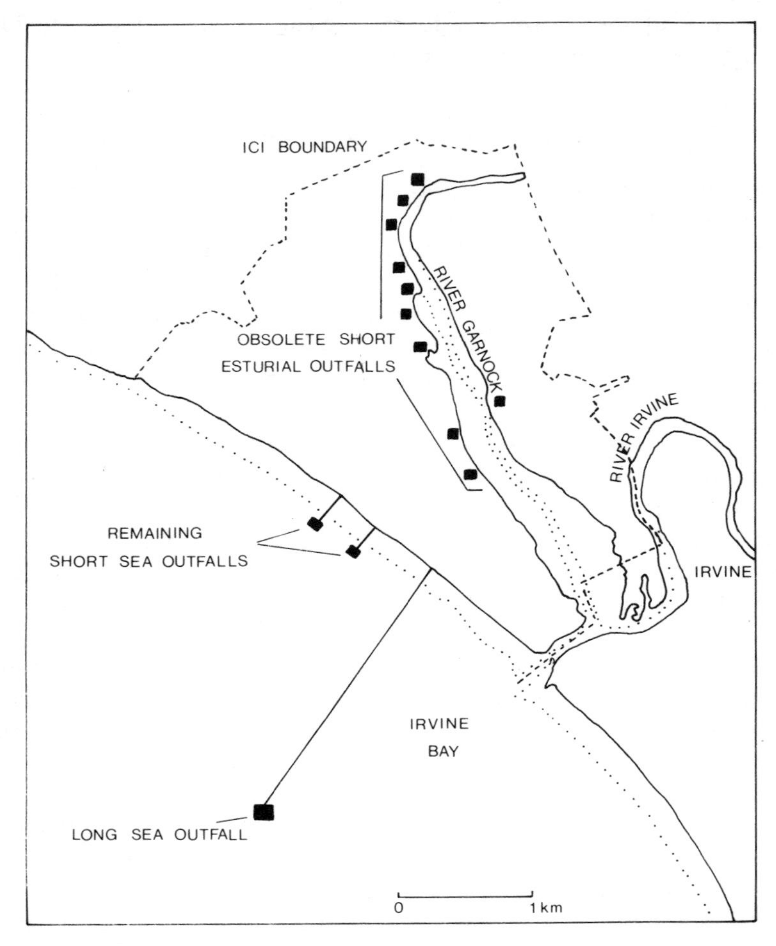

Fig. 7.14 Location of new ICI outfall, Irvine Bay.

passage of migratory fish, eliminated all macrofauna in the vicinity of the outfalls and produced a general biological impoverishment of the estuary. Particular problems resulted from the discharge of strong acids, toxic concentrations of ammonia and high levels of phosphate and nitrate. Furthermore the sediments near the outfall were contaminated with silicones, pigments and metals especially iron, copper, lead, and zinc.

In 1975 the CRPB began detailed monitoring of the area with a view to understanding the biological and chemical processes prior to revising the consent conditions. Shortly following this the company approached the board with firm proposals to collect all the existing discharges to the estuary into a single long sea outfall extending 2000 m into Irvine Bay. At the point of discharge a diffuser was to be provided which theoretically would give an initial dilution of ×200. The scheme required a new land interceptor sewer through which the combined effluents would be pumped into the outfall sewer. There were also to be trapping facilities for oils and immiscible solvents and screens to reduce suspended solids. The company also provided a suggested list of consent conditions which would provide for the present discharge together with some spare capacity for future growth.

The Board's task then was to evaluate this proposal and determine consent conditions which would fully safeguard the water quality of Irvine Bay without at the same time putting ICI to excessive expenditure.

(b) Irvine Bay studies

The first step was to determine whether Irvine Bay, which already receives various sewage and industrial discharges, had adequate spare capacity to receive the proposed discharge. Fortunately the Board had available the results of a £100 000 environmental survey which had jointly been carried out by CRPB and ICI Brixham on a contract basis for Strathclyde Regional Council in connection with a major municipal outfall. In addition following desk and pilot sewage, long-term routine monitoring was initiated and a literature survey was carried out.

A study of the various inputs to the Bay was essential and analysis of the river, sewage effluents and industrial effluents were carried out. Results from the four principal rivers showed that these were the main source to Irvine Bay of suspended solids, SiO_2, Mn, Ni, Hg and DDT/DDE.

Analysis of the principal sewage effluents showed these to be the main source of NH_3, PO_4, Cr, Zn, Cd, Dieldrin and alcohol. The major industrial effluents provided the main source of BOD, NO_3, Cu and Pb. Atmospheric inputs were assessed by means of a literature review but not determined directly.

Hydrographic surveys and routine monitoring showed that although the Bay was subject to only weak tidal currents with retention times sometimes in excess of four days the levels of pollution were generally low and for most determinations the bay had a spare capacity to take additional loads given proper siting and adequate dilution.

Having shown that Irvine Bay had adequate spare capacity the next step was to calculate the consent conditions by means of environmental quality objectives from which water quality standards could be determined.

However, the first essential was to ensure that the discharge point was correctly sited as the company needed early agreement on this in order that final designs could be approved and the contracts let. The principal factors here were (a) avoidance of existing outfalls to prevent 'hot spots', (b) the avoidance of river mouths because of the danger of tidal incursion, (c) the protection of beaches and (d) adequate initial dilution. The Board's view was that the company had chosen an outfall point which was satisfactory from all points of view and that no change would be required.

The final stage was the calculation of maximum permissible levels for each of 19 parameters. For Irvine Bay the following environmental quality objectives were adopted.

(1) To safeguard all legitimate uses and especially fishing interests:
 (i) No significant deterioration in water quality outside the mixing zone in respect of:
  (a) toxicity
  (b) tainting
  (c) bioaccumulation
  (d) eutrophication
 (ii) Protection of breeding grounds
(2) To safeguard aesthetic quality:
 (i) No surface slick or discolouration
 (ii) No sub-surface field
 (iii) No deposition of solids

The method used to calculate the maximum permissible level for each pollutant in the effluent was as follows: for a pollutant A in situation Z (i.e.

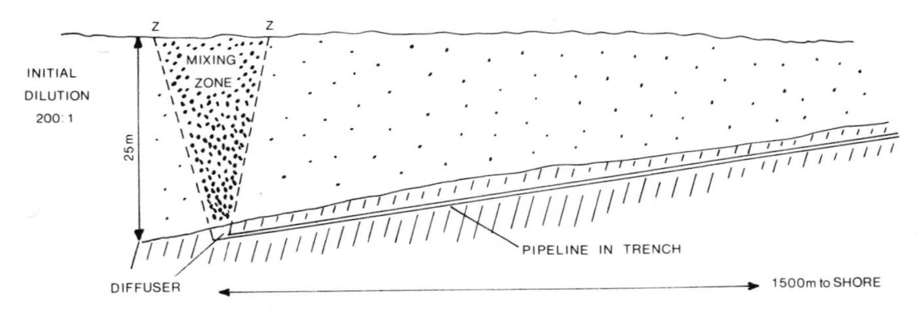

Fig. 7.15 Determination of consent conditions.

the open sea outside the mixing zone) (Fig. 7.15) an environmental quality standard (EQS) was set to accord with the desired environmental quality objective. If, for example, the environmental quality standard (EQS) for pollutant A is 2·5 mg l^{-1} and the maximum existing background level is 0·5 mg l^{-1} then the required consent condition – assuming an initial dilution in this case of 200 : 1 – becomes $(2{\cdot}5-0{\cdot}5)\times 200 = 400$ mg l^{-1}.

Fortunately for many of the components in the effluent the limits proposed by ICI were substantially below the levels at which harmful or toxic effects would be apparent and relatively few of the proposed conditions had to be tightened up by the Board.

It should also be noted that there are several safety factors at work here: in the first place the chosen EQS has an inbuilt safety factor; secondly the consent condition is an upper limit and a lower figure should normally be achieved while thirdly there will be progressive dilution as the effluent disperses away from the mixing zone.

Table 7.21 Consent conditions for ICI effluent

Parameter	*Consent condition*
Flow	max 5 MGD (dry weather basis)
Location of discharge point	NG co-ordinates at mid point of diffuser section: N636890 E228160
pH	1 to 10
Acid load (as $CaCO_3$)	< 80 tonnes/day
Dissolved solids	< 8000 g m^{-3}
Sodium sulphate	< 7000 g m^{-3} (additional to other dissolved solids)
Suspended solids	< 200 g m^{-3}
COD	< 4 500 g m^{-3}
BOD^5	< 1500 g m^{-3}
Nitrate (as N)	< 500 g m^{-3}
Ammonia (as NH_3)	< 200 g m^{-3}
Lead	< 2 g m^{-3}
Zinc	< 2 g m^{-3}
Copper	< 0·5 g m^{-3}
Chromium	< 0·4 g m^{-3}
Nickel	< 1·0 g m^{-3}
Mercury	< 0·1 g m^{-3}
Cadmium	< 0·02 g m^{-3}
Temperature	< 27 °C

The United States Environmental Protection Agency has issued a set of 'water quality criteria' for about 51 determinands but the CRPB follows a more flexible system which takes account of the following factors:

(i) Protection for mutagenicity/carcinogenicity, toxicity, persistence and bioaccumulation.
(ii) Existing levels in the receiving water and in unpolluted reference areas.
(iii) Degree of dilution, dispersal and removal in receiving area.

Toxicity values in this case were determined from the literature and also from static 'flow through' tests in the ICI laboratories using gobies, dabs and shrimps. These tests were carried out both with the whole effluent and with specific components such as nitroglycerine, ammonia and pH. These tests showed that fish could survive four days in the whole effluent with only

40× dilution compared with an initial natural dilution of 200 and between 11 000 and 28 000 after 4 days. Every care was taken to look at the effluent both in terms of its individual components and in terms of its net composition. The consent conditions finally approved by the Board and accepted by ICI were as shown in Table 7.21 (Hammerton *et al.*, 1980; 1981).

(c) River Clyde studies

During 1978–80 an industrial effluent treatment plant was constructed to treat the trade wastes from a large steelworks and discharge these to the river Clyde in admixture with domestic sewage from an adjacent municipal sewage works. In order to draw up a set of consent conditions which would fully protect the river (which is important as an amenity and as a trout fishery) a substantial laboratory evaluation of the effluent was carried out. In the event, because of legal problems, separate consents for the steelworks effluent and the sewage effluent have now been prepared but a brief account of the procedure for determining the original proposed joint consent conditions will serve to illustrate some of the problems involved.

In the setting up of consents for discharges to rivers the CRPB, in common with most River Boards and Water Authorities, seeks to achieve an EQO (environment quality objective) which will fully protect the rivers at what is termed the 95% exceedence flow. This is a much more scientific concept – based on long-term records – than the old method of using the so-called 'dry weather flow'. The 95% exceedence flow (Q95) is that flow which, on average, is exceeded on 95 days out of every 100.

In this particular instance the Q95 of the river Clyde is 6·1 $m^3 s^{-1}$ while the maximum dry weather flow of the combined effluent is 268 litres s^{-1}. Thus at the Q95 flow the effective dilution is 22·7 : 1.

From the outset it was evident that the concentrations of ammonia and cyanide in the effluent presented the main problem and if the treatment process could reduce these to an acceptable level the remaining component would be well within the Board's requirements.

In order to select the most suitable design of treatment plant a lengthy programme of laboratory and pilot-scale trials using effluent from similar works was carried out for the discharger and the results were made available to the Board. Simultaneously the Board's toxicity scientist carried out laboratory trials with the same effluent to determine its effect on rainbow trout. Regular meetings were held between the parties to ensure that realistic consent conditions could be designed, i.e. sufficiently stringent to fully protect the river under low flow conditions but not so stringent as to impose unreasonable costs on the discharger.

The difficulty in calculating a safe ammonia standard is that it is the proportion of ammonia in the unionized form rather than total ammonia or ammonium ion which is toxic to fish. According to EIFAC* standards and also the more recent EEC Directive on 'Quantity requirements for waters needing protection in order to support fish life' the maximum permissible level of unionized or 'free' ammonia is 0·025 mg l^{-1}, i.e. 0·025 p.p.m. The standard finally adopted for this effluent of 25 mg $litre^{-1}$ may, under certain circumstances, produce levels in the river in excess of this value. Thus, at Q95 and assuming the total ammonia upstream is 0·4 mg $litre^{-1}$ the theoretical downstream concentration when fully mixed will be 1·43 mg $litre^{-1}$. Theoretically this will provide the following concentrations of free ammonia:

At 15 °C and pH 7·5	0·85% of 1·43 = 0·012 mg $litre^{-1}$
at 20 °C and pH 8·0	3·83% of 1·43 = 0·054 mg $litre^{-1}$
at 24 °C and pH 8·75	24·9% of 1·43 = 0·356 mg $litre^{-1}$

At first this consent condition appears insufficiently stringent to fully protect the fishing interests. However, it is considered that no harm should result taking into account the following factors:

(1) The EIFAC limit of 0·025 mg $litre^{-1}$ has a tenfold safety factor and is based on a 96 hour exposure.
(2) The highest temperatures and pH values are likely to be experienced only for a few hours in the middle of the day in summer.
(3) Fish are mobile and are unlikely to remain for prolonged periods immediately downstream of the effluent.
(4) At high summer temperatures nitrification in the river will ensure a rapid decrease in ammonia levels.

With regard to cyanide the board imposed a standard which would produce a maximum concentration of toxic cyanide in the river below 0·01 mg $litre^{-1}$. A Royal Commission 20:30 standard for the sewage component of the effluent was considered fully adequate in view of the available dilution and the quality of the river upstream. The consent conditions proposed for the combined effluent were as shown in Table 7.22.

7.6.4 Discussion

There is a school of thought among some environmentalists which argues that controls on pollution must be so stringent that virtually no impact on the environment can be detected. This argument has reached its ultimate form in the 'zero discharge' principle now the subject of controversy in the USA. Although no doubt commendable as a utopian ideal, pollution control authorities would rapidly lose credibility if such a policy was adopted here. Enlightened pollution controls must aim to achieve a compromise between

* European Inland Fisheries Advisory Commission·

Table 7.22 Consent conditions for combined effluents

Parameter	*Consent condition*
5 day BOD	20 mg litre^{-1}
Permanganate value	30
Suspended solids	30
pH	6–8
Free and saline ammonia as N	25
Cyanide: ferro – and ferricyanide	0·75
other cyanides	0·075
total toxic cyanides	0·15
Thiocyanate (SCN)	5
Sulphate (S)	0·25
Total phenols	0·5
Zinc (Zn)	1·0
Lead (Pb)	1·0
Oil and tar	No visible traces
Standard Lc50 fish toxicity test on rainbow trout at 10 times dilution with river water	

unacceptable pollution and a completely unpolluted environment.

Thanks to the pioneer work of the Royal Commission on Sewage Disposal, which pointed us in the right direction, the cumulative experience of the water pollution control authorities and the expertise of the Water Research Centre, discharge standards can be tailor-made to the needs of individual situations with great confidence. However it must be conceded that the British method of controlling discharges by means of environmental quality objectives requires substantial skill and experience on the part of the controlling authority whereas fixed emission standards, as used in some of the EEC countries, are much simpler to operate and do not require highly qualified staff to evaluate each situation.

Although a system of fixed emission standards is relatively cheap to operate it is surely self-evident that such a system must cause an unacceptable degree of pollution where dilution is inadequate or impose excessive costs on the polluter where a lower standard would fully protect the environment.

During the last two decades enormous progress has been made in restoring the quality of our inland rivers and certain estuaries (notably the Thames and the Clyde) which had been blighted by the neglect of over 100 years. This progress has been largely achieved by persuasion and the willing co-operation of local authorities and industry, the powers of prosecution having had to be invoked on relatively few occasions. The rate of progress has, of course, been closely linked with the economic performance of the country and there can be no doubt that the deepening economic recession is severely reducing the rate of progress.

As this volume goes to press the government has announced a four year programme for implementing the *Control of Pollution Act* commencing in July 1983. The commencement of this Act has been severely delayed because of the economic recession and indeed has now been forced on the government because of the need to legalize discharges which otherwise would be in breach of various EEC directives and, in particular, the Directive on the Discharge of Dangerous Substances to the Aquatic Environment. In preparation for the application of this Directive the Department of Environment is now preparing Environmental Quality Objectives and Standards for a range of heavy metals in order to comply with the Directive on a uniform basis throughout the United Kingdom. The EQO approach is based on specific uses for water bodies so that Environmental Quality Standards will be defined for each particular usage such as potable water, recreation and amenity, support of game or coarse fisheries, irrigation and industrial supply. Where multiple use occurs obviously the most stringent EQSs would be applied. Although water authorities will still have the right to set the quality objectives and standards for waters for which they are responsible they will be strongly recommended to use these EQOs as guidelines. The author envisages that in the course of time nationally agreed EQOs will be available for most pollutants and that this will lead not only to a more uniform control but also to an overall improvement in the quality of our aquatic environment.

7.6.5. Conclusions

In conclusion there is always room for improving our system of environmental control and there can be little doubt that a more formalized method of environmental impact assessment, as proposed in the draft EEC directive, would have benefits even in the sphere of water pollution control where, as indicated in this paper, there has been a long history of development and refinement of control methods. Water pollution clearly only represents a part of the impact of any major industrial or urban development and therefore a fully integrated approach is highly desirable. While substantial progress in this direction has been made in our planning procedures there is, in the author's opinion, a strong case for adoption of new procedures on the lines of the EEC draft directive. The main danger is that of increased bureaucracy and of increased costs – although in some cases these might be reduced substantially, judging by the experience of British Gas and BP (see Syratt, Chapter seven).

The author therefore hopes that the Government will take a more positive approach to ensure that an EEC directive on environmental impact assessment, if adopted, would be sufficiently flexible to accommodate the best of our present procedures, strengthening these where necessary and allowing for various levels of impact assessment related to the scale and nature of the development.

7.7 ASSESSING THE IMPACT OF INDUSTRIAL EMISSIONS TO THE ATMOSPHERE

G. Diprose and N. W. J. Pumphrey

7.7.1 Introduction

In addition to the need to satisfy the statutory authorities the purpose of environmental evaluation is to enable decisions to be taken in a way that optimizes the use of resources. The resources involved are, on one hand, the capital which is used and the running costs which are incurred in controlling emissions to atmosphere; and on the other hand the atmosphere's capacity to disperse the emission and destroy its noxious components harmlessly, avoiding harm to humans or other parts of the environment.

When considering the impact of industry it must be recognized, that:

(a) It is not practicable to operate manufacturing processes without some emissions to atmosphere.
(b) These emissions may contain materials that would be noxious or offensive if they reached any target in too high a dose.
(c) Excessive costs in an industrial operation should be avoided by reasonable use of the atmosphere as a resource into which discharge can be made.

In return industry recognizes that it is not reasonable to discharge materials to atmosphere in a way which causes pollution where we define pollution as: 'The introduction by man into the environment of substances in such a way as to cause hazard to human health, harm to living resources and ecological systems, damage to structures or amenities or interference with legitimate uses of the environment.'

Thus proper assessment of the impact of industrial emissions enables manufacturing industry to steer a course between (a) 'Discharging to atmosphere in a manner which causes pollution', and (b) investing scarce and valuable resources needlessly in an effort to reduce a discharge well below the level where it can cause harm as judged by current knowledge.

The UK authorities responsible for controlling emissions to atmosphere have been, since their inception, aware of the need to balance *cost of control* with *effect of a reduction*. The Alkali Inspectorate is charged with the duty (among other duties) of ensuring that the 'best practicable means' are used to control emissions from a list of processes when noxious materials are handled and to ensure that 'such emissions of noxious or offensive materials as are necessary are made in an inoffensive manner'. The philosophy behind the UK law is reviewed by Nonhebel (1975) and the practice by Tunnicliffe (1975).

Having established some of the reasons why the impact of industrial emissions to atmosphere may need to be assessed, how is it to be done? In order to assess the impact of industrial emissions in a useful manner the overall problem is best divided into three main parts: (a) 'definition of the nature of emissions', (b) 'definition of the behaviour of the emissions in atmosphere', and (c) 'definition of the effect of the emission on any target; the human, animal or vegetable.' In the case of some emissions (such as SO_2, smoke, etc.) it may also be necessary to assess the magnitude of the background concentrations.

7.7.2 Defining the scope of the emissions to be considered

From a practical point of view there is likely to be this difference between the consideration of emissions from an existing plant and the likely emissions from a new plant. For a new plant all potentially noxious components in all emissions will have to be assessed. In contrast, for an existing plant the reason for undertaking an impact assessment is either that some specific emission is causing offence or a new appreciation of the toxicology of one material used on a works. Two examples of changing appreciation of the toxic effect of particular materials in the last decade have been vinyl chloride and mercury. Vinyl chloride is an example of a seemingly inoccuous material being shown to have subtle toxic effects. In the 1950s and early 1960s vinyl chloride was considered to be relatively non-toxic. Its acute toxicity is certainly fairly low, and it was at one time considered as a candidate for development as an aerosol propellant and even as an anaesthetic. Vinyl chloride's more subtle effect is its apparent capability of causing a particular form of cancer.

More interesting for its possible impact outside the work place is the heavy metal mercury. There is no doubt that the main anxiety concerning mercury has been its impact on the aquatic environment and its possible passage through the aquatic environment to man. But mercury is a uniquely volatile metal and the possible environmental impact of atmospheric emissions was a problem that faced ICI in the early 1970s and has to a greater or lesser extent occupied our attention since that time.

The background to the environmental mercury story is covered adequately elsewhere and there are some thousands of literature references to the environmental effects of mercury collated by Taylor (1976). Rather, ICI's experience with mercury at one site will be used to illustrate the difficulties and problems of the environmental impact assessment of atmospheric emissions of a pollutant whose behaviour in the environment has been closely studied.

7.7.3 Defining the emissions

One of the tools of environmental evaluation of emissions to atmosphere is mathematical modelling of plume dispersion, and for this a reasonable quantitative idea of the emissions themselves is required. In practice, there are many reasons why this is a much more difficult exercise than a casual glance at a flow sheet may suggest.

For a new plant it will be possible from the design flow sheet to obtain estimates of the composition and flow in all vents to atmosphere. However, even in seemingly quite simple processes subsequent measurement of flows and compositions are likely to show: (a) that there are large deviations from flow sheet values for normal operational reasons, i.e. load shedding, output variation to meet demand, etc., (b) that both flow and composition of material lost to atmosphere vary with time, and (c) commissioning and maintenance give rise to unexpected emissions.

On older operating plants there are likely to have been many changes, so that measurement of each emission is necessary. In addition, older plants may tend to have a multiplicity of discharge points. One very old process operated now for over 80 years is the electrolytic production of Cl_2 and caustic soda using the mercury cell. Mercury is used as the cathode in an electrolytic cell for producing chlorine from brine. A solution of sodium in mercury is also produced and this amalgam is decomposed with water in a separate cell to give (a) pure mercury which is recycled, (b) caustic soda for sale, and (c) hydrogen. Superficially this looks like a closed cycle for the mercury but there are ways in which mercury can reach the atmosphere.

The main sources of atmospheric emissions of mercury, accounting for perhaps 80–90% of the total were:

(a) The hydrogen. This is produced in the caustic cell or denuder and is saturated with mercury vapour. It is cooled to reduce the mercury concentration and then piped to a nearby power station where is it used for fuel. The flow of hydrogen depends on the cell room load which may vary over a 5:1 ratio during the day. The mercury content has proved very difficult to measure because it is near saturation at ambient temperature. It is discharged at very low concentrations in a buoyant plume from a 300 stack.

(b) Various sections at each end of the chlorine cell have to be flushed with air to prevent accumulation of hydrogen/air mixture. This stream, known as weak hydrogen, used to be vented to atmosphere through a short stack without treatment. This source is now treated to remove mercury.

(c) Ventilation air. In order to keep the atmosphere in the working area of the cell room tolerable, the cell room has a powerful ventilation system which moves many millions of cubic metres of air per hour. The air contains only a small concentration of mercury – less than the TLV of 50 $\mu g\ m^{-3}$. But because of the huge volume moved a significant proportion of the total loss goes by this route. Again it has proved difficult to define these losses due to variations in flow with wind speed, due to partial recycling of the air, and due to the large number of points of emission.

In addition to the main emissions of mercury, there were many more smaller emissions all of which need to be catalogued. These main gas streams have very different characteristics which effect the way they disperse. Ventilation air is emitted at low level, its buoyancy in the atmosphere is small. The weak hydrogen stream was similar but was emitted rather higher up. The strong hydrogen is mostly burnt in the power station and so the mercury in it is emitted to atmosphere in a buoyant plume at the top of a tall stack. However, some hydrogen used to be emitted at an intermediate level as an emergency vent. This should be very buoyant plume as it is almost pure hydrogen.

The mercury content of each stream was only established after very considerable effort and it is still not possible to obtain good measurements of gas streams which may be saturated.

I hope I have said enough to impress on you the difficulty of obtaining good data on which to base the next stage of the environment impact assessment.

The next question is: How do the emissions travel in the atmosphere from the sources to any possible target area? Once again there is a huge literature on this subject but there are some basic rules:

(a) The direction of travel is controlled by the wind.
(b) Whether the pollutant rises or falls in the atmosphere and hence the effect it may have at ground level depends to a large extent on its initial density relative to the surrounding air and to a lesser extent on its initial momentum.
(c) The rate at which the pollutant is diluted and spreads depends on the degree of turbulence arising from the buoyant movement of the plume through the air and from the thermal effects due to heat transfer from the ground to the atmosphere (due to sunlight warming the ground or works thermal losses) and due to airflow round obstructions, buildings etc.
(d) The rate of transfer of the pollutant to any surface depends on the degree of exposure of the surface and on the nature of the pollutant, whether it is gaseous or if particulate, on particle size.

From these basic ideas mathematical descriptions of the behaviour of plumes of contaminants in the atmosphere have been developed, for example by Pasquill (1974) in the UK.

These models now give a good description of the behaviour of plumes in simple cases but are as yet poor at following dispersion under low wind speed conditions or allowing for the effects of plume densities or of buildings near the source of the emission or of hills which could affect the path of the wind which carries the plume. Any feature which is large compared to the plume at that point in its travel will introduce an extra degree of turbulence or change of wind direction which will affect the plume in some way.

From the environmental evaluation view point an important aspect of the atmospheric transfer is to determine the way in which the dose arrives and over what period is it relevant to average the dose. This will, of course, depend on the target and how it is affected by whatever dose it receives.

Close to a source the mixing process within the overall envelope of a plume is incomplete and the atmospheric concentration recorded at a point with a fast response detector will show a high degree of variation of concentration with changes from peak to background within a few seconds. Further away from the source, turbulent mixing will have evened out many of the peaks and a more even concentration will occur with time. The ratio of peak concentration to average concentration will have reduced. This sort of effect is important for odour problems where the response time to a stimulus may be as short as a few seconds. More commonly in environmental evaluation longer periods are relevant. For a pollutant such as ethylene 1–8 hours is reasonable, whereas for fluoride one month to six months might be a sensible averaging period when the target is grazing animals. A 24 hour average period is usually taken to protect people with bronchitis/asthma for SO_2 and smoke.

For mercury, an average period of 30 days was considered sensible for measurements close to the works because (1) the potential problem was thought to be inhalation by humans, (2) the half-life of mercury in humans is believed to be 60–70 days. The Environmental Quality Standard for mercury outside the works was agreed to be a long-term average of $< 1\ \mu g\ m^{-3}$, which is now achieved.

Initially for modelling the dispersion of mercury around the source in Runcorn, ICI worked in conjunction with PPC Consultants Limited to validate their 24 average model. The input data for the calculation included the characteristics of all known mercury vapour sources and their relative positions and weather data recorded every hour on site. An example of the correlation between model and monitor is shown in Fig. 7.16 for a monitoring point near the work's boundary.

The points of interest are (a) that a reasonable overall correlation was obtained between PPC Consultants' predictions and our measurments, but (b) the measured minimum values of mercury never reach as low as calculation predicts. This illustrates a further difficulty with trace analysis.

Chemical contamination of environmental samples is often a problem.

Despite the protestations of analytical chemists it does still seem to be extremely difficult to measure environmental contaminants at a low concentration reliably and accurately on a routine basis. It was only by sampling and analysing in parallel with another laboratory and finally by changing the analytical method that we were able to eliminate spurious results. The combination of source quantification, mathematical modelling and monitoring enabled us to demonstrate which sources were contributing to the atmospheric concentration at any point around the works. The exercise

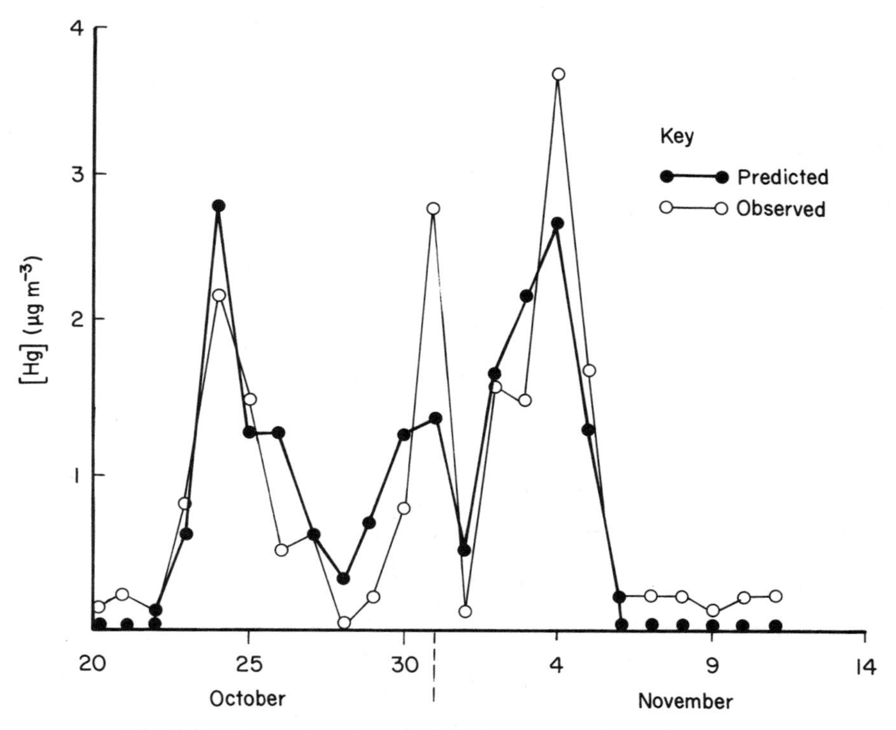

Fig. 7.16 Observed and predicted air concentrations of mercury.

showed that in our case the mercury discharged at a high level contributed virtually nothing to the ground level concentration near the works. Of the other two main sources of atmospheric mercury one, the weak hydrogen system, was scheduled for treatment because the PPC model suggested it contributed substantially to the mercury concentration outside the works and because it contained a relatively high concentration of mercury vapour. The other source, the ventilation air, was not treated directly. This was because the mercury concentration in the emission was low and the environmental impact assessment showed that this emission on its own would not cause the environmental quality standard to be exceeded.

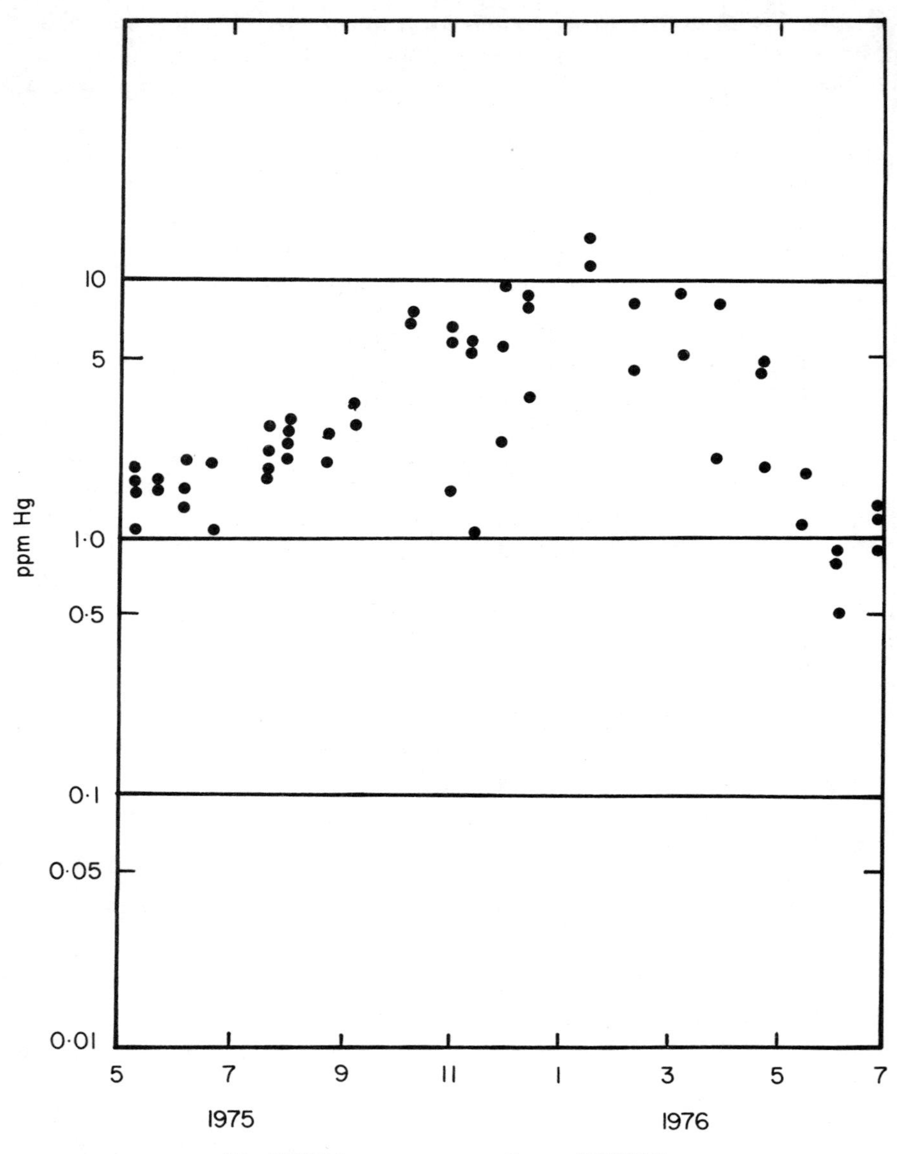

Fig. 7.17 Mercury content of grass (1975/76).

Because our significant sources were at a relatively low level, the long-term average mercury concentration fell off with the square of the distance from the source. At the mercury concentrations close to the works it is the direct inhalation of mercury vapour that is the main ingestion route of the primary target, Man. However, there are other possible indirect routes. It was

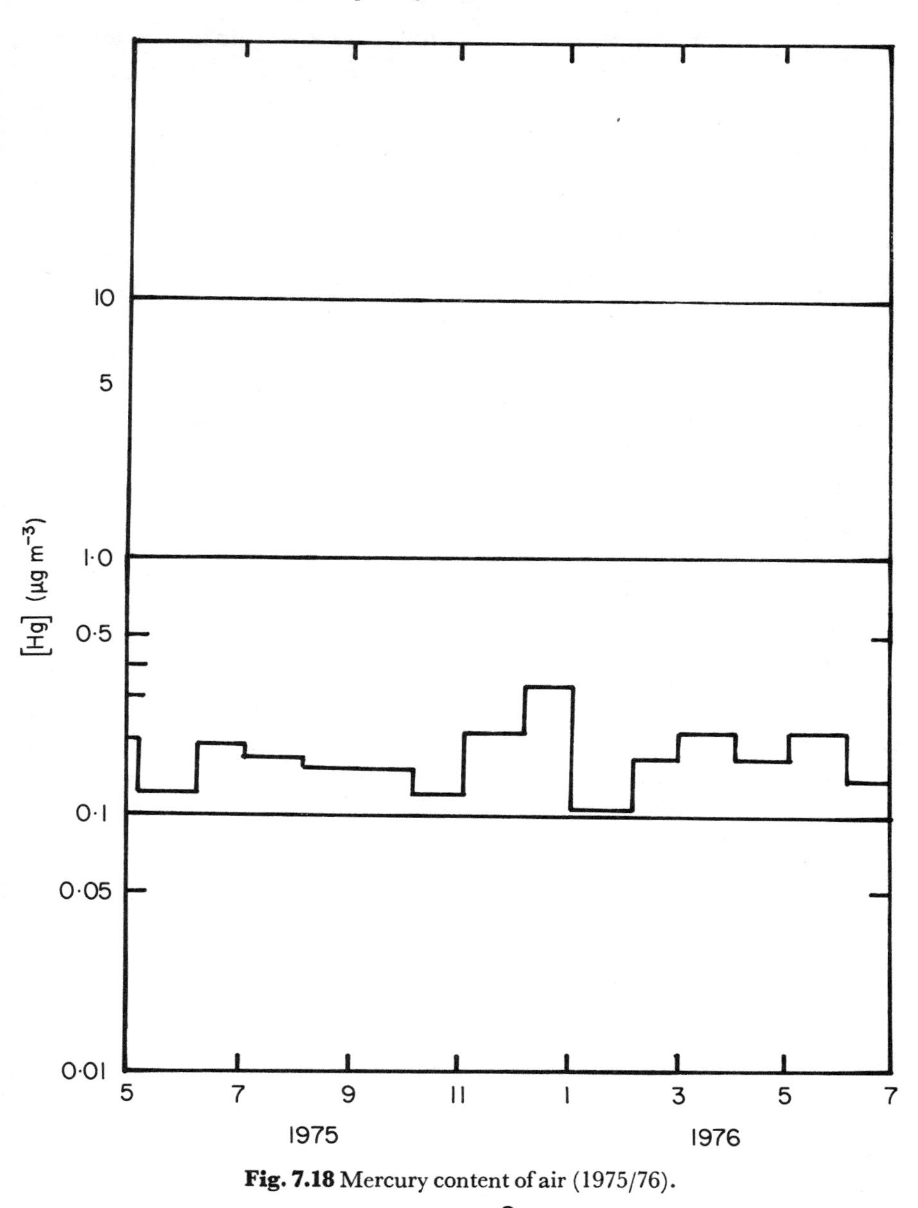

Fig. 7.18 Mercury content of air (1975/76).

anticipated at the impact assessment stage that these routes would be insignificant but there were no data on which this could be evaluated.

One we examined in some detail was the ingestion of mercury-contaminated forage by animals which might be affected themselves or which might later be eaten by humans (Edwards and Pumphrey, 1982). The conversion of the relatively innocent inorganic Hg into the more toxic methyl mercury in the rumen of grazing animals was a possibility we had to investigate.

Comparison of the atmospheric concentration and forage concentration for the same site showed that 0·2 $\mu g\ m^{-3}$ in atmosphere corresponds to 2 $\mu g\ m^{-3}$ mercury in dry grass (Figs 7.17 and 7.18). From this it can be deduced that grazing animals can ingest very much more of an atmospheric contaminant through their food than they will inhale. Making some assumptions about sheep – that they might breathe about 20 m^{-3}/day and ingest 2 kg dry wt. of grass the ratio between ingested and inhaled is 1000 to 1, and if the sheep retained the mercury they ingested in a couple of years they could have a flesh concentration of several hundred ppm of mercury.

In fact the area of grazing land which was affected to any measurable extent was very small, however, on one field close to the factory it was possible to establish a flock of 12 sheep with their lambs.

The objectives of this experiment were to assess the uptake and accumulation of mercury by the sheep during grazing, to see if the mercury was converted from inorganic into methyl mercury and to see if there would be any risk to human health from eating animals grazed on grass contaminated with low levels of inorganic mercury and, lastly, to see if there was any sign of a direct effect on the sheep as evidenced by their general health and by the breeding effectiveness of the flock.

The sheep were kept as nearly as possible under normal farming practice. However, to assess the mercury uptake it was necessary to take sheep at intervals for slaughter and analysis. Fig. 7.19 shows the measurements of mercury in the flesh of the sheep sampled before, during and after their period of exposure to mercury-contaminated grass. These measurements show that sheep do not accumulate mercury in the flesh to any significant extent.

The organ in the sheep that showed the highest mercury concentration was the kidney (Fig. 7.20) but even in the kidney the mercury concentration averaged only 1·4 $\mu g\ g^{-1}$. Measurement of the concentration of methyl mercury in kidney showed that there was at least an order of magnitude less of the methyl form than there was of the inorganic mercury. A similar ratio between methyl and inorganic mercury was found in the rumen contents showing that little if any methylation of mercury was occurring.

These measurements show that: (a) the sheep do not absorb much of the mercury in their diet; (b) the bacteria in sheeps' rumens do not methylate any of the mercury available to them to a significant extent; and (c) there would be no hazard to humans consuming their flesh.

The breeding achievements of the flock were comparable with that in normal agricultural practice. Our main trouble was worrying of the sheep by dogs rather that the effect of the mercury. However, the point of dwelling on the details of this investigation is not so much the results themselves which

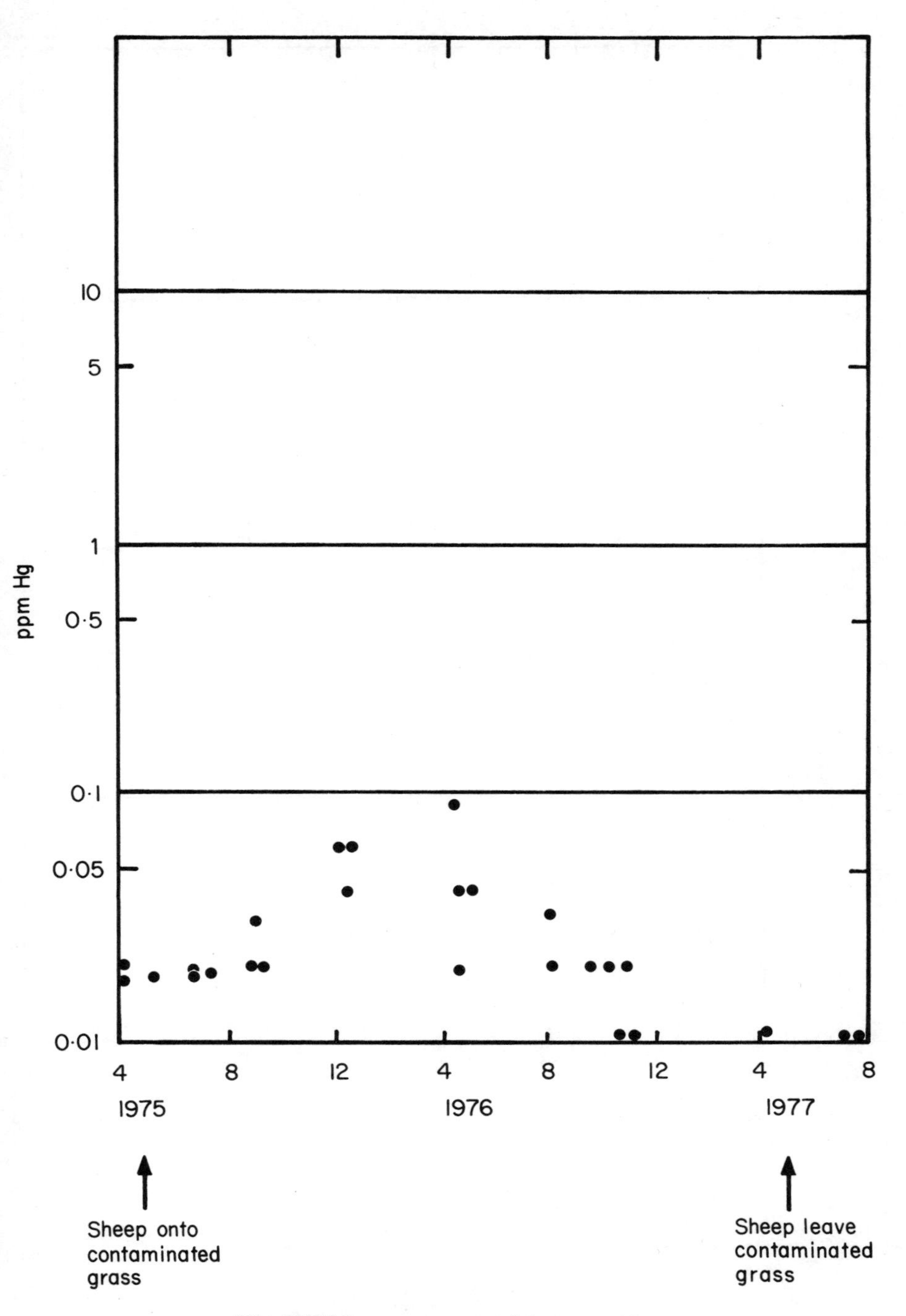

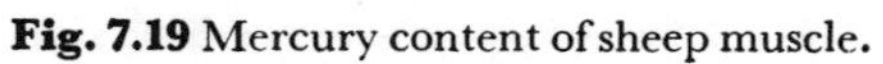
Fig. 7.19 Mercury content of sheep muscle.

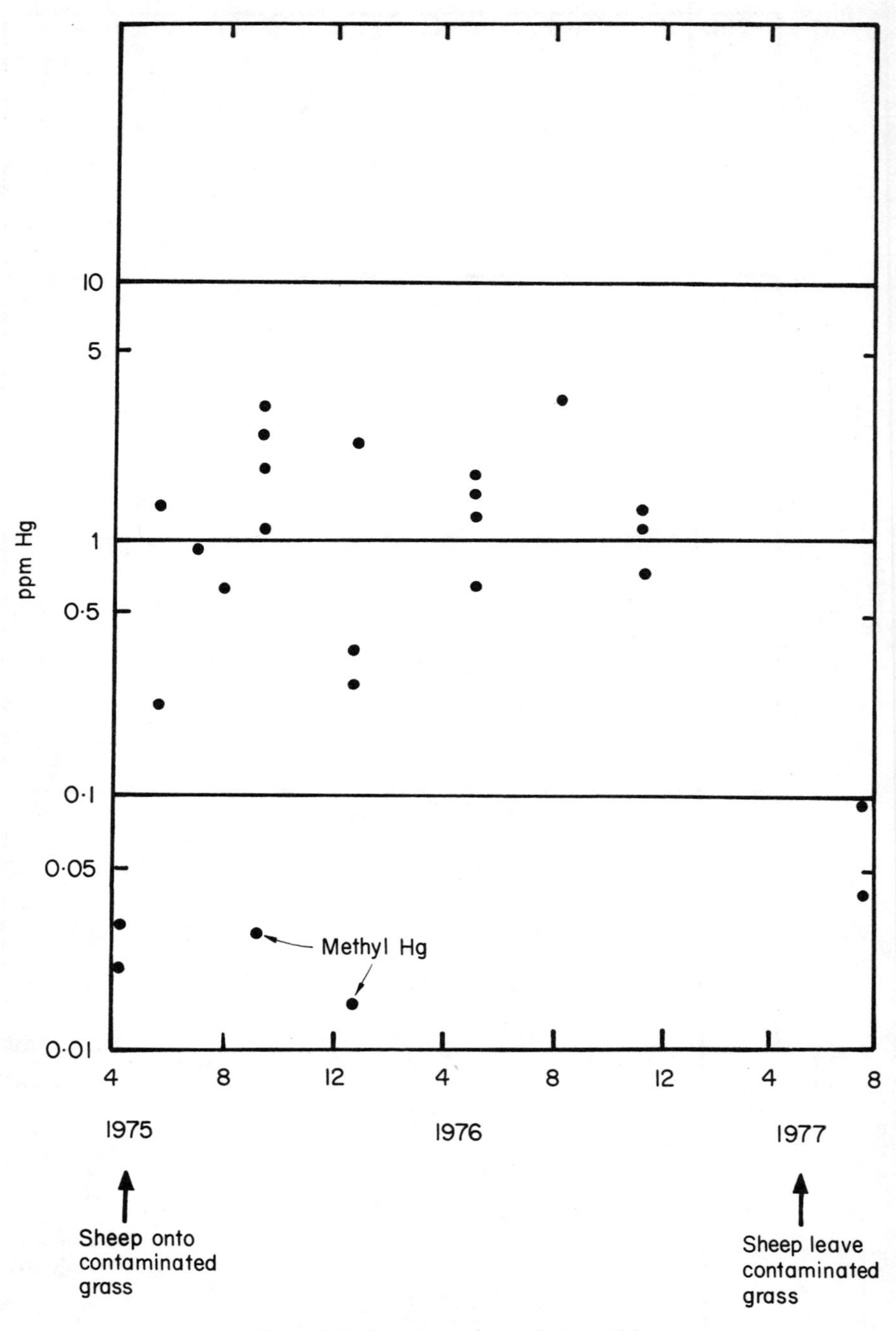

Fig. 7.20 Mercury content of sheep kidney.

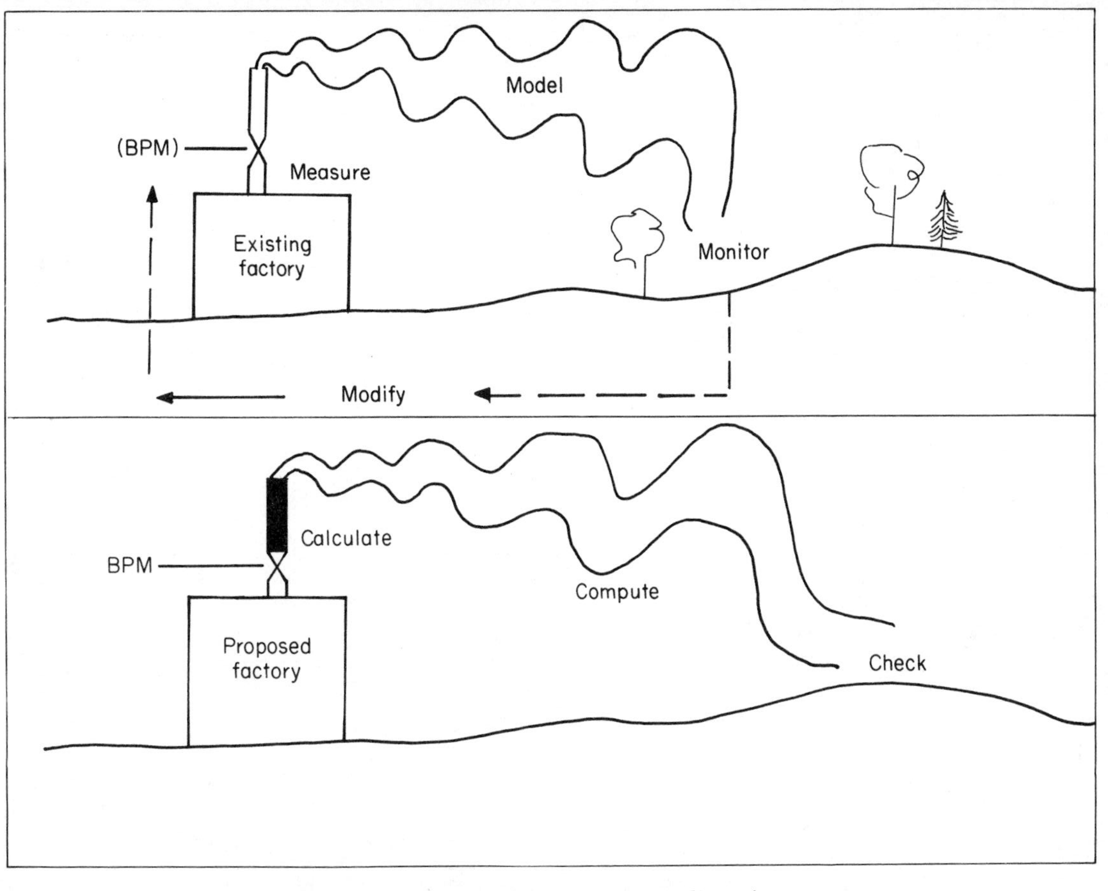

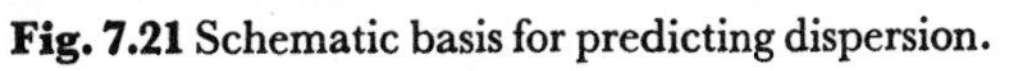

Fig. 7.21 Schematic basis for predicting dispersion.

have been published elsewhere but to illustrate some of the areas which may have to be covered for environmental evaluation of an existing plant where the path of a single component can be unambiguously followed in the environment.

7.7.4 Conclusions

Our experience of environmental evaluation of atmospheric emissions from existing factories shows that it can be a useful exercise in guiding the cost-effective use of capital. But for a new works in the design stage if there is more than one source of emission and if there are several components to be interrelated then the use of environmental evaluation in a predictive sense may not be very realistic. Therefore in no sense should environmental evaluation be imposed or undertaken 'because it is a good thing to do' but it should have a very carefully thought out set of objectives and constraints so that:

(a) The exercise itself does not waste resources.
(b) Any impact in the environment can be related to specific aspects of the project being assessed.
(c) A reasoned decision can be taken whether or not the existing or possible future impact should be alleviated and if it should, on the best practicable means to achieve this.

The techniques are summarized diagramatically in Fig. 7.21.

7.8 ECOLOGICAL ASSESSMENT OF THE EFFECTS OF ATMOSPHERIC EMISSIONS

A. W. Davison

7.8.1 Introduction

In trying to assess the effects of air pollutants, the biologist uses predicted ground-level concentrations of the pollutant in question, and dose–response relationships for relevant or important plant and animal species. It will be clear from the previous paper that predicting ground-level concentrations of a pollutant is fraught with difficulty but in most situations it is the biological data, the dose–response relationships, that limit the accuracy and usefulness of this facet of environmental impact assessment. It is important for non-biologists, legislators and decision-makers in general to appreciate the present limitations of the biological data that are used in impact assessment.

The following is an outline of an example of the way in which ground-level concentrations and dose–response curves are used. It illustrates both what can be done under favourable circumstances and the limitations of our knowledge.

7.8.2 Prediction of the effects of fluoride emissions – an example

The effects of fluoride on plants are well-documented and several good reviews are available (NAS, 1971; Weinstein, 1977; Suttie, 1977) but briefly, the principal effects on plants are chlorosis, necrosis, or distortion of leaves. It may also cause reduced growth or loss of yield. If grazing animals ingest excessive amounts of fluoride over long periods it may cause dental and osseous lesions, stiffness, lameness and loss of condition. It is therefore a pollutant of considerable economic importance so it is common practice for environmental impact studies to precede the building of an emission source. In this example, consultant engineers and the author were asked to provide a statement on the possible effects of emissions from a new aluminium smelter on native vegetation and domestic animals.

An initial survey of the semi-desert site showed that the area was largely flat and had a sparse xerophyte scrub vegetation. Examination of the literature showed that none of the wild species had been examined for susceptibility to fluoride and so nothing specific could be predicted for them without a major and lengthy research programme. This is in fact not an unusual situation because reserach has centred on western crop plants and very little is known about minor crops or wild species. However, to the south-east of the site there lay a large irrigated area that produced very good quality peaches and apricots as well as a variety of other crops (barley, garlic, tomato, etc.). This made a useful assessment possible because the two stone-fruits are among the most sensitive species known (Weinstein, 1977) and the dose–response curves have been as well researched as any. Exposure of the fruits to low concentrations of fluoride for short periods causes premature ripening of the suture and this affects appearance, keeping qualities and hence value. It was reasoned that if the effects of fluoride on these most sensitive species were assessed and they were protected, then a degree of protection would be given to other plants in the area.

In a temperate climate with western-type farming such as is practised in the UK it is possible to predict rates of deposition of fluoride to grass (Davison and Blakemore, 1976; Davison *et al.*, 1976) and therefore estimate uptake by grazing animals but in the study area, not only was the vegetation a type that has not been examined for fluoride deposition but also the grazing, by sheep and goats, was sporadic and completely unpredictable. Consequently no assessment of the effects on agricultural animals could be made.

Fig. 7.22 shows a series of dose–response curves for different plant species that were produced by McCune (1969) from a compilation of a large body of fumigation experiments. It is important to stress that the curves do not represent clear-cut divisions; the best way of interpreting them is to assume that the further above the curve a dose falls, the greater is the probability of there being visible injury. Conversely, the further below the curve, the lower is the probability of an effect.

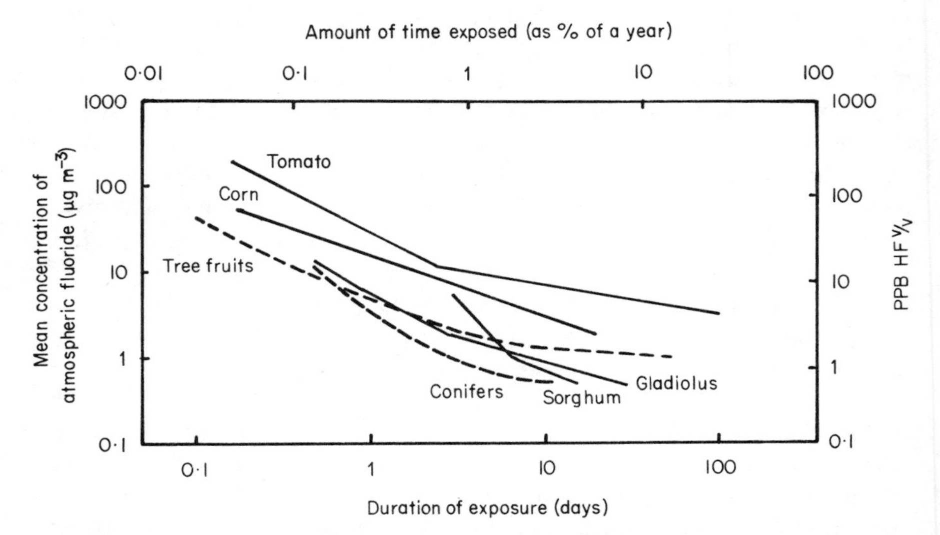

Fig. 7.22 Dose–visible response curves based on McCune's (1969) air quality criteria for protection of different species.

Examination of the current literature suggested that the curve for stone-fruits should be somewhat lower than that in Fig. 7.22 so a new curve based on the following doses was drawn up:

Concentration	*Time*
4·0 μgF m^{-3}	12 consecutive hours
2·2 μgF m^{-3}	24 consecutive hours
1·8 μgF m^{-3}	7 days

A consulting engineer, C. D. Weir, then produced predictions of ground-level concentrations for the time periods shown above, and for the months in which the fruits were in the critical stage of development. As several different designs of smelter were under consideration the predictions were produced for five emission rates (4–74 gF s^{-1}) and three stack heights (75, 100, 180 m). Examples are shown in Figs 7.23 and 7.24 for different emission rates.

The predicted concentrations were then used to estimate the risk of the critical doses being exceeded in the irrigated area. The results were summarized as in Table 7.23. This showed that at the lower emission rates and

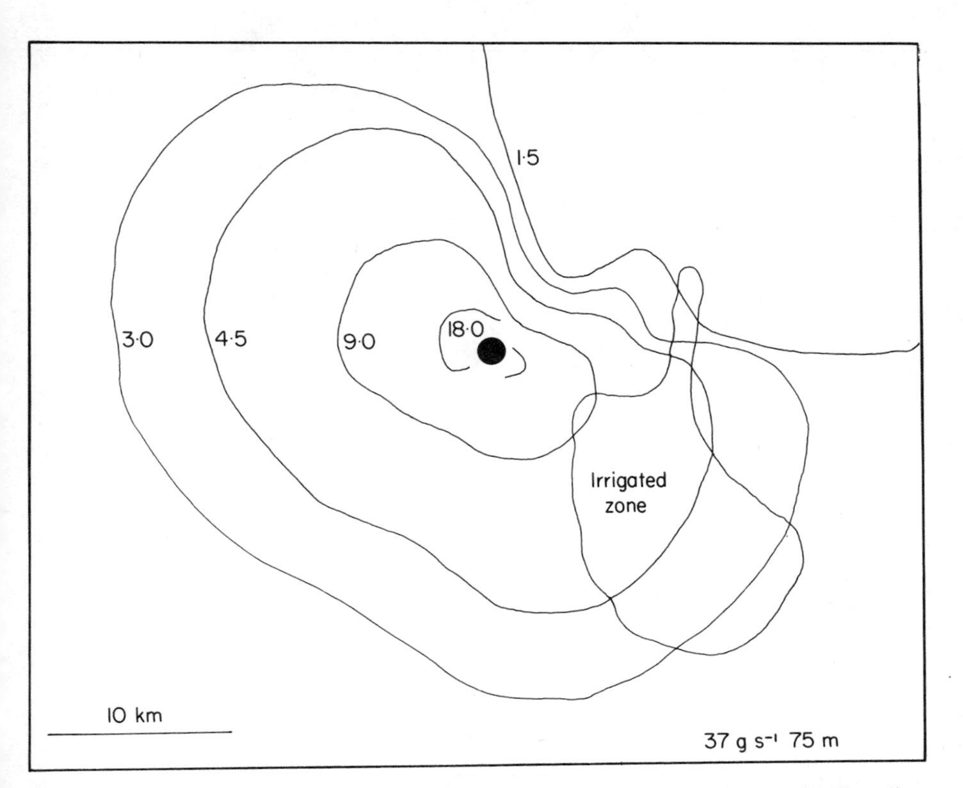

Fig. 7.23 Predicted 24 h mean ground-level concentrations of fluoride ($\mu gF\ m^{-3}$) for synoptic month. Emission rate 37 $gF\ s^{-1}$, stack 75 m. The irrigated zone is shown.

Table 7.23 Summary of the effects of different fluoride emission rates and stack heights on stone fruit

Emission rate	*Stack height (m)*		
($gF\ s^{-1}$)	75	100	180
4	A	A	A
12	B	A	A
24	C	B	A
37	C	B	A
74	C	C	B

A = low to very low probability of injury
B = indefinable but positive risk of injury
C = high to certain probability of injury

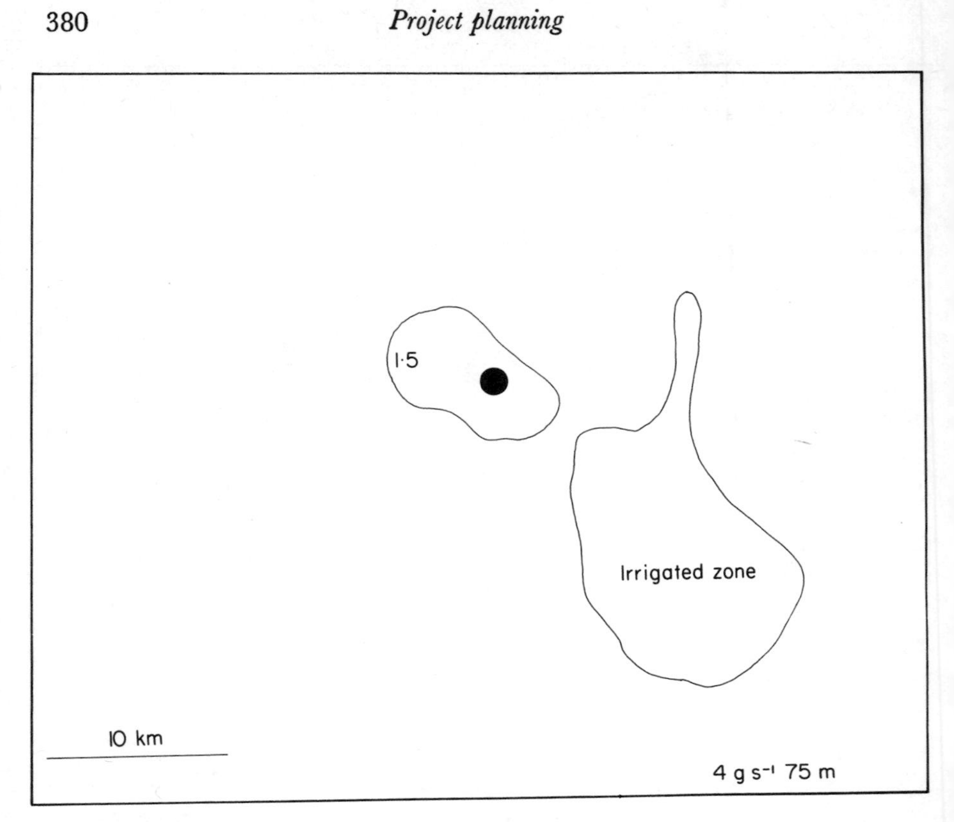

Fig. 7.24 As Fig. 7.23 except emission rate 4 gF s^{-1}, stack height 75 m.

with tall stacks there was little or no risk of injury, but with the highest rates and lowest stack there was near-certain risk of severe injury and therefore economic loss. There was also a 'grey' zone, a number of combinations of emission rates and stack heights where it was considered that there was a risk but an undefinable risk of injury. A similar study and summary was prepared for roof-level and fugitive emissions and these were used in the final report which recommended the optimum design parameters.

7.8.3 Analysis of the impact study

At this stage it is proposed to review the more important features of the study in order to demonstrate both the state and the limitations of the basic biological knowledge that is needed in order to predict successfully the effects of airborne pollutants.

The first aspect of this particular study that favoured assessment was the nature of the site. The topography was virtually flat and the proposed works was to stand on its own, with no large obstructions nearby. This, coupled with availability of adequate meteorological data gave optimum conditions for predictive modelling of ground-level fluoride concentrations. It need hardly be said that in most places in the UK, topographical variation and the almost inevitable presence of other buildings plus other sources of pollutants make prediction more difficult.

Just as important was the fact that there was in the area a readily identifiable and well-researched target in the form of stone-fruit. In many parts of the world there are crops such as rye-grass, *Lolium perenne*, ornamentals of high aesthetic value as well as the vast majority of wild species, that have not been researched for the effects of fluoride. Furthermore, in the case of the stone-fruit, it was possible to translate the visible effects of the pollutant into economic terms whereas in most cases it is impossible to place a monetary value on visible symptoms.

An essential feature of the assessment of the effects of the fluoride emissions was the use of simple dose–response curves. Fluoride is comparatively well researched in this respect but over the last few years there has been a tendency to revise the curves downwards and set higher air quality standards as research has proceeded. Where other pollutants are concerned either curves are not available for many species or dose–response models are a highly contentious area of discussion. For example, in the case of sulphur dioxide the first attempts at relating dose and response were made in 1922 (O'Gara) and then over the years simple models were accepted until in the 1940s and 1950s it appeared that dose–response data were adequate and reliable. In recent years models have become more sophisticated (e.g. Larsen and Heck, 1976) though for a limited range of species. In 1975 Mudd considered that the, 'dose–response considerations for SO_2 as a single pollutant are essentially complete'. However, this view is by no means universally held, especially in the UK. In the last few years it has been shown (Ashenden and Mansfield, 1977) that there must be considerable doubt about the results of earlier work because of the design of the fumigation chambers. In addition, a procession of workers (Bleasdale, 1952; Bell and Clough, 1973; Bell and Mudd, 1976; Ashenden, 1978, 1979; Ashenden and Mansfield, 1978; Horsman *et al.*, 1979; Black and Unsworth, 1979; Davies, 1980) have amply demonstrated that SO_2 may have considerable effects at concentrations lower than the previously supposed threshold. One group (Cowling and Lockyer, 1978) has shown growth stimulation, particularly when the soil is sulphur deficient. Bell (1980) stated that: 'It has proved most difficult, even for the most intensively studied species, *Lolium perenne*, to establish clear dose–response relationships and furthermore, different groups of workers have produced contradictory results with similar doses of SO_2'. In other words, even for the most researched and arguably the most important pollutant there is no consensus of opinion that would lead at present to universally acceptable dose–response curves or models.

Most dose–response curves and models are for visible effects on the plant, that is necrosis, chlorosis, premature ageing, discolouration and distortion. It is often assumed, and most air-quality standards are based on this assumption, that there is a relationship between these visible and non-visible growth/yield effects. That is, it is assumed that if a given dose does not cause any visible injury then there is no reduction in growth or yield. However, it has become clear in recent years that there is no clear-cut relationship

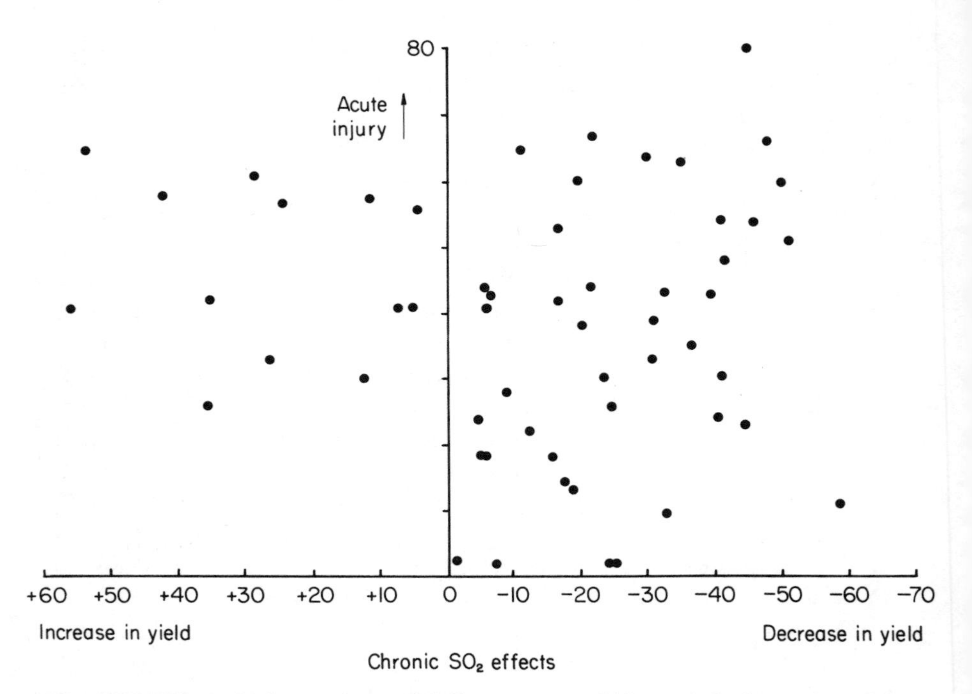

Fig. 7.25 Effects of SO_2 on clones of *Lolium perenne* cv S23. *x*-axis is change in weight of clone compared with clean air control when exposed to chronic (346 μg^{-3}, 60 days) dose of SO_2. *y*-axis is percentage of leaf length injured when exposed to an acute (5320 μg^{-3} 6 h) dose of SO_2. Data from Ayazloo (1979).

between the two types of effect (see Fig. 7.25). Conversely, and this applies particularly to fluoride (NAS, 1971), there may be considerable visible injury with no detectable reduction in growth or yield. We cannot use dose–visible response relationships to predict loss of yield and for no pollutant do we have reliable dose–growth or yield data.

Aluminium smelters produce pollutants other than fluorides (such as SO_2, hydrocarbons and NO_x) but fluoride is considered so much more important that the others are ignored. In industrialized areas it is extremely rare for there to be only one air pollutant and in fact many tend to occur together. For instance SO_2, NO_x and hydrocarbons often occur in constantly varying concentrations. Increasingly, research is showing (Mudd and Kozlowski, 1975; Wellburn *et al.*, 1976; Ashenden and Mansfield, 1978; Ashenden, 1979) that there are subtle interactions between pollutants, one modifying the effect of the other, and between plant/pollutant and pest or disease (Heagle, 1973), so even where dose–response relationships for single pollutants are available it is highly questionable whether they can be applied with any confidence to the field situation where the plant is bombarded by constantly varying mixtures of pollutants, and by pests and diseases that may themselves be affected by the pollutants.

Finally, the important target species around the proposed smelter were two long-lived crop trees that had little genetic variation and that were not subjected to many of the intense stresses and selective pressures that exist in pastures, meadows, hedgerows and forests. Other than in man-made monocultures, vegetation consists of complex populations of many species, each having a range of genetic material and therefore, the potential to respond to stress by natural selection. Most of the forage grasses grown in the UK are obligate outbreeders so even named cultivars contain a wide range of genetic potential (Snaydon, 1978). One of the best demonstrations of the range of variation that occurs within a species with respect to air pollution is provided by the work of Ayazloo (1979). She showed for several common pasture grasses that when named cultivars were subjected to either chronic or acute doses of SO_2, individual plants reacted in a different way. Fig. 7.25 shows some data for *Lolium perenne* cvS23. Each point represents an individual genotype. When subjected to a chronic dose of SO_2 most individuals showed reduced yield compared with controls (clonal material of the same individuals grown in clean air), and some showed 50–60% lower dry weight at harvest. On the other hand a proportion of individuals showed a greater dry weight than controls, some up to 50% greater. Where the same individuals were exposed to an acute dose of SO_2, the response again varied from individuals that showed 70–80% of the leaf length injured to a few that showed virtually no effect. Note that there was no relationship between response to acute and chronic doses.

Natural selection of genotypes starts as soon as the seeds are sown and, depending on the soil type, management, nutrient regime, weather and competition, the individuals that establish will represent only a proportion of the genotypes in the original material (Brougham and Harris, 1967; Charles, 1970; Snaydon, 1978). Where a pollutant stress is added, selection of the more tolerant genotypes would be expected (Snaydon, 1978; Horsman *et al.*, 1979) perhaps in a matter of weeks, perhaps over years. Just this one aspect of plant-pollutant relationships makes it impossible for the biologist to predict the effects of pollutants on yield; exposure of a pasture to SO_2 may reduce yield for a short time until selection occurs or it may have no effect or indeed it may lead to increased yield.

7.8.4 Conclusions

The current position on the prediction of the effects of air pollutants can be summarized by saying that the state of our knowledge and technical expertise is such that it should be possible to predict and therefore, prevent major visible injury caused by the main, well-researched pollutants but that it is not possible to do anything more subtle at present. It is currently impossible to predict the effects of multiple pollutants on growth and yield under field conditions and furthermore the technical problems are such that it will be some time before that is possible. If this biological aspect of environmental impact assessment is to make a real as opposed to cosmetic contribution to conservation of environmental quality, a much greater investment is needed in biological research on air pollution effects.

REFERENCES

Anon. (1971) Watch and Ward on the River Trent, *Midlands Power*, June 1971.

Anon. (1979) Chats Falls Nuclear Site Baseline Environmental Studies, Envirocon Ltd & Procter and Redfern Ltd, for Ontario Hydro.

Ashenden, T.W. (1978) *Environ. Pollut.*, **15**, 161–6.

Ashenden, T.W. (1979) *Environ. Pollut.*, **18**, 249–58.

Ashenden, T.W. & Mansfield, T.A. (1977) *J. Exp. Bot.*, **28**, 729–35.

Ashenden, T.W. & Mansfield, T.A. (1978) *Nature (London)*, **273**, 142–3.

Aston, R.J. (1973) *Hydrobiologia*, **42**, 225–43.

Aston, R.J. & Brown, D.J.A. (1975) *Hydrobiologia*, **47**, 347–66.

Aston, R.J. & Milner, A.G.P. (1980) *Freshwater Biol.*, **10**(1), 1–15.

Ayazloo, M. (1979) *Tolerance to Sulphur Dioxide in British Grass Species*, PhD thesis, University of London.

Bamber, R.N. & Henderson, P.A. (1981) Bradwell Biological Investigations: Analysis of Benthic Surveys of the River Blackwater up to 1975, CERL Report No. RD/L/2042R81.

Barber, F.R. & Martin, A. (1976) *Atmos. Env.* **10**, 281–4.

Bell, J.N.B. (1980) *Nature (London)*, **284**, 399–400.

Bell, J.N.B. & Clough, W.S. (1973) *Nature (London)*, **241**, 47–9.

Bell, J.N.B. & Mudd, C.H. (1976) in *Effects of Air Pollutants on Plants* (ed. T.A. Mansfield), Society for experimental biology seminar series, Vol. 1, Cambridge University Press, Cambridge, pp. 87–104.

Berry, R.J. & Johnston, L. (eds) (1980) *The Natural History of Shetland*, Collins, London.

Black, V.J. & Unsworth, M. (1979) *J. Exp. Bot.*, **116**, 473–83.

Bleasdale, J.K. (1952) *Atmospheric Pollution and Plant Growth*, PhD thesis, University of Manchester.

Blumer, M. (1972) *Proc. of FAO Tech. Cong. on Marine Pollution and its Effects on Living Resources and Fishing* (ed. M. Ruivo), Rome 1970, Fishing News (Books) Ltd, London.

Brougham, R.W. & Harris, W. (1967) *NZ J. Agric. Res.*, **10**, 56–65.

Brown, D.J.A. (1973) *Proc. 6th British Coarse Fish Conf.*, 191–202.

Buller, A.T., Charlton, J.A. & McManus, J. (1974) *Potential Movement of Oil Spillage and Pollutants: Scapa Flow Region*, Centre for Industrial Research and Consultancy, University of Dundee.

Charles, A.H. (1970) *J. Agric. Sci., Camb.*, **75**, 103–7.

Christie, A.E. (1979) *Ontario Hydro Report* 79382.

Clarke, A.J., Lucas, D.H. & Ross, F.F. (1971) *2nd Int. Clean Air Conference*, 1970.

Coughlan, J. (1966) CERL Report No. RD/L/R 1351.

Council of European Community (1976) *Directive on the Discharge of Dangerous Substances to the Aquatic Environment*, Official Journal L.129.

Cowling, D.W. & Lockyer, D.R. (1978) *J. Exp. Bot.*, **108**, 257–64.

Davies, T. (1980) *Nature*, **284**, 483–4.

Davison, A.W. & Blakemore, J. (1976) in *Effects of Air Pollutants on Plants* (ed. T.A. Mansfield), Society for experimental biology seminar series, Vol. I, Cambridge University Press, Cambridge, pp. 17–29.

Davison, A.W., Blakemore, J. & Wright, D.A. (1976) *Environ. Poll.*, **10**, 209–15.

Department of the Environment, Central Unit of Environmental Pollution (1976) *Pollution Control in Great Britain: How it works*, Pollution Paper, No. 9, HMSO, London.

Department of the Environment, Central Unit on Environmental Pollution (1977) *Environmental Standards: A description of United Kingdom Practice*, Pollution Paper, No. 11, HMSO, London.

Edwards, P.R. and Pumphrey, N.W.J. (1982) *J. Sci. Food Agri.*, **33**, 237–43.

Gammon, K.M. (1979) CEGB Newsletter No. 111 (and Nuclear Engineering International, September, 1979).

Goodier, R. (ed.) (1974) the *Natural Environment of Shetland, Proceedings of the NCC Symposium*, Nature Conservancy Council, Edinburgh.

Gurney, W.B. & Sons (1956) Minutes of Proceedings at a Public Inquiry 28 April 1956 (The Bradwell Inquiry).

Hammerton, D., Newton, A.J. & Allcock, R. (1980) *Effluent and Water Treatment*, **20**(6), 261–9.

Hammerton, D., Newton, A.J. & Leatherland, T.M. (1981) *Wat. Pollut. Control.* 189–203.

Hawes, F.B. (ed.) (1968) *Hydrobiological Studies in the River Blackwater in Relation to the Bradwell Nuclear Station*, CEGB, London.

Hawes, F.B., Coughlan, J. & Spencer, J.F. (1974) *J. Effluent and Water Treatment*, October, 1974, pp. 549–59.

Heagle, A.S. (1973) *Ann. Rev. Phytopathol.*, **II**, 365–88.

Henderson, P.A. & Whitehouse, J.W. (1980) CERL Report No. RD/L/R 2018.

Horsman, D.C., Roberts, T.M. & Bradshaw, A.D. (1979) *J. Exp. Bot.*, **30**, 495–502.

International Joint Commission (1976) Report of Water Quality Objectives Subcommittee, Great Lakes Water Quality Board, Canada.

Jones, A.M. (1980) Monitoring studies associated with an oil reception terminal. *Rapp. P-v. Reun. Cons. int. Explor. Mer.* (in press).

Jones, A.M. & Stewart, W.D.P. (1974) An *Environmental Assessment of Scapa Flow with Special Reference to Oil Developments*, Centre for Industrial Research and Consultancy, University of Dundee.

Klein, L. (1957) *Aspects of River Pollution*, Butterworths, London.

Knox, A.J. (1978) Ontario Hydro Report 79248.

Langford, T.E. (1978) CERL Note No. RD/L/N 145/78.

Larsen, R.I. & Heck, W.W. (1976) *J. Air Pollut. Control Assoc.*, **26**, 325–33.

Livesey & Henderson (1973) Sullom Voe and Swarbacks Minn Area Master Development Plans Related to Oil Industry Requirements: report to Zetland County Council.

McCune, D.C. (1969) *Establishment of Air Quality Criteria with References to the Effects of Atmospheric Fluorine on Vegetation*, The Aluminium Association, New York.

MacQueen, J.F. & Howells, G.D. (1968) *CEGB Research* **7**, 33–44.

Malins, D.C. (1977) *Effects of Petroleum on Arctic and Subarctic Marine Environments and Organisms*, volumes I and II, Academic Press, London and New York.

Martin, A. (1979) in *Proc. Int. Symp. Sulphur Emissions and the Environment*, London, May 1979, The Society of Chemical Industry, pp. 49–66.

Martin, A. & Barber, F.R. (1966) *J. Inst. Fuel*, **39**, 294–307.

Martin, A. & Barber, F.R. (1967) *Atmos. Env.*, **1**, 655–77.

Martin, A. & Barber, F.R. (1974) *Atmos. Env.*, **8**, 373–81.

Maul, P.R., Barber, F.R. & Martin, A. (1980) *Atmos. Env.*, **14**, 339–54.

Mudd, J.B. & Kozlowski, T.T. (1975) *Responses to Air Pollution*, Academic Press, New York, p. 383.

N.A.S. (1971) *Fluorides*. National Academy of Science, Washington.

National Academy of Engineering (1972) *Engineering for Resolution of the Energy–Environment Dilemma*, Committee on Power Plant Siting, Washington, DC.

Nature Conservancy Council (1976) *Shetland Localities of Geological and Geomorphological Importance*, NCC Geology and Physiology Section, Newbury, Bucks.

Nonhebel, G. (1975) *Atmos. Environ.* **9**, 709–15.

O'Gara, P.I. (1922) *J. Ind. Eng. Chem.*, **14**, 744.

Pasquill, F. (1974) *Atmospheric Diffusion*, 2nd edn, Ellis Horwood Ltd.

Sadler, K. (1980) *J. Appl. Ecol.*, **17**, 349–57.

Scriven, R.A. & Howells, G.D. (1977) *CEGB Research*, **5**, 28–40.

Snaydon, R.W. (1978) in *Plant Relations in Pastures* (ed. J.R. Wilson), CSIRO, Melbourne.

Suttie, J.W. (1977) *J. Occup. Med.*, **19**, 40–8.

Syratt, W.J. & Richardson, M.R. (1981) *Proc. Roy. Soc. Edin.*, Series B, **80**, 33–51.

Talbot, J.W. (1966) *Min. Ag. Fish and Food*, Fishery Invest. Series II, Vol. XXV No. 6, HMSO, London.

Taylor, D. (1976) *Mercury as an Environmental Pollutant. A Bibliography*, ICI, Brixham Lab.

The Sullom Voe Environmental Advisory Group (1976) *Oil Terminal at Sullom Voe: Environmental Impact Assessment*, Thuleprint, Sandwick, Shetland.

Tunnicliffe, M. (1975) *Industrial Air Pollution*, HMSO, London.

Weinstein, L.H. (1977) *J. Occup. Med.*, **19**, 49–78.

Wellburn, A.R., Capron, T.M., Chan, H-S. & Horsman, D.C. (1976) in *Effects of Air Pollutants on Plants* (ed. T.A. Mansfield), Society for Experimental Biology Seminar Series, Vol. 1, Cambridge University Press, Cambridge.

Whitehouse, J.W. & Foster, D.J. (1974) CERL RD/L/N 18/74 and RD/L/N 116/74.

US Environmental Protection Agency (1978) *Quality Criteria for Water*, US Government Printing Office.

Zetland County Council (1974) *Sullom Voe District Plan*, County Planning Department, Lerwick.

CHAPTER EIGHT

Environmental Audits and Research Needs

8.1 OVERVIEW

Retrospective analysis has been carried out at two levels (i) to assess the effectiveness of different planning procedures in addressing environmental issues and (ii) to assess the completeness and accuracy of predictions. Concern about the environmental effects of major developments and weighting of emphasis for environmental evaluations has led to the evolution of a variety of procedures for ensuring that due consideration is given to these in the planning process. Such procedures range from (i) no action prior to response to planning application by planning or statutory environmental authorities, (ii) orderly assessment of effects in an Environmental Evaluation or (iii) completion of a formal environmental impact statement. We are at present in the UK still in the evolutionary stages of an acceptable and generally applicable system of impact assessment. This is reflected in the variety of opinions expressed in this book on the advantages and disadvantages of the more formalized EIA systems. Those who stress the advantages suggest that EIA systems (a) improve the quality of environmental design, (b) reduce the costs of major projects, (c) reduce delays in the decision-making process, (d) ensure comprehensive consideration of the existing planning of land-use policies. In contrast, there are many who see disadvantages such as (a) the present 'best-practice' of environmental assessment is not improved upon, (b) increase in costs by expanding assessments beyond the 'critical factors', (c) increase delays in the decision-making process. These apparent conflicts can only be reconciled by a retrospective analysis of the effectiveness and benefits of the different procedures.

Dallas (Section 8.2) reviews the evolution of an EIA system in Ireland with special reference to mining development. It is clear that, in this context, EIA seen as an effective mechanism for ensuring early and orderly consideration of relevant issues identified by both ecological base-line surveys and public discussions to highlight 'perceived problems'. Unfortunately, very few analyses of this type have been carried out although experience is being gained in a number of areas. For example, a number of major industries now appoint trained ecologists to large developments to supervise base-line surveys, amelioration measures, reinstatement work and monitoring programmes. This experience should be extended more often to a post-audit of the developmental procedures.

Post-development monitoring does, however, provide a quantitative basis for clarifying uncertainties which inevitably exist in environmental evaluations and for analysing the accuracy and emphasis of predictions. This is a rational extension of the justification for ecological monitoring as a means of (a) meeting statutory obligations, (b) early recognition of problems, (c) a demonstration of no problems and (d) to provide a basis for informed debate with regulatory agencies. Bines *et al.* (Section 8.3) in analysing the predictions made during the Inquiry into the Cow Green Reservoir Development on a unique site for conservation, conclude that most impacts were predicted but the authors emphasize the difficulties encountered in quantifying the magnitude of the indirect impact (such as resulted from increased public access). Baker and Hiscock (Section 8.4) document the development of 'predictive ability' within the oil industry to the point where, at Sullom Voe, the effects of a large oil spill were largely anticipated. Nevertheless, although ecological monitoring is widely used in the industry there are still problems which arise from the relative sensitivity of biological monitors and our inability to predict effects at an 'ecosystem' level. An important point is made in this example, that prediction of ecological effects can only be based on monitoring experience and information from basic research.

A major constraint in publishing post-audits is that this may be seen as a criticism of the 'expert witness' who gave his 'best assessment' at the time of the Inquiry. Nevertheless, the considerable benefits in answering uncertainties for future considerations should outweigh such concerns. Impact assessment is a dynamic ongoing process and the objectives of good environmental planning cannot be satisfied by pre-project assessment alone. Applied ecological research must more often be interpreted in a planning context.

Retrospective analysis of the effectiveness of planning procedures or the accuracy of predictions is a major research tool for improving the validity of environmental impact assessments. The paper by Holdgate (Section 8.5) goes on to discuss the broader research needs under two headings (a) the methodologies for environmental impact assessments and (b) methodologies for evaluation of the resulting environmental impact statement. This paper illustrates the ways in which scientific information may limit the EIA process by reference to the possible consequences of mineral exploration and exploitation in the Antarctic. Holdgate concludes, however, that the EIA process is more often limited by poor communication between planners and ecologists than by deficiencies in ecological knowledge.

8.2 EXPERIENCES OF ENVIRONMENTAL IMPACT ASSESSMENT PROCEDURES IN IRELAND

W. G. Dallas

8.2.1 Introduction

It is probably true to say that EIA in Ireland began in 1971 with the approach and methodology adopted by Tara Mines Limited to their planning application for the development and operation of Europe's largest zinc/lead mine in an area of considerable social and environmental sensitivity. The Tara development is therefore a good case study in EIA effectiveness. Likewise it is fair to say that in 1971 we were unaware of the formal procedure of EIA. It is interesting that when eventually its missionaries brought the faith to Ireland there was already established a parallel doctrine. With time the doctrines merged and many are now devout adherents.

Environmental impact assessment was an integral part of the procedures leading up to the series of planning applications associated with this development namely the main mine project – the tailings dam project – the concentrate export terminal at Dublin Port and an alternative and ultimately abandoned export terminal on the River Boyne estuary.

A subsequent assessment, following operation of construction and commencement of production, of the accuracy of the impact prediction, was included as an integral part of EIA. The mine has been in operation since 1977 and the company, now at this assessment stage is confident that the initial impact assessment was accurate. If anything it was pessimistic.

This paper traces the development of the Tara project and provides a retrospective look at the EIA procedures used.

8.2.2 The Tara project

The ore reserves at Navan are estimated at 52.9 million tonnes, with a zinc–lead ratio of 5 : 1. The ore also contains approximately 0·04 % cadmium and between 12 and 20 g/t silver. The ore body dips from immediately sub-surface to 500 m and lies within 500 m of the town of Navan which had a population of 6000 in 1971 (12 000 in 1980). The mine is surrounded on three sides by rich agricultural land. On the immediate mine boundary there are three dairy herds and also livestock fattening, grain-producing and potato-producing units. The River Blackwater flows over the ore body and joins the River Boyne in the centre of Navan. These rivers support substantial salmon, trout and eel fisheries. Both the orebody and the tailings dam are situated in areas of high amenity and archaeological interest. Apart from the mine area abutting on Navan's urban fringe there are over 40 residences located on the mine property perimeter.

When the orebody was discovered in 1971 Navan was just beginning to luxuriate in the respectable status of a developing dormitory town for Dublin 48 km away. The unemployment level was less than 1 %. A mining operation, especially as envisaged then, was the last thing the residents wanted.

As a result of the use of our concept of an EIA strategy the planning process of the mine and tailings projects from application to consent, including the appeal proceedings, took only 14 months. This still stands as an Irish record for such a potentially controversial development.

8.2.3 The EIA process

It seemed to us in 1971, as we put together our own EIA formula that an environmental impact assessment should consist of seven parts.

(a) Appraising of the pre-development 'pristine' environment in quantitative and qualitative terms, alias the baseline study.
(b) Determining the extent and detail of each sector of the proposed development and its process in specific terms.
(c) Examining the existing and relevant laws of the land, relevant impending legislation (EEC) and appropriate recommendations and standards.
(d) Predicting the effect of superimposing (b), above, on (a).
(e) Where impact was obvious at (c) and (d), modifying the development and its processes until an acceptable impact was achieved (otherwise the proposal must be abandoned and alternatives sought).
(f) Use of (a), (b), (d) and (e) as a basis for public discussion of the development with a view to ensuring a better understanding and acceptance of it.
(g) Subsequently assessing post-construction operation and production performance to determine the accuracy of the original prediction.

The baseline study began in 1971, a few months after the decision to develop the orebody and prepare the EIA format. The first environmental sector examined was the Boyne and Blackwater River system. This commenced 8 km upstream of the mine on the River Blackwater, 2 km upstream of the confluence on the River Boyne and included the estuary, approximately 32 km away. An initial water chemistry survey was carried out on contract and continued by Tara's own environmental team until production commenced in 1977. It was supplemented in 1972 by an aquatic flora and fauna study which was repeated in greater detail in 1975 when it included two extensive electro-fishing surveys which established not only population information but also heavy metal status of the principal species, i.e. salmon, trout and eels. A detailed hydrological investigation was also carried out. Since the original objective was to mine the orebody by an open-cast method and recognizing the pollution potential of such a decision an extensive soil, herbage and domestic stock survey was also commenced in 1971, to determine anomalous heavy metal locations and the reflection of these in herbage and domestic stock uptake. In 1972 a network of dust deposit gauges was established to determine pre-development ambient heavy metal levels in the atmosphere. While this information was accumulating predictive exercises based on similar scale quarrying operations were under way to establish possible dust emissions, noise levels and vibration effect.

Concurrently the first results of our Community relations programme, which is dealt with later, were indicating specific people problems and apprehensions. The predicted ecological and social impacts of open-cast mining were found to be unacceptable to the company and to a large section of the community. Supported by a favourable economic evaluation, a decision was made in late 1972 to abandon the open-cast proposal and develop an underground mine.

This was the first management decision based on environmental impact assessment and it illustrates that the EIA process need not necessarily be completed and an impressively bound work of art before decisions are made. Indeed throughout the planning stage the prospective process was constantly considered in relation to the accumulating environmental information and where adverse effect was obvious or suspected the process was continually modified until the predicted impact was considered acceptable.

As information accumulated on each environmental component and on the community, areas of concern became apparent. As process details (e.g. effluent quality) became evident from test plant work on actual ore samples and as the hydrological history of the local rivers was uncovered possible limitations on available fresh water and dilution capacity were obvious. Ultimately a decision had to be made to abstract fresh water from and return effluent to the less accessible River Boyne, a very expensive alternative because of the additional pipe work necessary. Also when the possible public reaction to the location of the tailings pond adjacent to the mine site was considered it was felt prudent to locate it three miles north of the urban area.

When the planning application was made in 1973 a comprehensive environmental study had been carried out. The more obvious components had been examined in detail and we had also completed a check study for rare vegetation species as well as an ornithological study of the river and tailings pond areas.

The Tara case study also spotlights another important issue which cannot be overemphasized. This is that the sociological impact must be assessed as meticulously as the ecological impact. When it becomes apparent that attitudes are so entrenched that in the lead time prior to planning application desensitizing is impossible, objectives may have to be altered and alternative methods and processes adopted. With mining there is not the luxury of the alternative site. The action must be where the orebody is. I do not wish to labour the social issue but the developer who prepares his community strategy well in advance of the first visible local activity is a long way towards his initial objective which is planning consent in minimum time which means with minimum, ideally zero, objections.

By the time the planning application was submitted the first part of the Community relations plan had been completed. Tara had by then identified interest groups, pressure groups, decision-makers and influential people and had convinced them that while the presence of the operation would be obvious the community and individual life style would not be disimproved. A total of 1500 school children were also involved in educational projects on mining and processing. Joint working groups were established with the Local Authority and with the National Trust and regular updating and discussion meetings arranged with them and with fishery interests. As a point of interest the joint liaison meeting with the Local Authority is still a monthly fixture at which monitoring data and public complaints are reviewed.

During the 14 months between submission and final approval the baseline programme continued, recording cyclic and seasonal changes and establishing further detail where this was required. In fact, baseline activities continued right up to the time production commenced in 1977. This extended baseline study has been extremely valuable and is strongly advocated.

I have not yet mentioned operational noise and blasting vibration and yet both are major and obvious problems in a mining area. Baseline noise surveys were carried out and limits imposed in the consent conditions. These conditions insisted on three fixed continual monitoring stations for noise and vibration. Each station has a microphone and a seismometer relaying information by cable to a central recording and interpretation unit and operational performance audits are carried out frequently.

As already stated, the EIA approach to project planning was further used in the separate planning sequences for Tara's 242 ha tailings pond and also in relation to the development of its port facility in Dublin. It was also used in relation to a shipping terminal proposal on the Boyne estuary, abandoned in favour of Dublin. In this instance the Company, thankfully, had the luxury of an alternative site.

The location of the Dublin ship loading facility is interesting in that it is exactly 100 metres across the harbour from a large grain off-loading terminal which in turn is beside a large milling operation. The planning application for this facility was again processed in record time and with three objectors only. The decision to opt for the Dublin terminal and to rail the concentrate from Navan substantially extended the baseline programme and necessitated soil and herbage sampling for farm land and urban gardens along almost 64 km of track, together with an extensive atmospheric monitoring programme around the terminal area, monitoring harbour water and an estuary mud study.

I have mentioned the tailings pond planning application. This particular development led us into an unusual environmental field when an unrecorded prehistoric settlement was identified right on the dam wall line and also the location of an ancient Church, adjacent to another section of the dam wall was confirmed. Extensive excavations had to be undertaken on both sites and experience emphasizes the advisability of making provision for archaeological parameters in any EIA methodology.

While baseline monitoring was on a fairly extensive scale, Tara's planning consent conditions insist on an equally extensive and intensive operations monitoring programme. Compliance with these conditions necessitates the collection and assaying of 50 water samples on daily, weekly and monthly frequencies. A total of 30 well water samples are taken quarterly and 36 dust samples are taken monthly from 9 deposit samplers. Sixteen directional high volume samplers are operated on a weekly sampling frequency. Five emission sources including drier stacks are monitored weekly and five others monitored on a monthly basis. Soil is sampled annually at 120 locations and herbage twice per year at 120 stations. In addition the Company operated, for a five year period, a series of farm units around its property by the monitoring of which it was intended to establish a relationship between air pollution and the end product in the grazing animal. Stock was weighed and blood sampled monthly and tissue samples were taken from all animals slaughtered on disposal. It was therefore possible to determine live weight gain and evaluate on-going physiological parameters. An additional benefit of the farm enterprise was that knowing of its existence, adjoining farmers were and hopefully still are reluctant to pursue spurious claims for adverse effect. This project has recently been terminated after two year's baseline and three year's operational experience since it was considered that sufficient data had been accumulated for the present. It is possible that the project will be reactivated in the future should circumstances demand it.

8.2.4 Retrospective analysis

There can be no doubt that Tara's approach to EIA was of positive benefit in obtaining rapid planning permission and ensuring favourable community response throughout the development and operational phases.

The most important decision of the entire strategy was the initial recognition, that there were two fundamental problems to be tackled, in parallel, one ecological and one psychosociological. In the early stages of the development the potential aesthetic impact of the mine, with its coal mining and slag heap connotation, shared first place in people's minds with the fear of 'lead poisoning'. Tara put considerable effort into softening the visual impact of the buildings with both the geometry of their various profiles and the colour scheme. Due to the pressures experienced at a very early stage in the pre-planning application discussions with the Local Authority, the

Company agreed to lower the profile of the concentrator by excavating 10 m into bed rock.

It was landscaping that gave Tara its first break through the very considerable public apprehension barrier in that it established a large tree nursery, in 1972, containing 50 000 trees which were eventually grown to semi-mature size prior to using them on the site. This was seen as a major and early commitment to public consideration.

The attention paid to community views and the progressive educational approach was extremely beneficial. Tara's community relations strategy avoided the potentially explosive public meeting and concentrated on a small group and individual relationship. The size of each group and the communication method used was geared to the identity characteristics of those involved.

In order to assess the success of the original community relations strategy Tara audited its progress using a public opinion survey. This indicated that some changes in direction and emphases were required to achieve the objectives set. This was a most valuable tool and one which is recommended to those in a similar position.

8.2.5 EIA requirements in Ireland

In the Republic of Ireland EIA is already provided for in Irish planning legislation, namely *Local Government (Planning and Development) Act*, 1976 which introduces the concept. Local Government (Planning and Development) regulations, 1977 subsequently set a £5 million threshold on the requirement for EIA. The 1976 Act empowers the Minister of the Environment to make regulations providing for the furnishing to Planning Authorities, in cases where development to which a planning application relates, 'of a written study of what, if any, effect the proposed development, if carried out, would have on the environment relative to the place where that development is to take place'. The situation in Ireland, however, has shortcomings of which two of the most obvious are:

(a) It applies to developments for which planning application is made and not to – exempted developments – such as works of National and Local Government, e.g. Arterial Drainage and Forestry are not subjected to it.
(b) The £5 million limit ignores the small industry with a high pollution potential.

8.2.6 Conclusions

Following experience of EIA in Ireland I must confess that I can muster little support for those in vehement opposition to it. Certainly it can be overdone but I cannot regard it as the obstacle that many see it. I see it, properly defined and wisely used, as an advantage rather than a deterrent. It is a mechanism for ensuring the early and orderly consideration of all relevant issues and for the involvement of affected communities. It is in this last area that its true benefit lies. We have entered an era when the people decide. It is therefore in the interests of developers to ensure that they the people, are equipped to so so with the confidence that their concern is recognized and their future life-style protected.

I am confident that the future industrial, agricultural and infrastructural development of Ireland can proceed more efficiently and economically within an EIA concept than in its absence.

8.3 A RETROSPECTIVE VIEW OF THE ENVIRONMENTAL IMPACT ON UPPER TEESDALE OF THE COW GREEN RESERVOIR

T. J. Bines, J. P. Doody, I. H. Findlay and M. J. Hudson

8.3.1 Introduction

The controversy and debate prior to the construction in the North Pennines of the Cow Green reservoir in Upper Teesdale (Fig. 8.1) has been adequately described elsewhere (Gregory, 1971; Whitby and Willis, 1978) and it is not the intention of this paper to concern itself with the decision-making processes involved. The reservoir proposal, presented as a Private Members Bill, was examined by the Select Committees of both Houses of Parliament and subsequently debated. In this paper we examine the predictions made at the time, their accuracy and general emphasis.

8.3.2. The Cow Green development

The requirements in the late 1960s for increasing the summer water flow in the River Tees, came from the needs of the growing industries on Teesside. This growth in demand was mainly caused by the rapid development by ICI of an alternative and more economic process for producing ammoniacal fertilizers. The new process required large volumes of water and when taken together with the water needs of the other Teesside industries the estimated new demand for water invalidated the water requirements of industrial Teesside predicted by the Tees Valley and Cleveland Water Board.* In 1959 the Water Board estimated that 295×10^3 m^3 day^{-1} would be adequate to meet requirements up to 1983 and had planned accordingly,

*Following the 1973 *Water Act*, nine Regional Water Authorities were established in England – the TVCWB became part of the Northumbria Water Authority.

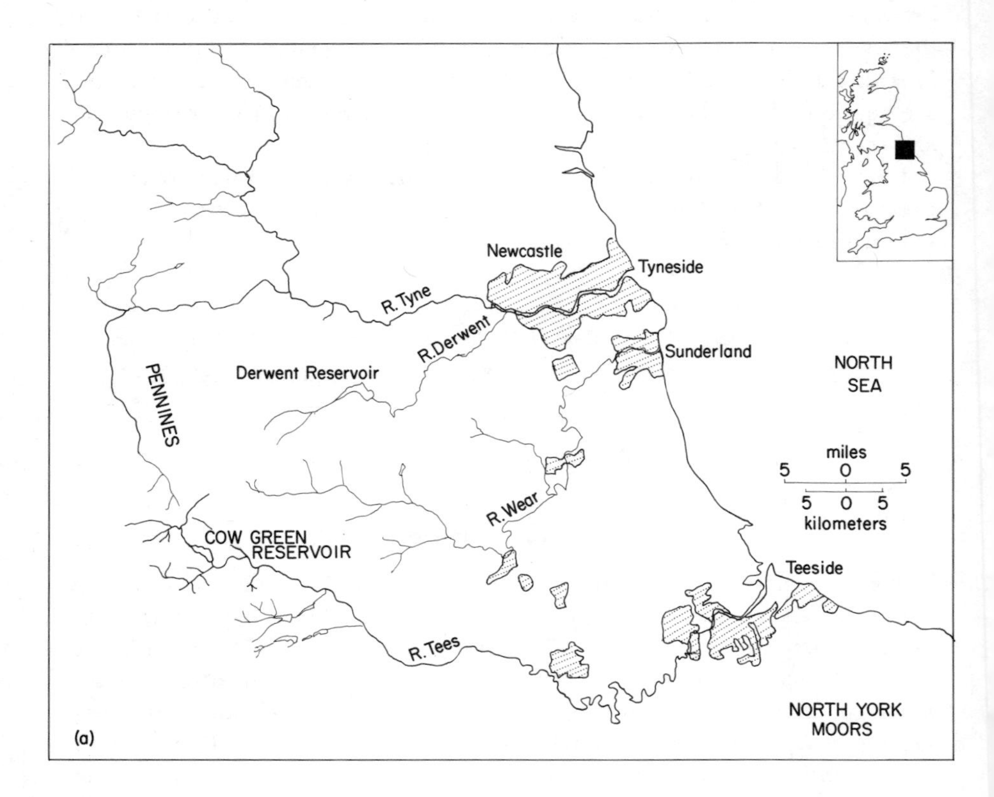

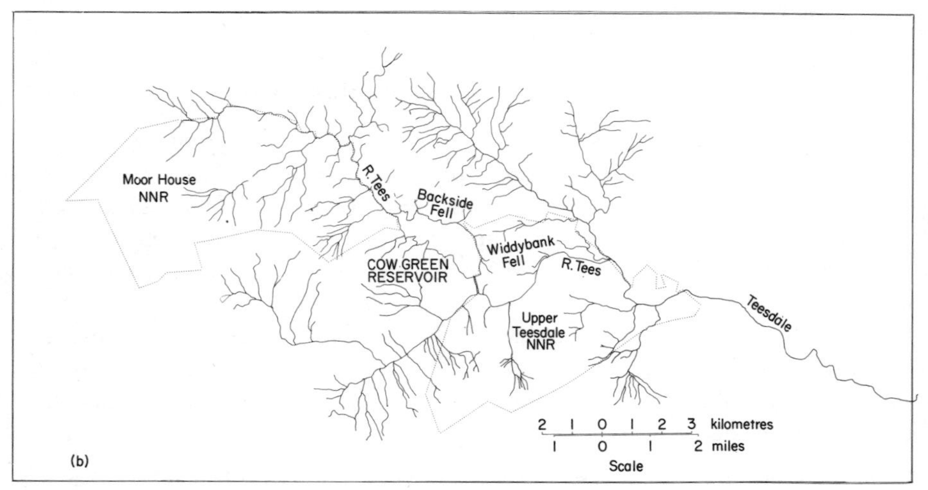

Fig. 8.1 (a) and (b) Location of Cow Green Reservoir.

whereas in 1964 the picture had changed dramatically with a predicted requirement of 455×10^3 m^3 day^{-1} by 1970. In order to meet the new demand the Water Board elected, with industrial support, to press for a river regulating reservoir at Cow Green in Upper Teesdale. This proposal involved the construction of a dam across the river Tees above Cauldron Snout which would impound some 312 hectares with a top water level at 488·5 m. The reservoir would fill in winter from the large catchment area of the Upper Tees, any excess water flowing over the dam. During winter the major streams feeding into the River Tees below Cauldron Snout would, it was considered, together with any surplus reservoir water, meet Teeside's needs. In summer the water held in the reservoir would be released in quantities appropriate to guarantee the required supply. At all times a statutory daily minimal flow of $38{\cdot}6 \times 10^3$ m^3 day^{-1} would be met.

The opposition to this proposal from the ecological community was almost unanimous and it created an international controversy at the time. To put the opposition into perspective the reasons for considering Upper Teesdale – and the proposed reservoir basin in particular – of outstanding ecological importance must be considered.

Since the eighteenth century, Upper Teesdale has been recognized as of outstanding botanical and scenic interest (Clapham, 1978). About 3500 ha of the Upper Teesdale area are now designated as a National Nature Reserve. A combination of features – climatic, altitudinal, geological and pedological – have enabled an unusual complex of plant communities to persist since the last glaciation (Command 7122, Ministry of Town and Country Planning 1947; Ratcliffe, 1977). The communities range from those associated with the River Tees through herb-rich meadows and pastures to birch and juniper woodland and, on the higher ground, dwarf shrub heath and blanket bog communities (Pigott, 1956; Godwin and Walters, 1967; Bradshaw and Jones, 1976; Clapham, 1978). These are interspersed with species-rich limestone grassland and associated flush communities which give place to sub-montane vegetation on the summit of Mickle Fell (790 m).

The most important feature of Upper Teesdale, part of which unhappily coincided with the proposed reservoir site, is the 'sugar' limestone produced metamorphically as a result of contact between the Melmerby Scar limestone and the intrusive Whin Sill, which latter forms impressive inland escarpments. Mineralization occurred and the mineral veins have been extensively worked for lead and barytes. Indeed at the time of the controversy, the Cow Green Mine had only recently closed down (1954) and the buildings were a conspicuous feature in the proposed reservoir basin.

The formation and character of the 'sugar' limestone has been described in Johnson and Dunham (1963), Johnson *et al.* (1971), Robinson (1971) and Bines (1977). The exposures of this rare crystalline rock with its rapid draining and peculiar nutrient properties (Jeffrey and Pigott, 1973) support grass swards, close-grazed by sheep and dominated by *Sesleria albicans* and *Festuca ovina* in which *Agrostis tenuis* and *Festuca rubra* are conspicuous by their absence, and extensive systems of species-rich calcareous flushes. The 'sugar' limestone exposures, and the associated soils and plant communities occur in two main centres – Cronkley Fell to the south of the River Tees and Widdybank Fell to the north, the latter forming the eastern part of the proposed Cow Green reservoir basin. In both areas the swards have 'an extraordinary number and abundance of rare or local montane plants' (Ratcliffe, 1977) at surprisingly low altitude.

What then were the interests in the proposed reservoir basin which aroused such opposition? The prime concern was the direct loss of examples of the species-rich 'sugar' limestone grassland and the related flush communities. These included such rarities as *Gentiana verna*, *Kobresia simpliciuscula*, *Carex capillaris*, *Minuartia verna*. *Tofieldia pusilla*, *Juncus alpino-articulatus* and *Viola rupestris*. The vegetation of the calcareous springs was made up of *Cratoneuron commutatum* hummocks and smaller cushions of a range of interesting bryophytes. It was clear that significant losses of such species as *Saxifraga aizoides*, *Juncus triglumis*, *J. alpino-articulatus*, *Selaginella selaginoides*, *Carex capillaris*, *Equisetum variegatum*, *Tofieldia pusilla*, *Primula farinosa*, *Thalictrum alpinum* and *Trollius europaeus*, would occur, as communities containing these species, in varying abundance were found below top water line. There were also sedge-dominated calcareous soligenous mires in the basin of the proposed reservoir, in some of which the rare alpine *Kobresia simpliciuscula* was the major species. These too supported a rich flora including *P. farinosa*, *S. selaginoides* and *Cinclidium stygium*. These communities were principally associated with two major streams or sikes arising from springs on Widdybank Fell. In both of these rich algal communities containing *Rivularia biasolettiana* and *Phormidium* species occurred abundantly.

The bogs in the proposed basin were also of interest ranging from Callunetum on drier ground through Trichophoreto-Callunetum (certain examples of which were rich in *Sphagnum* spp.) to true Pennine blanket-bog on the wettest ground. There were three discrete areas of blanket bog in the basin which lay below the proposed top water line and it was anticipated that a fourth would be damaged during dam construction. The *Sphagnum* dominated faces of blanket peat which is found in the Pennines is very local as a consequence of drainage operations and one of the bog areas was especially important since the rare *S. imbricatum* and *S. fuscum* were important components of its flora.

Intermediate between the blanket peat and truly calcareous communities a range of flushed bog communities were found including Sphagneto-Juncetum effusi, Sphagneto-Caricetum and Trichophoreto-Eriophoretum caricetosum. These too have been inundated.

The Tees immediately above the planned dam site was known as 'The

Weel' and was here a slow flowing river some 35 m wide and up to 2 m deep. Because of its sluggish nature the river bottom was generally overlain by a few centimetres of mud and a high proportion of the adjoining area consisted of alluvial flats over which the river flowed, changing its course periodically and leaving pools of standing water in abandoned channels. The tributary streams were also slow-flowing where they crossed the alluvial flats and aquatic macrophytes had established themselves in both the Weel and the tributaries. In contrast both above and below this section the Tees assumed the normal upland river characteristics of fast movement, strong scouring action and stony bottom.

The Reverend Proctor (in Clapham, 1978) records that there were 26 macrophytes to be found in the reservoir basin and an interesting and diverse vegetation had become established on the alluvial flats. The two most interesting species were perhaps *Carex aquatilis* and *C. paupercula*. However, at the time of the proposal these were not known to occur in the basin and were only found during later survey work.

The opposers of the Cow Green scheme gave evidence of the existence in the reservoir basin of unique plant communities and hence the inescapable direct loss of scientific interest and conservation value which would result from inundation. Following the decision to proceed with the construction of the reservoir (Royal Assent, 22 March 1967) certain arrangements were made in an attempt to restrict damage to the natural history of the area to that which was unavoidable.

At this time ICI offered £100 000 for research to be centered in the Cow Green basin and upon the effect of the reservoir on the ecology of the area. This money was allocated on behalf of the Trustees by a scientific committee and enabled a crash research programme to be carried out. The results of the work have been published in numerous papers and a concise account is available in Clapham (1978).

8.3.3 The ecological consequences of construction

The effects of the reservoir proposal are now considered in relation to both the predicted effects and those recorded following completion of construction and inundation in 1970.

(a) Direct loss

The most obvious and important of the direct effects of the proposal was the submergence of some eight hectares containing the 'sugar' limestone communities briefly described above. This amounted to 12 % of the total resource on Widdybank Fell and some 7 % of the 'sugar' limestone communities of Upper Teesdale as a whole. These figures do not, however, convey the true loss since much of the interest centred on the unusual complex of communities in the reservoir basin. For instance the gradation from limestone grassland through herb and grass-rich stands of Callunetum transitional between these and the bogs in the basin is a very rare occurrence in Britain and of very great ecological, as opposed to floristic, interest and importance. It proved difficult for the opposition to the reservoir proposal to convey this point when presenting evidence – much of the examination being taken up with the details of the threatened loss of individual species – i.e. the proportion of the populations of some species which would be lost on inundation. The loss of the diverse aquatic vegetation and blanket bogs – cores from which have assisted in the piecing together of the vegetation history of the area – was also of great importance. It is unfortunately true to say that the Cow Green reservoir has destroyed an association of plant communities unique in Britain.

(b) Effects of construction

One of the fears expressed by the opponents of the scheme concerned the damage which would be caused:

(1) To the reservoir basin prior to inundation, which would interfere with the extensive operation that was to be launched to rescue species of highest scientific value which would otherwise be lost. It was also argued, that inundation having offered the unique opportunity of undertaking what would otherwise have been regarded as unacceptably destructive research (Jeffrey and Pigott, 1973), the experiments would be severely hampered by ongoing construction work.

(2) To the areas above top water-line where damage particularly from movement of machinery was envisaged. In fact the construction was carried out (by the contractors and other agencies concerned) very sympathetically. The appointment of a Site Research Officer (SRO), whose prime function was to liaise between those engaged on construction and those involved in scientific work, – a post unique at that time – was of considerable assistance and significance in this.

The details of the siting of roads (both above and below top water line) were especially important if damage was to be minimized. The main access road was sited within the reservoir basin and skirted the areas of highest botanical importance, so that research work might proceed unhindered. This, coupled with the careful siting, again below top water line, of the works infrastructure and machinery parks enabled both the scientific and construction teams to carry out their work programmes with remarkably little friction (Fig. 8.2).

The development (other than the actual inundation of 311·8 hectares) which has caused the greatest disturbance to the area resulted from publicity

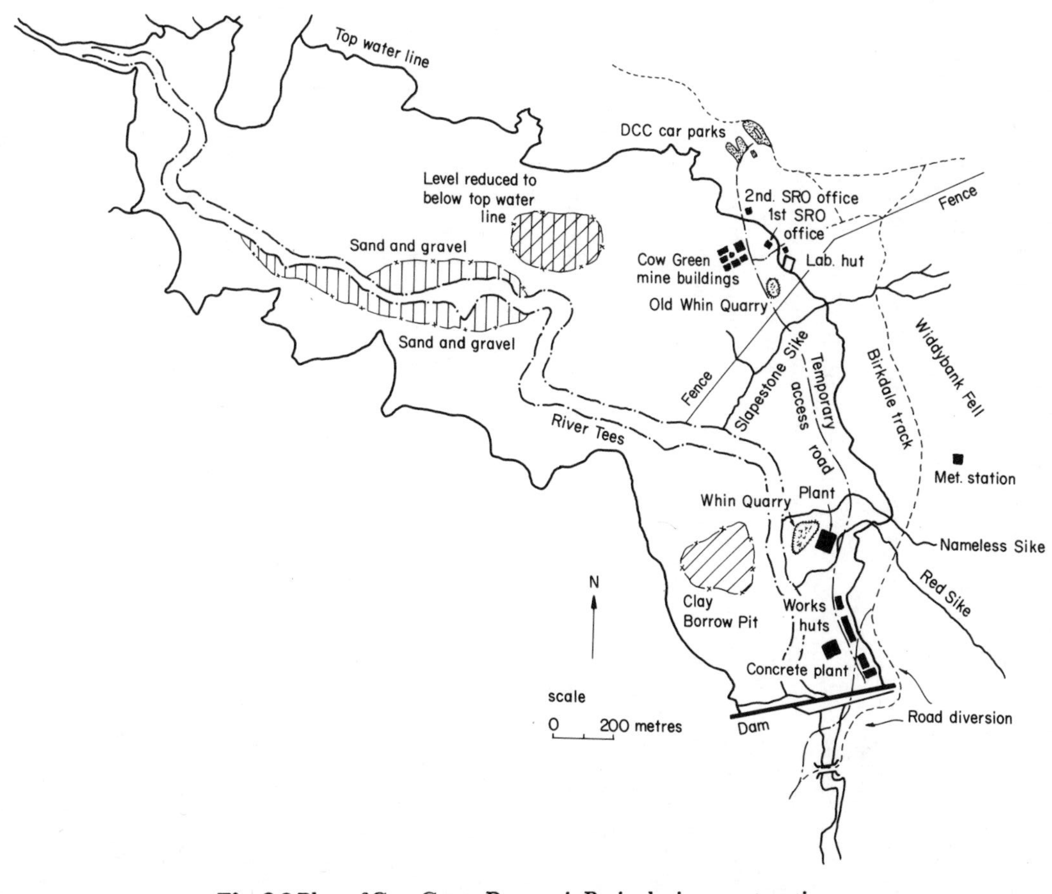

Fig. 8.2 Plan of Cow Green Reservoir Basin during construction.

over the controversy and the improvement of access to Widdybank Fell. These effects were anticipated by the opposers of the scheme and are described elsewhere. There was an obvious need to improve road access to the site for construction purposes. The metalled track which led to the site had to be resurfaced and passing bays provided to enable machinery, supply lorries and indeed the work-force itself, to gain easy access to the construction site. This improvement, by virtue of the careful attention that was paid to conservation requirements during construction and surfacing work, did not affect the biological interest of the site directly, but its indirect effect upon visitor pressures was foreseen and has fully lived up to expectations (see below).

The reservoir basin itself furnished much of the material needed for construction of the dam – clay, gravel and whin sill were all won from within the basin.

In summary, the direct damage to the botanical interest of the site during the construction period was relatively minor, both above and below the top water line, and enabled the intensive research programme to proceed successfully (Fig. 8.2). However, access improvements and more especially, publicity concerning the controversy brought about considerable increases in visitor pressure, which has in turn induced change in the flora of Widdybank Fell.

(c) Leaks and grouting

In the early stages of the Cow Green debate it was feared by the opposition that a reservoir impounded over Carboniferous limestone rocks would leak. This would necessitate extensive grouting operations (insertion of a concrete curtain) causing considerable damage to the flora above top water line. The proposers of the scheme dismissed this and Knill has published the details of his consideration of this matter (Kennard and Knill, 1969). The proponents were indeed so firmly convinced that leaking would not occur that they gave binding undertakings that they would not grout. The situation has been monitored by consultant engineers, and to date we have no evidence that the reservoir is leaking.

(d) Aquatic effects

The opponents of the reservoir scheme were concerned over the effect that river regulation would have on the vegetation downstream of the dam through a reduction in the flash floods so characteristic of the River Tees. The scouring action of the river would be minimal at periods when this could be very important biologically. Pigott (1956) argued that certain sites along the River Tees, below Cauldron Snout, consisting of steep banks supporting plant communities of national significance, were dependent on continuing instability to maintain the open communities. It was considered that the instability was, at least in part, due to the action of flood waters. Similarly, populations of the rare shrubbery cinquefoil (*Potentilla fruticosa*) which are closely associated with the river banks in Teesdale, were thought to require some periodic flooding or scouring action if the colonies were to survive. This situation may be compared with that of the Burren, Co. Clare, where Praeger (1934) has described the association of *P. fruticosa* with seasonal flooding along the margins of turloughs (Summer-dry lakes).

Reduction in the amount of flooding by the River Tees, especially in the spring to autumn periods, was thought therefore to put at risk these features. A monitoring programme has been established on the most important eroding riverside bank and the *P. fruticosa* colonies are being closely studied for any signs of change. As yet, we have detected no changes since construction, as a consequence of the lack of flooding outside the winter period. It should be pointed out, however, that scouring still occurs in winter and this may be sufficient to meet the needs of plant communities adapted to unstable river habitats. The Water Board undertook to provide artificial spates when so requested by NCC, but to date this has not been considered necessary.

There have, however, been profound changes in the aquatic environment, which were not discussed at the time of the controversy. These are concerned with the populations of aquatic macrophytes (Proctor, in Clapham, 1978) and with changes in the invertebrate and vertebrate populations which latter have been thoroughly described by Crisp (in Clapham, 1978).

Carex aquatilis – not known to be in the reservoir basin at the time of the controversy, has been lost from the Cow Green area, but has since been found at two stations near Middleton-in-Teesdale. Of the 26 aquatic species listed as present in the basin prior to inundation, only 12 have been subsequently recorded from the reservoir, though a further eight have been found growing in the vicinity. Apart from *C. aquatilis*, *Ranunculus peltatus*, *Potamogeton alpinus* and *Carex paupercula* have all disappeared. However, species only found at intervals, for example *Sparganium angustifolium* which was recorded in the 1930s, but not found subsequently prior to inundation, have since been rediscovered.

Crisp *et al.* (1974) described the distribution of the three fish species – trout (*Salmo trutta*), minnow (*Phoxinus phoxinus*) and bullhead (*Cottus gobio*) – in the Cow Green area. They found that minnows were scarce below Cauldron Snout and never occurred upstream of the reservoir basin. Trout were found throughout most of the length of nearly every stream, though not above the waterfalls upstream of the confluence of Lodgegill and Rowantree Sikes, while bullheads, not abundant in the main river, occurred throughout the year in the lower reaches of most of the afferent streams, but in no case above top water level of the proposed reservoir.

In regions of sluggish flow with *Callitriche* spp., *Potamogeton* spp. and *Carex* spp. the lower reaches of Dubby Sike and the Whitespot/Dead Crook/Furness Lodge Sike systems held minnows but they too were never found above the future top water level.

The trout populations of the Tees prior to impoundment were characterized by slow growth, low population density and low annual production. Both trout and bullhead enjoyed a high annual survival rate which Crisp (in Clapham, 1978) regarded as a reflection of low metabolic rate, with the species benefiting from spreading increases in weight over a number of year classes so facilitating efficient utilization of available resources, and enabling several year classes to participate in reproduction each year, thus providing an 'insurance' against the effects of a succession of poor breeding seasons.

Impoundment has resulted in a marked reduction in the extent and rapidity of the environmental fluctuations of the natural river. Crisp (in Clapham, 1978) recorded the following effects on the river downstream of the dam:

(1) A delay of the rise in water temperature in Spring by one month and of its fall in Autumn by two weeks. Moreover the Summer peak of water temperature is reduced by almost 2 °C and the seasonal range of daily temperature changes is almost entirely suppressed.
(2) The variations in concentration of the common elements in solution were much reduced. Calcium concentrations in the unregulated river varied from 37 ppm in dry weather to 3·5 ppm during spates whereas following impoundment rates have varied between 6·4 and 8·1 ppm.
(3) Small flash spates and very low dry weather flows are eliminated permitting the development of a permanent cover of vegetation affording additional habitats for the aquatic fauna.
(4) Impoundment markedly reduces the burden of suspended solids carried by the Tees since most brought by the river into the reservoir settle out there. This has the effect of significantly reducing the amount of potential food available for detritivores in the river.

In the reservoir basin itself, impoundment led to a twenty-fold increase in the surface area of water with the following consequences (Crisp, in Clapham, 1978):

(1) The submergence of large areas of terrestrial vegetation which decomposes only slowly providing in the meantime mineral nutrients and a food supply to the aquatic flora and fauna developing in the reservoir. Much terrestrial animal life was also submerged.
(2) A gradual increase in the size of fish populations to exploit the greatly extended, although changed, habitat.

(3) It is estimated that of the order of 4 million kg of suspended solids will settle out in the reservoir each year providing an abundant food supply for detritivorous invertebrates.

(4) Thermal stratification occurs comparatively rarely and the reservoir is a body of water in which ionic composition, temperature and dissolved oxygen concentrations are largely uniform at all depths throughout the year.

The changes in the invertebrate fauna were recorded by Armitage (1976). He found that upland river species, chiefly Plecoptera (stone flies) such as *Leuctra moselyi* and *L. inermis* and Ephemeroptera (may-flies) such as *Rhithrogena semicolorata* and *Baetis rhodani* were largely replaced by aquatic worms, the water-shrimp *Gammarus pulex*, the leech *Glossiphonia complanata* and by Chironomid midges. The population density and biomass of the bottom fauna increased greatly as compared with that of the river, and a zooplankton dominated by *Daphnia hyalina* quickly developed within the reservoir. Armitage and Capper (1976) estimated that 150 kg dry weight of zooplankton, chiefly *Daphnia hyalina*, was being discharged annually in water from the reservoir, and represented a significant food source for detritus feeders for several kilometres downstream of the dam. One species of may-fly *Ephemera danica*, which occurred in the lower reaches of Dubby Sike has not been recorded since the basin was flooded.

Below the dam Baetid may-flies and non-biting midges became much more numerous while Ecdyonurid may-flies and stone-flies decreased in numbers. *Hydra vulgaris* and small worms of the genus *Nais* became significant components of the fauna and eleven other species of various groups were recorded in the river for the first time following impoundment.

The limestone grasslands and other vegetation submerged by the filling of the reservoir amounting to over 300 ha contained large numbers of earthworms and craneflies, together with other invertebrates. Estimates of 95 000 kg of earthworms and 35 000 kg of craneflies have been advanced and from late 1970 to the early spring of 1973 terrestrial invertebrates, particularly earthworms, were the principal stomach contents of trout in the reservoir. Following the gradual development of the planktonic and benthic fauna, aquatic organisms have since formed the major proportion of the trout diet.

As might be anticipated, the major alteration of the aquatic environment at Cow Green brought about a significant change in growth of the trout. As early as the Autumn of 1970 a marked increase in the rate of growth and average length of the reservoir trout was apparent. This was followed, however, by a slight decrease between 1972 and 1975. Crisp found that four year old reservoir fish reached an average length of 34 cm and a weight of 454 g, while river trout were less than 20 cm long and only about 100 g weight at the same age. By virtue of their larger size the reservoir trout lay more and larger eggs. Numbers of large trout (over 15 cm in length) increased approximately five-fold between July and September 1975. These increases have made the Cow Green reservoir a viable sporting fishery without stocking – the average trout caught weighing almost one pound (446 g) and an average catch of 0·72 fish per angler visit.

The average weight of trout taken by anglers has shown little sign of a decrease in growth rate, but some reduction is anticipated in due course in response to increasing population density.

There have also been changes in the distribution and movements of fish. Before impoundment, bullheads were found in the afferent streams only below the future top water level. Within about a year of the reservoir filling, bullheads were found from time to time in most of the sikes and a small breeding population is established in Near Hole Sike. This latter population is exceptional and it is thought that the majority form part of the population breeding in the reservoir in Spring, entering the sikes in Summer and returning to the reservoir again in Autumn. The species has thus extended its distribution and appears to have developed a previously unrecorded migratory habit. The Cow Green trout have in contrast always exhibited migratory behaviour, most of those in the Tees spawning in the afferent streams or in the headwaters. Recruitment to the river population was chiefly through the stream fish entering the main river, the streams holding their own resident populations of sexually mature fish. Interchanges of stream and river fish have continued since impoundment. However, the average age at entry into the reservoir has fallen. Whilst before impoundment fewer than 10 % moved into the Tees in their first year and over 80 % in the second to fourth years, by 1973 30 % entered in the first year and only a negligible proportion after their third year.

(e) Climatic effects

The climate in Upper Teesdale is best described as sub-arctic (see Table 8.1) with a short vegetative growing period lasting approximately from mid-April to mid-October (Manley, 1952). This late start to growth and short season not only impedes vegetative growth, but affects the length of time available for flowering and seed ripening (Pigott in Clapham, 1978). This may well be one of the main factors (the others being related to nutrient status and management) which prevent many lowland plant species from competing successfully with the less vigorous of the arctic/alpine species found on Widdybank Fell.

At the time of the controversy it was considered by the opponents of the scheme that the Cow Green reservoir, if built, would significantly alter the micro-climate of Widdybank Fell. The effect was variously predicted either

Table 8.1 Summary of climatological data for Widdybank Fell (July 1968–June 1978)

	Jan	*Feb*	*Mar*	*Apr*	*May*	*June*	*July*	*Aug*	*Sept*	*Oct*	*Nov*	*Dec*	*Year*
Rainfall													
Average	196·4	134·6	120·5	99·2	84·5	79·8	88·1	100·1	139·6	123·6	170·7	150·1	1487·2
Max. fall	323·0	313·2	240·1	166·8	208·8	153·9	118·9	184·1	237·7	236·7	295·2	253·0	
Min. fall	91·7	34·5	49·9	17·1	28·6	38·1	53·9	22·1	24·3	24·9	86·7	69·9	
Most in a day	35·6	60·5	42·7	56·1	39·1	30·2	38·2	55·9	66·5	47·0	52·6	66·8	
Air, temperature °C													
Average max.	3·3	2·5	4·3	7·1	11·5	14·3	16·0	16·2	12·7	10·2	5·2	4·0	8·9
Average min.	−1·3	−2·5	−1·4	−0·1	3·1	6·0	8·3	8·5	6·5	4·5	0·1	−0·9	2·6
Mean	1·0	0·0	1·5	3·5	7·3	10·1	12·2	12·3	9·6	7·3	2·7	1·5	5·7
Extreme max.	12·7	10·0	14·4	16·7	21·7	26·3	26·4	25·8	21·2	20·8	12·5	13·1	
Extreme min.	−13·7	−14·6	−10·1	−8·2	−2·7	−1·3	−0·1	0·7	0·1	−4·8	−8·9	−8·3	
Average no. of days of													
Rain days 0·2 mm	25·5	21·4	22·1	18·7	17·9	15·5	16·6	14·4	18·2	20·2	23·3	21·8	235·6
Air frost	19·3	21·3	21·0	15·0	4·5	0·6	0·1	0·0	0·0	2·8	14·6	17·7	116·9
Ground frost	24·6	24·2	26·0	21·4	14·1	6·3	1·9	1·5	4·8	9·0	20·0	23·5	177·3
Snow lying at 0900 GMT	12·4	13·0	10·8	3·7	0·1	0·1	0·0	0·0	0·0	0·1	6·1	9·6	55·9

to be most pronounced in Autumn when the reservoir would act as a heat source producing a slight but significant amelioration in temperature close to the margin or that the mean monthly temperatures would be consistently higher.

Few data have been obtained on climate amelioration by large inland water bodies and Gregory and Smith (1967) examined the effect of one of the then existing Teesdale reservoirs on temperature amelioration. They predicted that changes in air temperature would amount at most, to an increase of 2·5 °C (and more often than not, of less than 1 °C) at the reservoir margin on the Widdybank Fell side and decrease progressively further inland for about 400 m when a 0·5 °C change might be detected. However, these effects would be short lived, lasting a few hours only and would be dependent on winds blowing across the resevoir surface towards Widdybank Fell.

The effect of this potential reduction in climatic severity, even if short lived, on the plant communities containing species with diverse geographical distributions posed an unquantifiable threat. It was thought that the competitive balance between the species in the communities could well be upset and that critical species growing close to their climatic limit might well not be able to survive.

As part of the programme of investigation into the ecology of the area prior to flooding, and to monitor the effects of the reservoir, a climatological station was established on Widdybank Fell, some 200 m from the reservoir margin and daily standard observations of temperature, humidity, wind speed, rainfall and sunshine have been recorded since late 1967. As the reservoir filled in 1970–71, only three years pre-impoundment data are available for comparison with post-impoundment data and the assessment of possible effects of reservoir construction.

The collected data have been compared with that from the neighbouring Moor House climatological station, 7·5 km further up the Tees Valley, where continuous recording dates from 1952 (Harding, 1978). This station, is in a generally similar location but far enough away from the reservoir not to be affected by it. Harding found (Fig. 8.3) that there has been little effect on the overall climate following impoundment but, as predicted, by Gregory and Smith (1967) small temperature changes have apparently occurred. These it was anticipated would be of short duration and perhaps occur only at certain times of the year. However, Harding has also found great seasonal variation. The fact that the climatological station is so far from the reservoir margin implies that changes much greater than those predicted by Gregory and Smith have arisen. Thus, the lowest air minimum temperatures are moderated by about 1 °C and the lowest grass minimum (or ground) temperatures are ameliorated by between 2 °C and 3 °C. This has therefore reduced the severity of the climate somewhat with a 15 % reduction in the number of frosts. The major part (70 %) of this effect occurs between April and September when frosts are generally of low intensity.

Thus the only observed effect to date has been a moderating influence on a small proportion of nights when there is low wind speed and intense nocturnal cooling. In these conditions it would seem that air is drawn eastwards across the reservoir surface, warmed in its passage and then, in turn warms Widdybank Fell. The climate is so severe that many consider these effects to be inconsequential. The studies of the population dynamics of selected species undertaken by Bradshaw in Clapham (1978) and Bradshaw and Doody (1978) in the first years after impoundment have not indicated that these

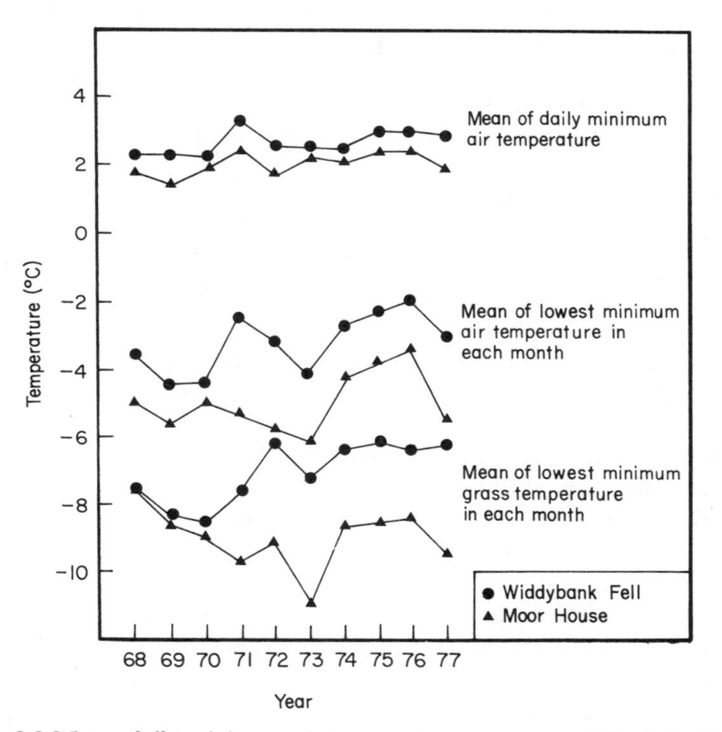

Fig. 8.3 Means daily minimum air temperature on an annual basis for Widdybank Fell and Moor House climatological stations (from Harding, 1978).

changes have as yet had any significant effect. This does not rule out the possibility of changes in plant community composition in the longer term.

(f) The gene bank

The decision to proceed with the construction of the Cow Green dam offered the opportunity to establish a gene bank of those species, which would otherwise be lost due to inundation. Thus, during the Summer of 1970, specimens of individual species together with turves and soil were carefully removed from the area below the future top water line.

The material consisting of 85 species, and including the majority of the rarities was sent to a number of workers largely in Universities who requested it. The major part, with a portion set aside for the Northumberland and Durham Naturalists' Trust, was taken by Durham and Manchester Universities.

In Durham the plants were kept in concrete tubs as single species stands and although attempts were made to retain turves and create an artificial stream with associated species, this was not entirely satisfactory. It was found that regular weeding was essential to prevent the smothering of plants by more competitive species. This was not always successful and it became clear that different species reacted in different ways to the change of environment. Thus, for example, *Draba incana* normally a short-lived monocarpic perennial which does not flower in Upper Teesdale until three years or older and then dies, flowered in the same year following Spring germination and some plants survived to flower a second time the following year (Doody, 1975). *Primula farinosa* reacted to the amelioration of the climate by becoming much more robust and producing a great many more flowers than in Teesdale. By contrast *Gentiana verna* and *Viola rupestris* showed few changes in growth habit, although the latter did produce open flowers in both Spring and late Summer, a phenomenon not observed on Widdybank Fell. Some of the species, set aside for the Trust, survived and have been transported back to Teesdale where they have been used in a display outside the Durham County Conservation Trust Information Centre at Bowlees. The fact that this material was available ensured that further pressure was not placed on the population within the NNR and has enabled easy viewing of selected rarities, as well as a number of the more common species.

In Manchester similar problems were encountered in tending species and again it was found that monoculture was the most effective technique (Valentine and Cranston, in preparation). Even with care, changes occurred; *Gentiana verna*, *Primula farinosa*, and *Armeria maritima* survived well for the first three years, but had died out by 1979. In the first two cases the overproduction of flowers, fruits and seeds is considered to have caused the failure, whilst *A. maritima* was persistently attacked by moth larvae. Success was achieved with other species, notably *Antennaria dioica*, *Galium boreale* and *Carex capillaris*, although in the case of *A. dioica* the male plants died out. On the whole the majority of the transplanted species survived more successfully at Durham than at Manchester.

In summary the maintainance of populations of rare species in cultivation has not proved easy, not least because of their different life strategies and their apparent inability to adapt to new climatic regimes. This may not be without relevance, for the populations remaining at Cow Green should significant climatic changes occur. If such changes were to arise then it is difficult to forsee what action could be taken to rectify the situation and thus it is vitally important to understand the detail of the life strategies of the species potentially at risk in any similar future situation.

In addition to the collection of plant specimens for the gene bank experiments described above, a collection of seed from selected species was made and sent to the Royal Botanic Gardens, for storage. Further material was collected and some 5000 sheets of 139 species were prepared and distributed to major reference herbaria.

(g) Visitor pressure

It has proved very difficult to obtain adequate figures indicative of the levels of visitor pressure. Indeed neither proposers nor opponents of the Bill at the time of the controversy were able with confidence to indicate the numbers of people then visiting Widdybank Fell. Such figures as have been obtained fall into four main categories:

(i) *Fishermen*

There appears to have been an approximate fivefold increase (from 10 to 50 per day) in the number of fishermen visiting the area. In general terms the fishing interest has had little effect on the ecology of the area but it should be noted that fishing is not permitted within the reserve. In fact fishing has assisted in the development of an understanding of the changing fish populations in the reservoir.

(ii) *Walkers*

The stretch of the Pennine Way which passes through Upper Teesdale is considered by many to be the most scenically attractive of the whole route. As a consequence the Cow Green reservoir proposal was strongly opposed by amenity bodies. The number of walkers in this country continues to grow annually and it would appear that the Pennine Way, in particular, is being used by more walkers each year with the local youth hostel returning 4000–5000 bed nights/annum.

The publicity over the Cow Green controversy undoubtedly stimulated use of the Teesdale stretch of the Pennine Way. However, it is extremely difficult to disentangle the effects of the publicity in stimulating usage from that arising from general growth in countryside recreation. At the present time it is estimated that total usage of the Pennine Way is of the order of 10 000 walkers per annum. This has necessitated maintenance work to prevent excessive path erosion, especially in blanket peat areas where the potential damage is most serious. This in turn may stimulate increased usage by making this section of the Pennine Way increasingly accessible to a range of walkers – from the experienced long-distance walker to the day visitor.

> 'Finally, there is the whole question of human disturbance. The presence of a water body in such a well-known area, adjacent to the Pennine Way and accessible by road, would be sure to create considerable visitor pressure, even without the provision of any specific attractions. Unless exceptionally restrictive measures were laid down to confine visitors to areas where they would do little harm, it is inevitable that the scientific interest would suffer. The delicacy of the habitat and the marginal conditions in which these communities survive would lead to the loss of much that is of scientific value through trampling, soil compaction, flower picking, collecting of plants by unscrupulous gardeners and botanists, pollution, fires and other disturbances, which could not be wholly prevented.' (NERC, 1966)

This was countered by the protagonists with the view that 'people have been walking there for decades and centuries and they are not likely to do any more damage in the future than they have in the past.' (Boydell, 1966). This last proposition overlooks entirely potential growth in visitor pressure. However, it is interesting to note that local residents recall playing football and cricket on the close cropped sugar limestone grasslands in their breaks from mining activities.

Prior to the Cow Green controversy, day visitors to Cauldron Snout and Widdybank Fell were relatively few in number, the majority making for the more easily accessible High Force waterfall lower down the valley.

In Spring and Summer visitors went to Cow Green, especially when the Spring gentian (*G. verna*) was in flower. The pressure on this species – one of the most conspicuous and attractive of the Teesdale rarities – in the form of picking and uprooting was probably greatest in the period immediately prior to dam construction. Indeed the situation then necessitated wardening by volunteers and even earlier in the 1930s the land owner had erected notices requesting visitors to leave the flowers for others to enjoy.

The number of visitors prior to the controversy is very difficult to ascertain, accounts varying from 'none at all' to 'great numbers'. The national coverage of the controversy by the media must certainly have increased the number of visitors and a survey during 1965 (Bradshaw, personal communication) indicates the numbers involved at weekends (Table 8.2). It is noticeable that the numbers reach a peak during the flowering time of *G. verna*. The figures are contrasted with figures for the same period in 1979,

Table 8.2 Comparison of visitors to Cauldron Snout 1965/1979

Date	1965	1979*	*Change*	*% increase*
Whit Sunday	169	497	+328	194
Whit Monday	249	302	+53	21
1st Sunday after Whit	177	390	+213	120
7th Sunday after Whit	136	259	+123	90
8th Sunday after Whit	81	302	+221	273
9th Sunday after Whit	97	324	+227	234
Total	909	2074	+1165	128

* Assumes 60 % of the car visiting population visits Cauldron Snout

which indicate the magnitude of change which has occurred following construction and this is explored further below.

The stimulus to the number of visitors by the construction of the Cow Green dam necessitated the provision of visitor car parks and careful control and guidance of visitors, both during and after construction to prevent the pressures outlined above from getting totally out of control. Following

Table 8.3 Car numbers visiting Cow Green, April–September, 1974–1979

	Year					
Month	1974	1975	1976	1977	1978	1979
April	2033	1198	2216	2250	1478	1553
May	2585	2358	2300	2279	2914	2078
June	2010	2183	2717	2606	2333	2063
July	2455	2465	2976	3226	3205	2498
August	3570	3671	4799	4419	3770	3326
September	1832	1536	1947	1919	1637	1773
Total	14 485	13 411	16 955	16 699	15 337	13 291

completion of the dam a regular monitoring of car numbers was carried out and Table 8.3 gives the total number of cars on a monthly basis for the last six years.

It can be seen that despite adverse weather conditions and petrol price increases the number of visitors to Cow Green has increased very substantially, but now appears to have reached equilibrium, with an estimated 60 000–70 000 people per annum visiting the site.

(iii) *Educational parties*

The Upper Teesdale area has been a training ground and source of inspiration for ecological study since the eighteenth century. The development of biological education in this country has enabled formal studies to be undertaken in areas such as this – the level of teaching and research accommodated ranging from that appropriate to young school children to the highest levels of botanical and ecological research (Clapham, 1978).

In the early fifties the then fast-growing interest in field studies brought many parties to Teesdale and the local Universities made frequent visits. It is thought that prior to reservoir construction some 12 Universities used the Upper Teesdale area. School parties have been using Teesdale for educational purposes since the 1920s, and whilst the number of parties prior to construction has not been precisely ascertained, it was certainly nothing like so great as at the present time.

In 1978 – a typical year – a total of 120 parties (6000 individuals) visited Widdybank Fell for various educational purposes. This represents a considerable increase, exactly as predicted by the Natural Environment Research Council in the evidence to the House of Lords.

In the original Private Members Bill it was envisaged that recreational facilities would be provided at Cow Green (as part of the justification for the costs of reservoir construction), but this was strongly opposed by the conservation and amenity interests and as a consequence such provisions were later removed from the Bill.

The increasing number of visitors to Cow Green, however, made it necessary to provide limited facilities to prevent damage to the delicate and vulnerable plant communities of Widdybank Fell, including car park and toilet facilities. As many of the visitors walk to Cauldron Snout there has arisen an opportunity to encourage an awareness of the natural history of the site – a process which itself may stimulate further visits, but perhaps by people with a greater understanding of the features of the area. A Nature Trail was therefore originated between Cow Green and Cauldron Snout by the Northumberland and Durham Naturalists' Trust. Since the extension of the National Nature Reserve this has been maintained by the Nature Conservancy Council.

In the case of Cow Green the prime functions of the nature trail were: to interpret the countryside to visitors; to educate school parties and others seeking information about the countryside and its ecology and to act as a management tool by channelling visitors along prescribed routes (Bayfield and Burrow, 1976). The trail has proved very successful and elicited an excellent response from the visiting public. Further development has taken place recently with the installation of an interpretative display and perhaps as a consequence of the educational initiatives, damage to the site is now largely contained.

The large influx of people into the countryside in recent years (Countryside Commission, 1978) makes the assessment of what would have happened at Cow Green if the reservoir had not been built, very difficult. The number of visitors to National Trust and DOE Ancient Monuments is perhaps one of the statistics that can be used to obtain some insight into the probable growth of visitor pressure in the absence of a reservoir, and they show a 7–10%/annum compound growth rate, but this picture differs substantially from that observed at Cow Green (Table 8.4). At the present time the annual day visitor population to the site averages some 60 000 compared with a pre-construction 'guestimate' of 2000 (Aponso, 1976).

The associated returns for nature trail leaflets sold are included in Table 8.4 and indicate a turnover of one leaflet per seven cars up to 1978, when a new leaflet was issued and interpretive facilities expanded. This appears to have had the effect of increasing the sale of booklets to a level of one for every four cars.

Table 8.4 Total car numbers, estimated day visitors and nature trail sales at Cow Green 1974–1979

	Year					
	1974†	1975	1976	1977	1978	1979
Total car numbers	14 485	15 837	20 572	19 792	20 072	15 147
Total day visitors*	46 342	53 846	72 632	71 251	72 259	54 529
Increase % on 1974	—	15%	57%	54%	56%	18%
Total nature trail leaflets sold	2543	2636	2829	2919	2854	3476
Ratio of leaflets sold : cars	1:5·7	1:6·0	1:7·3	1:6·8	1:7·0	1:4·4

* Durham CC (personal communication) annual conversion factor for car occupancy
† Figure only available for April–September inclusive

The majority of the visitors to Cow Green (62%) are from the Tyne/Tees conurbations and a further 7% from other parts of the North-East. The 'average' visitor to Cow Green stays some two hours and a minimum of 60% walk along at least part of the nature trail. The majority of these visitors walk to Cauldron Snout, 2·4 km and return, whilst a surprising 25% walk even further afield. Approximately one third of the total visitor population do not leave their vehicles at all, simply using their vehicle as a means of viewing the countryside – and this perhaps reflects the sub-arctic climate!

The visitor who uses the Nature Trail as requested will have little effect on the very high botanical interest. There is, however, a direct effect on sheep grazing which has caused a broad sheep-free corridor to develop on either side of the trail when visitors are present. The sheep are only present in the mornings and evenings when the visitor (with his dog) is at home. However, no change in the overall balance of grazing pressure has been detected.

Where the visitor 'cuts corners' on the trail, taking the shortest route or preferring to walk on the close cropped sugar limestone grassland, very marked botanical changes occur associated with soil compaction and erosion of vegetation. The number of visitors who trespass in this way, is, we believe, now much smaller than it was a few years ago, largely perhaps because of the quality of the surface and the educational effect of the nature trail. However, once changes in the sward have been produced, the process is extremely slow to reverse.

At the present time a detailed study of the effects of trampling on the vegetation is in progress which we hope will enable us to evaluate fully the nature and extent of the floristic changes which occur. The dry sugar limestone grassland falls into two broad phytosociological alliances – the pure Seslerio-Mesobromion with a high proportion of rare herbs and the Ranunculo-Anthoxanthion which whilst still containing many of the species in the Seslerio-Mesobromion also contains weedy species such as *Prunella vulgaris* and *Bellis perennis* (Bradshaw and Jones, 1976). This latter community is distributed along the Nature Trail and there is evidence to suggest this community reflects past disturbance.

From the data obtained to date it is clear that *Bellis perennis*, *Polygonum viviparum* and *Plantago maritima* are more numerous adjacent to the trail, whilst the converse is true for *Briza media*. The reason for this response is not clear and may be related to the destruction of apical growing points by trampling, with a resulting stimulus to vegetative growth. Some caution must be used in considering these data, since the purpose of the quadrats is to monitor change rather than determine changes that have already occurred, or distinctions which may be inherent in the vegetation alongside the trail. Evidence from other work suggests that trampling has already caused changes similar to those described elsewhere. Thus moderate trampling on neutral grassland causes an increase in *Poa annua*, *Plantago major* and *Lolium perenne* (Crawford and Liddle, 1977), and similarly in calcareous grassland there was an overall reduction in species richness with an increase in *Carex flacca*, *Plantago lanceolata* and *Bellis perennis* (Chappell *et al.*, 1971).

The general observation of a number of workers of an increase in rosette-forming and low-growing species under trampling conditions appears to apply also to the upland species-rich 'sugar' limestone communities in Teesdale. It is perhaps significant that *Gentiana verna* – a short-lived perennial – which might be considered most susceptible to trampling is still relatively frequent in those areas which have suffered greatest trampling pressure. This may relate to the preference of many of the rarities for open conditions with *G. verna* able in addition to withstand at least some trampling. Supporting evidence for this can be found in the Burren (Co. Clare) where *G. verna* occurs in a heavily trampled sand dune area adjoining a popular amenity development.

(h) Erosion

When the reservoir was first proposed it was apparent to the opponents that not only would there be direct loss of habitat by inundation but that wave action could have a serious effect. Evidence presented in support of this appears in retrospect not to have been considered sufficiently seriously. It was argued by the proponents of the scheme that the water level would only rise 3 ft above top water line once every 100 to 150 years in 'catrastrophic' storm conditions and it was with this in mind that the height of the dam spillway was determined. Similarly waves were predicted not to reach a height greater than seven feet above top water level. The wave action was seen by opponents to the scheme as one of a number of consequences of inundation – spray was anticipated in stormy weather and linked to possible detrimental effects of raising the open water level to 488·5 m.

We have found that the rate and extent of wave erosion on the Widdybank Fell margin has been greater than was anticipated at the time of the public inquiry. This has necessitated the drawing down of the reservoir by the Water Authority during winter in a pragmatic attempt to reduce the rate of erosion. Marshall (1977) has shown that erosion proceeds apace as a consequence of:

(1) The prevailing wind direction (south-west to west) (see Fig. 8.4) and strength – winds may often reach 60 kph in winter.
(2) The friable nature of the sugar limestone resulting from sub-surface weathering which extends to a depth of 3 to 12 metres, with the joint planes facing into the wind.
(3) The fact that as a regulating reservoir it is filled during the winter period when winds are at their strongest.
(4) The severe winters when frost-heave occurs, which coupled to the weight of snow will cause any areas undercut by waves to collapse.

The erosion extends along 1500 m of the fell side and follows a pattern of steep bank formation with subsequent undercutting. This occurs both in the sugar limestone communities, and in acidic grassland and peat communities on drift overlying the limestone. The situation is probably made worse by the drying out of the exposed peat faces in Summer which destroys the otherwise 'semi-elastic' peat structure. The material that is removed from the 'cliffs' is roughly sorted by the moving reservoir water and the sugar limestone is transported by 'longshore drift' to form white 'sandy' beaches when the reservoir is drawn down in summer. On drying, the finer fractions of the sugar limestone are blown on to the fell immediately adjacent to the shore. This places a physical selective pressure on the grasslands akin to that normally encountered in a sand dune system. The rate of erosion has been found to approximate to 1 m per annum since construction, but is highly

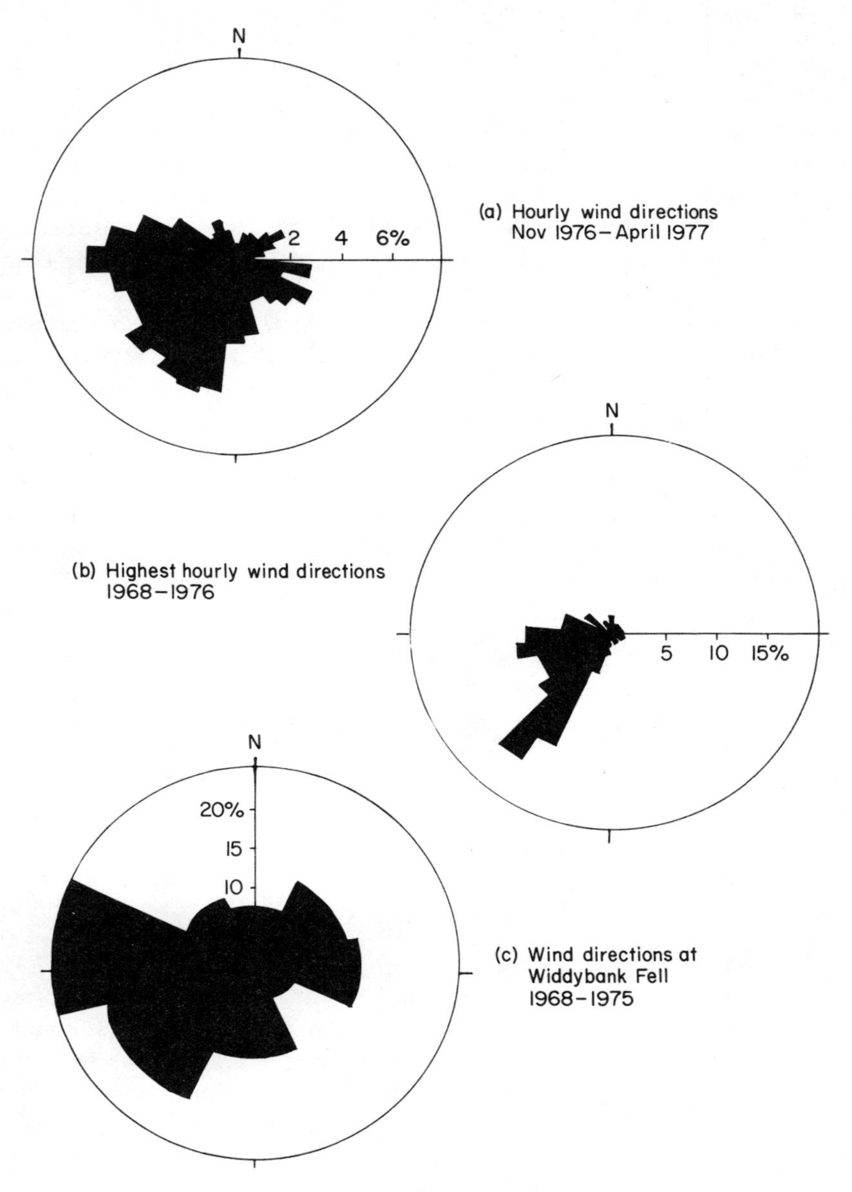

Fig. 8.4 Widdybank Fell wind statistics.

Table 8.5 The rate of erosion of 'sugar' limestone grassland at selected points along the Widdybank Fell margin of Cow Green reservoir

Date established	*Post No.*	*Loss of turf in (m) to June 79*	
November 1974	1	2·5	Mean 5 yr loss = 3·45 m ± 0·378
	2	4·1	
	3	4·6	
	4	3·2	
	5	3·9	
	6	2·4	
November 1975	7	2·5	Mean 4 yr loss = 3·77 m ± 0·378
	8	4·5	
	9	2·5	
	10	3·2	
	11	4·5	
	12	4·6	
	13	4·6	

irregular (Table 8.5). Thus, new 'cliff' faces are formed and successively destroyed with a maximum recorded height of 2 m.

The general slope of Widdybank Fell prior to impoundment was between 6 and 8 ° and wave action has cut back this angle to the more gentle profile of 4–6 ° (see Fig. 8.5). Marshall (1977) estimated that in total some 4000 m³ of sugar limestone has been eroded with an annual loss along the most important 300 m of 225 m³ per 100 m. This amounts to a loss up to 1977 of 0·25 ha above top water line in the most valuable sugar limestone grassland area around Slapestone Sike.

It is uncertain how far this process can continue, but as the sugar limestone extends to a depth of 12 m it is possible that, in the absence of appropriate

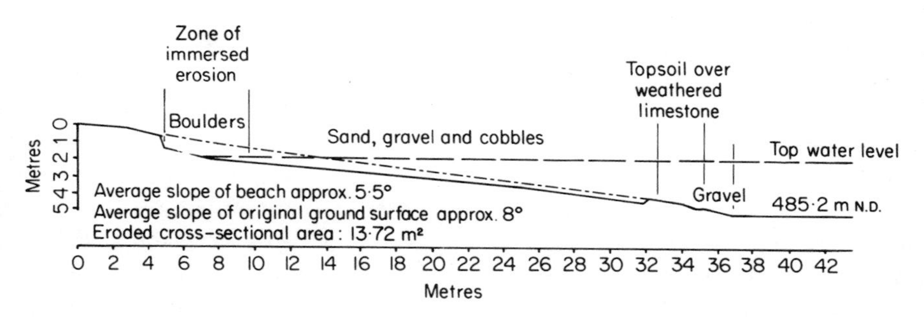

Fig. 8.5 Typical profile through Widdybank Fell showing wave action (from Marshall 1977).

control measures, all of this might be eroded in time. Meanwhile the cliff face moves back across the fell each year with a cumulative loss for conservation. Attempts are presently being made to prevent further loss and it is hoped that remedial action will be possible with the financial and technical assistance of the Water Authority.

8.3.4 Land management considerations

(a) Loss of grazing

Much detailed consideration has been given to the significance of inundation from the viewpoint of the botanist, but it was also of agricultural significance. The limestone and acidic grasslands, moorland and bog which were submerged beneath the reservoir provided grazing for substantial numbers of sheep. Much of the vegetation was relatively poor in nutritive value, in common with that of hill land generally, but was nevertheless of importance, depending on the prevailing wind, in offering grazing in relatively sheltered conditions. One tenant farmer and five or six stint holders were affected by these losses. Average stocking rates prevailing in the area are one sheep to 0·70 hectares so that grazing for approximately 430 animals was inundated. (Evidence presented to the House of Commons Committee suggested that grazing for 450–500 sheep would be lost.)

(b) Recreational developments

The appeal of the 'new landscape' of the reservoir is clearly quite considerable and as indicated earlier this may be seen as an addition to the attractions of the region and thus of interest as the destination for a weekend drive – to this extent it therefore serves as an informal recreational amenity. During the course of the Committee enquiries the recreational potential of the reservoir was considered but there were objections to the introduction of intensive recreational facilities and it was agreed that it was in any case unsuitable for sailing by virtue of water temperatures and vulnerability to gale-force winds, and that provision would be made only for passive pursuits such as walking, picnicking and fishing.

The creation of the reservoir has had a profound effect upon the numbers, size and growth rate of trout in the reservoir such that even without the artificial introduction of fish a viable sporting fishery was soon established. Usage of the fishery during April to September 1971 amounted to 357 angler visits by 265 anglers who took 121·0 kg of fish (Crisp and Mann, 1977*a*). By 1974 usage had increased to 1083 visits by 830 anglers taking 371·0 kg of fish. The mean catch per angler visit rose from 0·74 fish in 1971 to 0·94 in 1975. In 1976, 1557 visits produced 2329 fish (an average of 1·50 per angler visit) weighing a total of 833 kg (Crisp and Mann, 1977 *b*).

8.3.5 Conclusions

By virtue of the botanical, geological, and scenic importance of the area which would be affected, the potential impact of the reservoir on Upper Teesdale was examined in considerable detail before the decision to proceed with the construction was finally taken. In consequence few of the changes which were observed following inundation had not been predicted.

The inundation caused both direct losses and a number of changes resulting from the creation of the water body which were more difficult to quantify. There was little problem in predicting the floristic and vegetation changes in the Basin except that a full inventory was only completed some time after the decision to proceed had been made. Similarly the general nature of the effects upon the fauna of the Basin were largely predictable although at the time of the debate little attention was paid to them. The value of the juxtaposition of discrete communities and the complex matrix they formed proved difficult to convey adequately in the inquiry. The argument that representatives of each species and community present in the basin could be found elsewhere does not mean that their loss is unimportant. Indeed, a unique complex of communities forming an irreplaceable resource was destroyed and this must be regarded as the major impact of Cow Green.

Great problems were experienced in predicting the extent of the indirect effects and their relative significance. In fact, as has been shown here, all the major consequences were largely anticipated but the stress laid on the various aspects was perhaps not correct – in retrospect greater emphasis should have been placed on both wave erosion and visitor pressure.

The wave erosion of sugar limestone turf has been even more damaging than was expected. The publicity, the improved access and the attraction of the man-made landscape, have led to much increased visitor pressures, which have been the apparent cause of significant changes in the vegetation of areas – some of them of out-standing importance for conservation – adjacent to footpaths.

Opponents of the scheme were pleasantly surprised by the extent to which damage during construction was confined within the reservoir basin. So far the worst fears of botanists that climatic amelioration and reduction in the incidence of river floods would have adverse effects upon the Teesdale assemblage of rarities show no signs of being realized.

Much valuable experience was gained in the assessment and determination of the consequences of such developments and in the making of an effective nature conservation case. This gain cannot compensate for the loss of a unique area such as Cow Green. There remains a need for the acceptance of the general principle that sites of national and international conservation importance should be sacrosanct.

8.4 PREDICTING THE IMPACT OF OIL TERMINAL DEVELOPMENT ON THE IN-SHORE MARINE ENVIRONMENT: RETROSPECTIVE ANALYSIS

J. M. Baker and K. Hiscock

8.4.1 Introduction

The aim of this paper is to critically discuss the process of ecological impact assessment at oil terminals in the British Isles, drawing upon case history examples. These range from the Fawley Terminal and refinery in Southampton Water, opened in 1951, to the Sullom Voe Terminal in Shetland which started operating in 1978. In all cases there has been some assessment of ecological impact, though this has often been retrospective rather than predictive.

The case-history section of this paper is intended as a straightforward

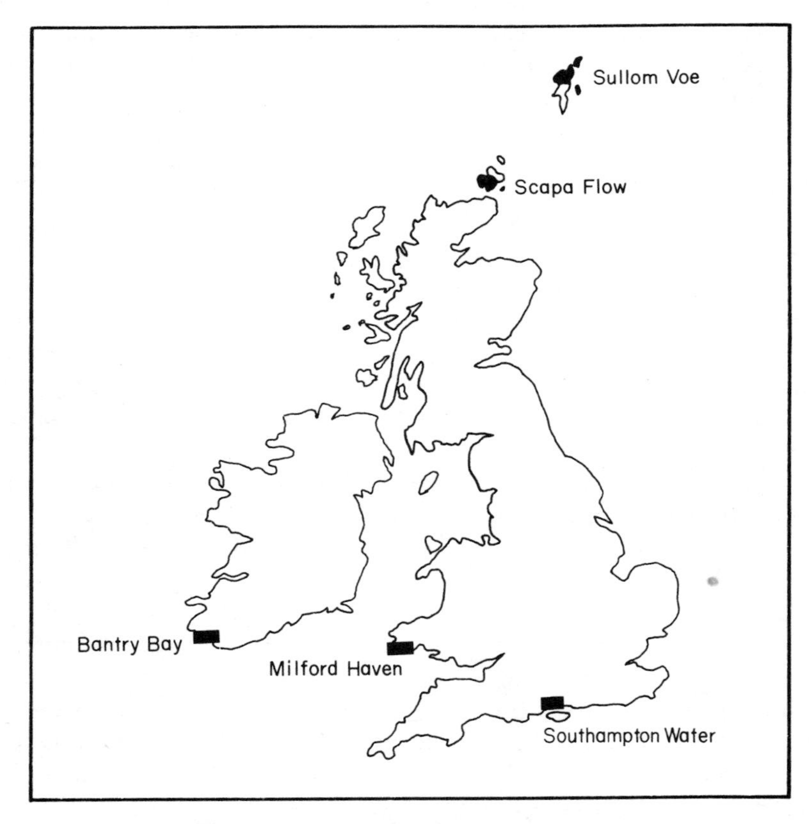

Fig. 8.6 Location of areas cited in the text.

summary of the work carried out at various terminals, the locations of which are shown in Fig. 8.6. Detailed reports are available and are referenced, but the intention for the purposes of this paper is to describe the general approaches used and major effects noted.

The concluding section of the paper examines these approaches with particular reference to their usefulness in identifying and predicting the impact of oil terminals on the in-shore marine environment.

8.4.2 Potential oil-terminal disturbance of in-shore marine environments

In-shore marine communities may be affected by both accidental and normal operational aspects of oil-terminal activities. The former include oil spills and subsequent clean-up, and the latter include discharges of treated ballast water and refinery effluents.

(a) Oil spills

The following categories of oil spills in oil ports are slightly modified from those of Dudley (1976):

(1) Fairly frequent minor spillages involving a few gallons, resulting from tank overflows, valve failures etc. They may occur anywhere within a port and are not restricted to operations at terminals, or specifically to oil tankers.
(2) Infrequent medium spillages involving up to five tonnes of oil, usually resulting from damage or mechanical failure. They are only likely to occur during loading or discharging operations in the vicinity of terminals.
(3) Rare serious spillages of up to 500 tonnes due to human error, damage, grounding, etc. They may, therefore, occur anywhere within a port, for instance on the edges of channels and anchorages or at terminals.
(4) Catastrophic spillages. Pollutions of the Torrey Canyon type are very unlikely to occur inside a port. However, in any port handling large numbers of tankers, collision or grounding could occur, from which a pollution of more than 500 tonnes might result.

(b) Clean-up

Much time, effort and money has been spent on developing physical methods of containing and recovering spilt oil floating on water, but these operations are still difficult to organize effectively if the weather is bad or the current strong. For this reason dispersants still play an important role in clean-up. Dispersants may be used routinely in oil ports for the rapid treatment of small floating slicks, and this approach, used for example in Milford Haven since its inception as an oil port, helps to reduce the oiling of amenity beaches, other areas of shore, boats and other structures. It may also, according to time of year, help to minimize bird casualties. Following larger spillages, treatment of both large floating slicks and stranded oil may be necessary. Dispersant spraying of floating slicks has traditionally been from boats fitted with spray booms, but recent progress with aerial application of dispersant concentrates suggests that this could be an efficient and attractive treatment method. On the shore, dispersant treatment of the thinner oils is commonly carried out using knapsack sprayers or specially designed light vehicles. Thicker oils and mousse do not disperse effectively through the use of chemicals, and physical removal is necessary.

(c) Ballast water discharges

All tankers coming to a terminal to collect cargo will be carrying ballast water. When this water is carried in tanks that have previously carried oil, it is necessary to treat it before disposal to the sea. The main contaminant is dispersed oil and this is usually removed in shore-based skimming and emulsion-breaker tanks. Ballast water treatment should ensure that the oil content of the final discharge will not normally exceed 25 ppm, with an average of 10–15 ppm (SVEAG, 1976).

(d) Refinery effluents

Refinery effluents are considered here because refineries are commonly associated with oil terminals and because ballast water may be treated through the refinery waste-water system. In most cases, primary treatment (gravity separation) is used to remove oil and coarse particulate matter. Some refineries in addition have some form of further treatment for removal of oil (e.g. activated sludge). Final effluents comprise varying proportions of process, cooling, storm and ballast water and may contain oil, phenols, sulphides, mercaptans, cyanides, ammonia, some heavy metals and suspended solids, and possibly other compounds. Variation in concentration of these components occurs both between refineries and within any particular refinery over a period of time.

8.4.3 Case histories

(a) Southampton Water

Southampton Water is fringed with mudflats and *Spartina*-dominated saltmarshes. Sub-littorally it has fine black muds along the south-western side of the estuary and more heterogeneous deposits of fine muds between large shells and stones along the north-eastern side of the estuary. A map of the area, with locations of places mentioned in the text, is given in Fig. 8.7. Apart from any possible effects of the large Fawley oil terminal and refinery, the picture is complicated by other industrial discharges, municipal discharges, and the heavy use of Southampton Water by ships of all kinds.

Fawley – Britain's first major post-war oil refinery – was constructed at a site where a small-scale refinery had been operating since the early 1920s.

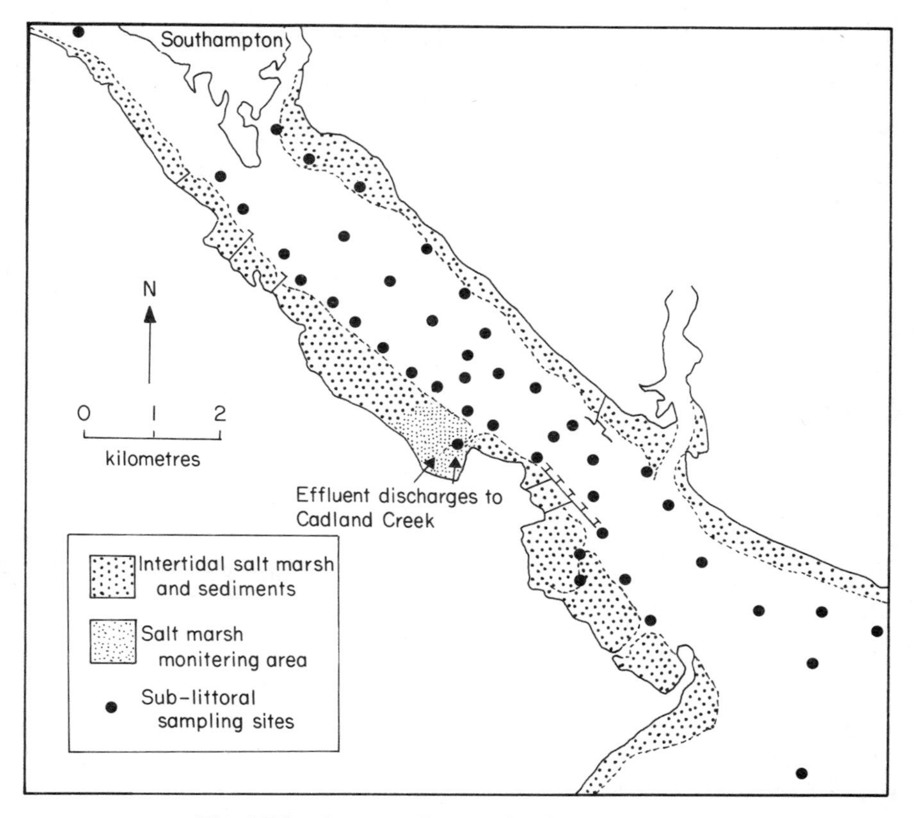

Fig. 8.7 Southampton Water, showing survey sites.

The new, greatly extended refinery and terminal were opened in 1951 and development has continued until the present time, the current capacity of the refinery being $19{\cdot}5 \times 10^6$ tonnes per year. Liquid-waste sources are cooling water (20 000 $m^3\ h^{-1}$), ballast water (100 $m^3\ h^{-1}$) and rainwater (average 100 $m^3\ h^{-1}$). Treatment is by API gravity separators.

In recent years, over £15 million has been spent on equipment specifically aimed at reducing the environmental impact of the refinery. This includes £4 million expenditure on reducing the quantity of cooling water used and improving the quality of the effluent water. At the marine terminal, which has up to 3000 ship movements a year, detailed procedures and check-lists have been drawn up over the years in an attempt to prevent spills, and booms and dispersant spraying vessels are available if pollution should nevertheless occur. These precautions are now standard practice at all oil terminals.

Ecological aspects of environmental impact did not form part of the original planning but have been considered later in connection with the salt-marsh and sub-littoral studies described below. In addition, hydrocarbon survey work has been carried out by the Ministry of Agriculture, Fisheries and Food (Burnham-on-Crouch Laboratories) and the University of Southampton has carried out hydrocarbon analysis of water and sediments as part of their long-term investigations into the carbon budget of the estuary.

The salt-marsh, which has had refinery effluent discharged through its creek system since 1951, was surveyed in 1969 and 1970 to assess the extent of ecological damage. The salt-marsh has been re-surveyed twice a year since

Table 8.6 Effluent improvements at Fawley

1963–1970	Average oil content 31 ppm Total flow 28 000 $m^3 h^{-1} = 4{\cdot}7 \times 10^5$ l min^{-1}
1972	Average oil content 25 ppm Total flow 28 000 $m^3 h^{-1} = 4{\cdot}7 \times 10^5$ l min^{-1}
1974	Average oil content 14 ppm Total flow 23 000 $m^3 h^{-1} = 3{\cdot}9 \times 10^5$ l min^{-1}
1976	Average oil content 10 ppm Total flow 21 000 $m^3 h^{-1} = 3{\cdot}5 \times 10^5$ l min^{-1}

Note: oil contents measured by infra-red spectrometry

1972 to monitor any changes in the distribution of plant species in association with the refinery effluent improvement programme. The results of these surveys are presented in detail and discussed by Baker (1971), Dicks (1976) and Dicks and Iball (1981).

A limited amount of information (mainly aerial photographs) is available on the condition of the salt-marshes from 1950 up to the start of the recent series of surveys. This information and the monitoring programme results provide a picture of, firstly, severe damage to an area of salt-marsh vegetation resulting in bare mud, followed in recent years by recolonization of some parts of this previously denuded area by common salt-marsh species, especially the annual plants *Salicornia* spp. and *Suaeda maritima*.

Baker (1971) concluded that the initial damage to the marsh was mainly a result of repeated light oiling of the shoots. This occurred because oil in the effluent rose to the surface during the calm water conditions at high tide, and was stranded on the marsh as the tide went down. Oil spills from the jetty, and effluent constituents other than oil, may have contributed to the damage.

Since the time of the initial survey, the oil content and volume of the effluent have been progressively reduced as is shown in Table 8.6, and recolonization of the marsh is proceeding. The monitoring programme is now being extended to include marsh fauna and sea-bed fauna in the effluent discharge area.

(b) Milford Haven

Milford Haven is a large drowned valley at the south-west tip of Wales. A map of the area is shown in Fig. 8.8. There are five oil terminals and four refineries on the shores of the Haven, established between 1960 and 1974.

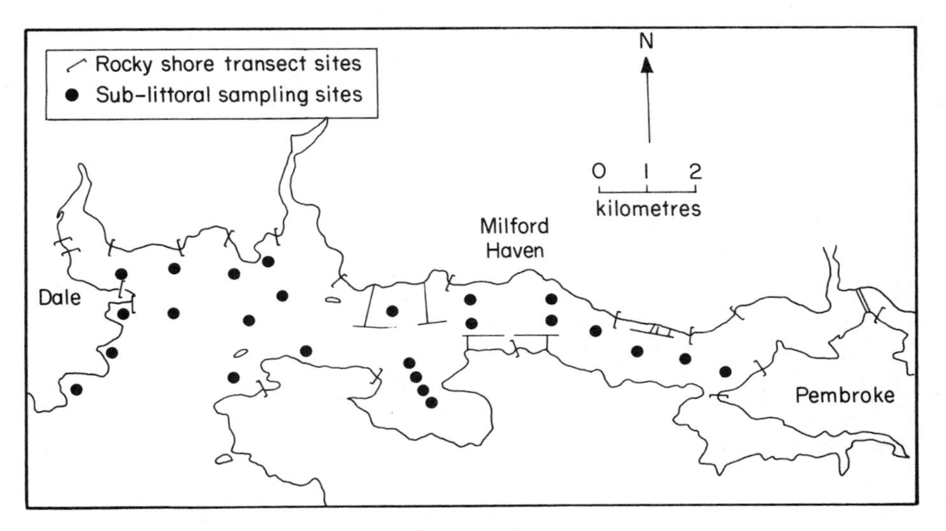

Fig. 8.8 Milford Haven, showing survey sites.

The terminals receive crude oil from tankers of up to 270 000 tons and export products in smaller tankers, with a total of 3338 tankers entering the port in 1979. All of the refineries are modern, air-cooled installations and, as an example of effluents produced, one refinery produces a dry weather flow of about 360 $m^3 h^{-1}$ with a mean total oil content of about 15 $mg l^{-1}$ through the year. There is also a large oil-fired power station which discharges heated cooling water into the Haven.

The shores of Milford Haven are predominantly rocky with some shingle beaches, some extensive areas of mudflat on the south side, and some muddy bays on the north side. The sub-littoral seabed includes bedrock, pebbles, sand and shell in the outer regions, merging to more muddy sediments with variable amounts of shell and stone present in the middle reaches of the Haven to a predominantly rocky seabed east of the terminal developments.

There is no formal environmental impact assessment for Milford Haven, though the rocky shore communities were described by Nelson-Smith (1964, 1967) before the first oil terminal started operating. There has been no formal monitoring programme to establish the effects of the oil-related developments on marine life or hydrocarbon levels. However, a wide range of studies has been carried out since 1967 by the Field Studies Council's Oil Pollution Research Unit aimed at re-surveying Nelson-Smith's sites, describing communities in the other main habitats, studying the effects of spills and refinery effluents, and analysing sediments for hydrocarbon content. Possible spill and clean-up effects have been further investigated through field experiments involving the application of selected oils and dispersants to different types of intertidal community.

Fig. 8.8 shows the location of rocky shore monitoring sites and sub-littoral sediment biological survey sites in Milford Haven. The Welsh Water Authority have a statutory obligation to set effluent quality standards and have carried out studies of water exchange in the Haven to assist in setting those criteria. The Central Electricity Generating Board have carried out a monitoring programme aimed at assessing the effects of warm water discharge from the Pembroke Power Station on mudflats.

The anti-oil pollution measures in Milford Haven began with the methods then in existence in the most oil pollution conscious port of the day, i.e. Southampton. These took the form of regular patrolling and of prosecution where possible. It quickly became obvious, however, that to minimize shore pollution there was a need to deal with floating slicks quickly, before those responsible had been discovered and had accepted their responsibility. By 1964 the oil companies operating in Milford Haven, together with the Milford Haven Conservancy Board (The Harbour Authority) decided to try to solve this problem. The result was an oil spill contingency plan, unique in that it was necessary for the oil companies and the Conservancy Board to accept responsibilities and expense which they were not legally bound to do. The latest edition of the plan provides information on the location of sensitive areas, the authorities to contact if, for instance, oil is likely to strand at a Site of Special Scientific Interest, and other information to ensure as far as possible that spills are dealt with efficiently with as little damage to marine life as possible.

A summary of ecological changes in Milford Haven since the establishment of the terminals is given in Baker (1976) and this summary is still applicable to the impact of the developments up to 1980. This information, together with recent re-surveys of rocky shore sites, indicates that there are no obvious overall changes which can be attributed to the oil industry, though short-term localized effects occur. The most severe effect was noted following a spill of petrol from the 'Dona Marika' in 1973 which resulted in changes over about one mile of coastline. There was a large reduction in numbers of limpets (*Patella* spp.) from over 50 m^{-2} to 1 or less m^{-2}. This was followed by abundant growth of the green alga *Enteromorpha* spp. and subsequently by growth of fucoid algae. Reduction in numbers of limpets and other molluscs is also evident in a small bay receiving refinery effluent and constitutes a chronic but very localized impact on some species.

Recent studies of the amount of petroleum-derived hydrocarbons in sediments in Milford Haven indicate levels of contamination ranging from 4 to 480 ppm (total aliphatics), with a general trend from low levels in the exposed, wave-washed mouth of the Haven (<30 ppm) to highest levels in the inner Haven of up to 480 ppm. There was no clear pattern of elevation of levels around industrial developments or their discharges.

(c) Bantry Bay

In 1968 Gulf Oil Terminals (Ireland) Limited opened an oil trans-shipment terminal on Whiddy Island in Bantry Bay, south-west Ireland. The installation was designed to receive and store crude oil from 300 000 ton tankers for subsequent shipment to European ports (including Milford Haven) in smaller vessels. The Bay, shown in Fig. 8.9, has predominantly rocky shores and there is in general a zone of sub-littoral rock between the shores and the offshore bottom sediments. By 1968, surveys designed to provide information on the flora and fauna of Bantry Bay before possible exposure to oil pollution had been carried out by workers from the Department of Agriculture and Fisheries (Dublin), the Lancashire and Western Sea Fisheries Joint Committee, the University of Durham, and the Field Studies Council's Oil Pollution Research Unit. An initial survey of the rocky shores of Bantry Bay was described by Crapp (1970). There was, however, no formal ecological impact assessment and no plan for regular comprehensive monitoring of Bantry Bay.

Work carried out since the terminal started operating includes:

(1) A more detailed survey of the rocky shores (Crapp, 1973). The abundance of littoral animals and plants was assessed at regular intervals. The species were described in relation to two major environmental variables: emersion and exposure to wave action. Crapp intended that this work should be a useful survey in its own right and also serve as a 'baseline' for the detecting of future changes.

(2) Re-surveys of transects established under (1) in 1975, 1978 and 1979. These re-surveys, carried out by the Field Studies Council Oil Pollution Research Unit, were intended to find out if particular oil spillages had caused large changes in intertidal flora and fauna against a background of natural changes.

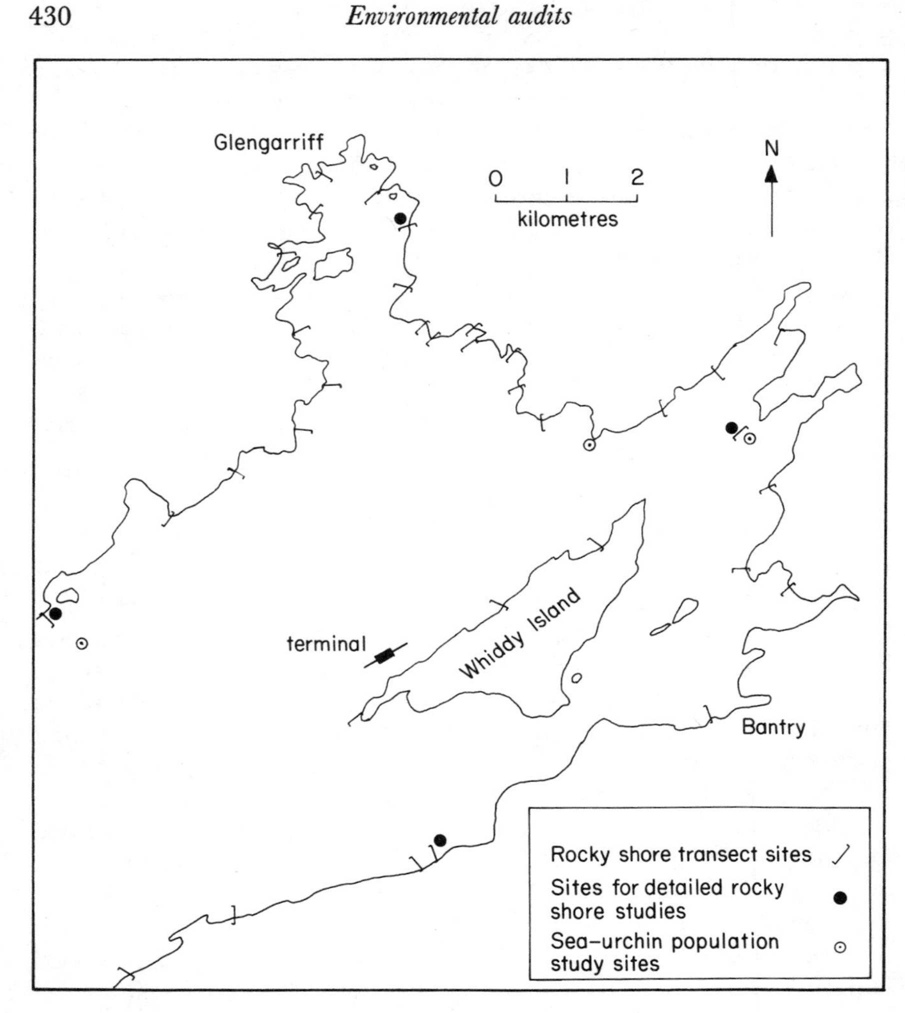

Fig. 8.9 Bantry Bay, showing survey sites.

(3) Post-spill observations, e.g. the reports of O'Sullivan (1975), Cullinane *et al.* (1975), Cross *et al.* (1979) and Myers *et al.* (1979).

(4) Work on sea urchin (*Echinus esculentus*) populations carried out in 1977, 1978 and 1979 by the Cambridge University Underwater Exploration Group.

(5) Detailed work designed to quantify seasonal and longer-term fluctuations at five rocky shore sites, work on the fauna of algal turfs; and rocky sublittoral studies. This programme was undertaken from University College, Cork, with the financial support of the National Board of Science and Technology. Details are given by Myers *et al.* (1978, 1979).

Table 8.7 Major incidents in Bantry Bay

Date	*Incident*	*Effects*	*References*
21/22 October 1974	Spillage of Kuwait crude oil, estimate 2600 tonnes, at Whiddy Island terminal from the 'Universe Leader'. Some dispersant spraying of floating oil, plus some physical removal from shores	Immediate damage to intertidal rocky areas not generally severe. *Fucus serratus* reported as the most badly damaged alga; several lichen species discoloured/eroded, some recovered by December 1974. Some detachment of limpets (*Patella* spp.) and damage to saddle oysters (*Anomia ephippum*)	O'Sullivan (1975) Cullinane *et al.* (1975)
10 January 1975	Spillage of bunker oil (heavy fuel oil), estimate 460 tonnes, from tanker 'Afran Zodiac' being towed out of Bantry Bay. Some dispersant spraying of floating slicks and on shore, plus physical removal from shores	Small salt marsh oiled, most plants showing signs of recovery by May, 1975. Winkles (*Littorina littorea*) and top shells (*Monodonta lineata*) seen crawling over oil stranded on rocks	Baker *et al.* (1981)
		The 1975 transect re-survey showed many changes in abundance from the 1970/71 survey, but these could not be consistently related to either the 'Universe Leader' or the 'Afran Zodiac' incidents	
8 January 1979	Tanker 'Betelgeuse' exploded at Whiddy island terminal, killing 50 people. Some of the cargo burnt, some leaked slowly for several months. Accurate estimates of amounts not available. Dispersant spraying of floating oil slicks (including aerial spraying of concentrates) plus dispersant spraying and physical removal on shore	Little acute or chronic effects at mainland sites. 1 km of shore on Whiddy Island severely affected. Asphalt-like residue from the fire coated the shore from MHWN to MTL, killing everything beneath it. Many deshelled *Patella* and dead dog-whelks (*Nucella lapillus*) below MTL, probably affected by heat. Possible chronic oiling effects at some sites	Cross *et al.* (1979) Myers *et al.* (1979)

A summary of major incidents in Bantry Bay and their effects is given in Table 8.7.

(d) Sullom Voe

Sullom Voe is a large sheltered inlet on the north east of the Shetland mainland. A map of the area is shown in Fig. 8.10. The terminal for the stabilization and transshipment of North Sea oil has a planned capacity of 1·5 m+ barrels of oil per day. Construction commenced in 1975 and operations started in 1978.

The shores of the Voe are predominantly bedrock in the outer region but of sand, shingle or boulder with some areas of bedrock in the inner Voe. The sub-littoral sea bed includes areas of mud near the head of the Voe grading into sand and coarse sand with stones near to the entrance to the Voe. There are extensive beds of the horse mussel *Modiolus modiolus* on the level sea bed within and at the entrance to the Voe.

The Sullom Voe Environmental Advisory Group (SVEAG) was established in 1974 to advise on the environmental aspects of development associated with the oil terminal and, in 1976, published an Environmental impact assessment (SVEAG, 1976). The assessment provided a comprehensive overview of the Shetland environment, the terminal development, the sources of disturbance, the probable effects of disturbance and pollution, and also described proposals for environmental protection. The main potential sources of disturbance to marine communities following construction and during operation were seen as ballast water discharge and oil spills.

Although some biological surveys were carried out prior to 1976, studies aimed at providing a description of the marine communities present in the region of Sullom Voe as a basis for monitoring change were mainly commissioned in 1976 by SVEAG on behalf of the terminal operators. The aims of the monitoring programme as stated in SVEAG (1976) were:

(1) To determine the nature and scale of the environmental effects of the terminal, especially to provide early warning of any effects that could lead to permanent and irretrievable damage if unchecked.
(2) To provide a continuing check on the adequacy of measures taken to control the impact of the development, and to rehabilitate areas temporarily affected in the process of construction.
(3) To assess the extent of recovery of such damaged sites following treatment.
(4) As a result of these studies, to provide for updating and improvement of the control measures to be taken should there be further development at Sullom Voe.

Routine operational monitoring was intended to include the detection of

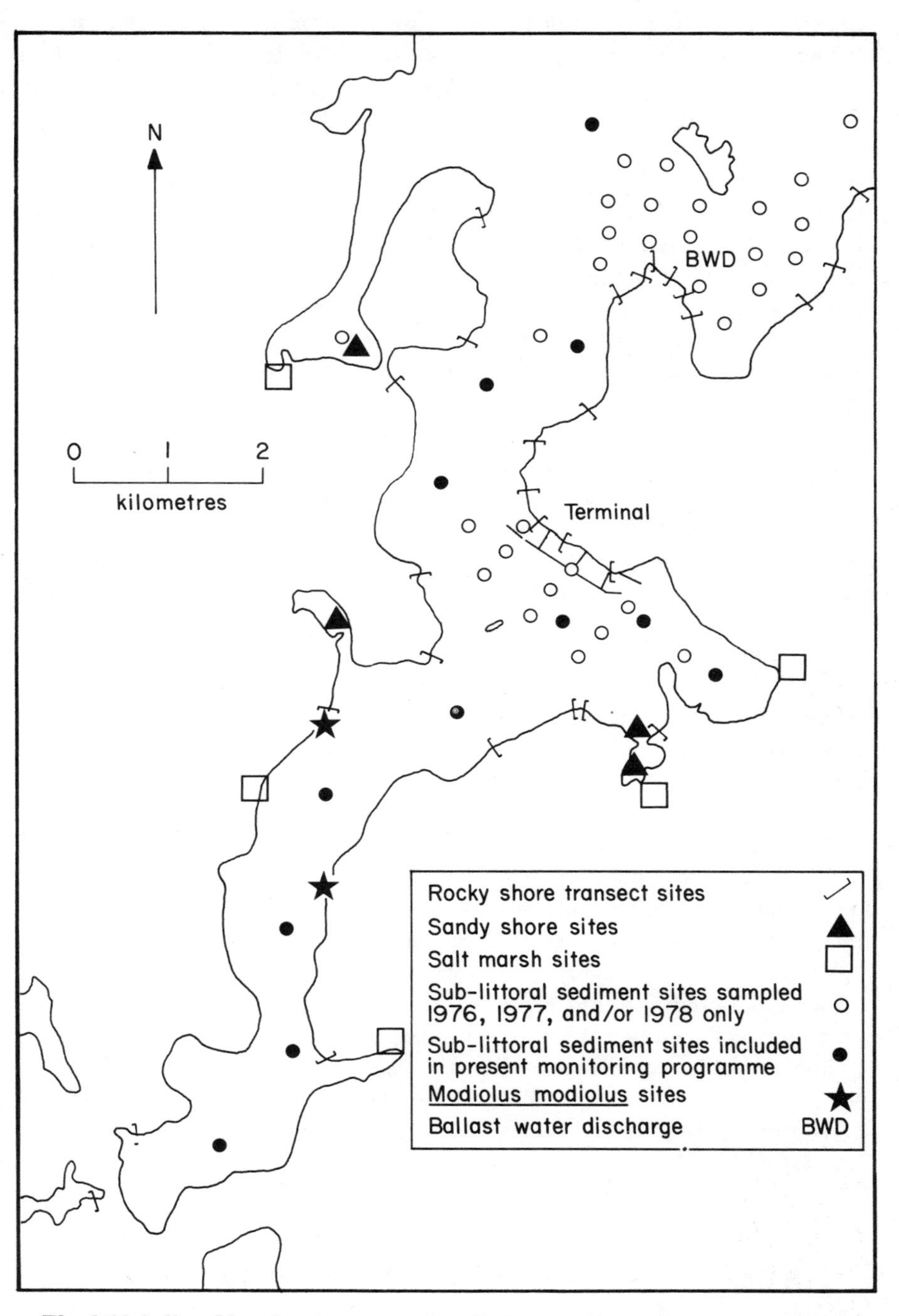

Fig. 8.10 Sullom Voe, showing survey sites. Reference sites established at locations remote from Sullom Voe are not shown.

increases in hydrocarbons in the water and sediments and the detection of biological change amongst other features. The biological surveys and monitoring commissioned by SVEAG included salt-marsh, rocky shore, sandy shore, sub-littoral sediment and sub-littoral rock communities. Analysis of hydrocarbon levels in sediments and in *Modiolus modiolus* were carried out and physical and chemical parameters in the water column and sediments were made. In addition, the Department of Agriculture and Fisheries for Scotland (DAFS) carried out hydrographic surveys and biological surveys of benthic communities and plankton as a part of their statutory requirement to consider the implications of the terminal development to fisheries. Also the Nature Conservancy Council in association with SVEAG instituted counts of cliff nesting and other birds. The work initiated by SVEAG was continued under a new group which was established in 1977, the Shetland Oil Terminal Environmental Advisory Group (SOTEAG). Since 1976, the chemical monitoring, particularly of hydrocarbons, has been increased and biological monitoring rationalized to ensure as little overlap between the various studies as possible. However, the number of sites in the region of the ballast water discharge and jetties has been reduced, providing a much reduced capability for observing effects in the latter area. In 1979, the terminal operators initiated a biological and chemical monitoring programme in the region of the ballast water discharge as required by the Local Authority Works Licence Condition. Initial survey and present monitoring sites are shown in Fig. 8.10.

The results of the biological surveys up until 1979 were summarized in a symposium on the 'Marine Environment of Sullom Voe and the Implications of Oil Development' (Royal Society of Edinburgh, 1981).

The Sullom Voe terminal has, at the time of writing, suffered one major oil spill, the 'Esso Bernicea' spill (December 1978/January 1979), and several minor spills. The effects of the 'Esso Bernicea' spill on rocky shore communities were much as predicted in the environmental impact assessment (SVEAG, 1976). However, the problems relating to clean-up of heavy fuel oil were not predicted and, since neither dispersants nor skimmers could be used on the oil, considerable ecological damage was caused through the use of bulldozers and drag lines at sites where oil was mechanically cleared from the shore. Also, the escape of oil from failed retaining booms led to large mortality of several bird species and to mortality of molluscs, particularly *Cerastoderma edule*, in a sandy inlet. Eleven otters are also known to have been killed.

8.4.4 The role of monitoring schemes in impact analysis

Monitoring, whether chemical, toxicological or ecological, is an essential element in the impact assessment scheme shown in Fig. 8.11. Monitoring

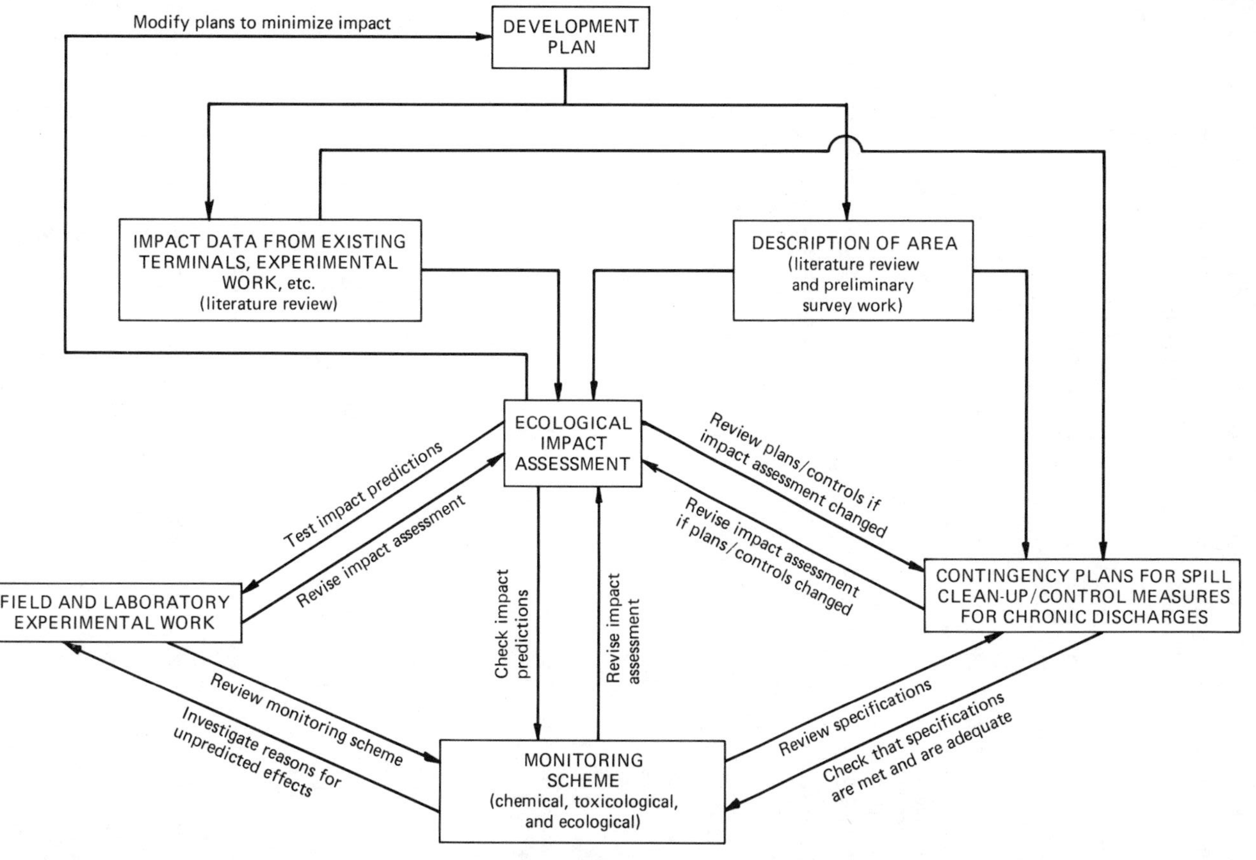

Fig. 8.11 An ecological impact assessment (depicted as being initially derived from literature review and preliminary survey work, but subsequently subjected to a continuing process of revision through input from experimental work and monitoring schemes).

makes possible a continuing process of revision of the initial impact assessment and of spill contingency plans and control measures for chronic discharges. Initially unpredicted effects can only come to light through monitoring, which in such cases stimulates field and laboratory research into the reasons for the effects. A variety of complementary monitoring objectives and approaches are given in Table 8.8 and at least three of these approaches have been used at each of the case-history sites described, though not necessarily on a regular basis.

Table 8.8 Examples of monitoring objectives and approaches

Objective	*Monitoring approach*
Comparison of effluents with statutory chemical specifications	Chemical analyses (hydrocarbons, phenols, sulphide, ammonia, etc.)
Information on toxicity of effluents or discharge water	LC_{50} or sub-lethal tests
Detection of physiological and genetic changes in organisms	Several approaches are possible, for example 'scope for growth' (Bayne *et al.*, 1976); screening of extracts from accumulator organisms for mutagenic activity (Parry and Al-Mossawi, 1979)
Detection of changes in species composition and species abundance; delimitation of area so affected	Ecological surveys, for example of the types outlined in the case histories in this paper. Reliable detection of small changes is likely to require intensive quantitative sampling
Information on the movements and fate of pollutants	Chemical analyses of water and sediments
Detection of uptake of pollutants by organisms and possible biomagnification	Chemical analyses of tissues. Decisions must be made on selection of species, trophic levels, tissue types and size classes

A particularly fruitful approach has been quantitative biological sampling from localized grids of sites around chronic discharges or jetties. In combination with physical and chemical measurements, for example sediment particle size and hydrocarbon content, this approach allows comparison with changes occurring at reference sites at the edge of grids, the identification of gradients of effect, and correlation between biological, chemical and physical data. Ecological monitoring which is not specific to an effluent discharge or other point-source of disturbance presents greater problems in selection of reference or 'control' sites and interpretation of change. The

Milford Haven, Bantry Bay and Sullom Voe rocky shore schemes have, however, been useful in allowing comparison between otherwise similar oiled and non-oiled shores following accidental spillages.

The monitoring carried out at oil terminals so far leaves many questions unanswered. One group of questions concerns long-term effects, natural fluctuations, the concepts of recovery, and indicator or key species. There are seasonal and secular variations in many intertidal organisms which are difficult to separate from the effects of oil spills. Although the seasons of annual recruitment for most of the common shore animals and plants are known, fluctuations in the strength of recruitment from year to year are not understood, though work is in progress (Lewis, 1977; Jones *et al.*, 1979). There are also secular fluctuations in distribution and abundance, some of which have been related to climatic changes, but they complicate the interpretation of post-spill surveys.

In many cases following single oil spillages, 'recovery' of conspicuous shore organisms appears to be good, if not complete, in periods ranging from one to ten years. The term 'recovery' has however, to be used with caution as we know little about the inconspicuous shore organisms and we are not yet sure if certain conspicuous organisms ('indicator species' or 'key species') can be used to provide measures of ecological change which are valid for the whole system. Moreover, as shores are subject to large and often patchy natural fluctuations it is difficult to define a norm and therefore difficult to say when 'recovery' is complete.

The effects of oil spills and of refinery and other oil-bearing effluents on near-shore subtidal rock and sediment communities are poorly known, though the work in progress at the four terminals described and elsewhere, should provide information on changes in seabed communities in relation to disturbance and petroleum-derived hydrocarbons. However, interpretation of the effects of oil pollution is hampered by a lack of information on the scale of seasonal and long-term fluctuations in the components of the seabed communities, though some idea of the natural changes in sedimentary benthos can be obtained from work such as Buchanan *et al.* (1978), Pearson (1971) and Rees *et al.* (1977). Far less is known about natural fluctuations in subtidal rock communities though several programmes of study are under way in Scandinavia.

A second group of questions concerns the use of dispersants. Modern dispersants approved by the Ministry of Agriculture, Fisheries and Food are far less toxic than those used to clean up the 'Torrey Canyon' oil. Laboratory tests used for measuring toxicity cannot, however, be used for predicting complex ecological interactions following cleaning. Recent incidents have highlighted the continuing lack of critical information about the effects of dispersed oil. For example, there was much discussion regarding possible use of dispersants in cleaning 'Christos Bitas' oil on Skomer Island (off the mouth of Milford Haven), where a marine reserve was at risk. The communities involved included subtidal algae and animals, and sea-grass (*Zostera*) beds – there is hardly any quantitative information on the effects of dispersed oil on such areas.

It is often recommended that dispersants should not be used in shallow bays, on mud flats or on salt marshes. The reasoning behind this is that relatively toxic dispersant/oil mixtures would not be diluted quickly enough in shallow water to prevent damage (e.g. to shellfish beds) and that penetration of oil into sediments might be promoted by dispersant use. The effects of dispersants on the interaction between oils, water, sediments and organisms has never, however, been adequately tested in field experiments, though work is now in progress both in Britain and in North America.

The third group of questions concerns chronic discharges, in particular refinery effluents. Many refinery effluents which have received primary treatment only, are toxic at their discharge-point to marine organisms. In some cases numbers of species and/or numbers of individuals decrease near the discharge point and in other cases no such effect has been observed, probably because the effluent is rapidly diluted to below acute toxicity levels by the receiving water. It is not known which of the effluent components contributes most to the overall toxicity. The large number of hydrocarbons and other chemical compounds contained in oils and in refinery effluents yield such a high number of possible combinations that the application of some general principles applicable to the discharge of mixtures of oil and other substances into receiving waters would be preferable to an infinity of ad hoc tests of individual effluents. Furthermore, it is not known whether ecological changes, where present, result from above-specification peaks of effluent components, or from continuous low-level discharge.

8.4.5 Conclusions

Ecological impact assessment for oil terminals has been, and should continue to be, an evolutionary process. Retrospective survey data from older terminals have provided a useful input to predictive assessment for new terminals, and monitoring scheme data should be used regularly to up-date current ecological impact assessments.

The intermittent surveys or monitoring schemes now used at terminals do not, however, adequately answer a range of questions on long-term effects versus natural fluctuations, effects of dispersant use, and reasons for effects in chronic discharge areas. These questions are best answered by field and laboratory experimental work and basic ecological research. Such work is relevant to ecological impact assessment at all terminals but need not be carried out at all of them.

8.5 THE NEED FOR RESEARCH ON ENVIRONMENTAL IMPACT ASSESSMENT

*M. W. Holdgate**

8.5.1 Introduction

Environmental impact assessment has been defined by Munn (1979) as 'an activity designed to identify and predict the impact on the biogeophysical environment and on mans' health and well-being, of legislative proposals, policies, programmes, projects and operational procedures, and to interpret and communicate information about the impacts.'

Clearly therefore EIA is a process of analysis. It is designed to make assumptions explicit. As Munn says, 'the task is not to prepare a scientific treatise on the environment, but rather to help the decision-maker select

Table 8.9 Judgements in the EIA process

Objective	*Subjective*
1 Definition of proposed developments (a) Nature of the technology (b) Nature of the impacts	1 Value of proposed development (a) Employment, value of product
2 Definition of the environment affected (a) Components likely to be affected (b) Likely changes in spatial and temporal patterns: probability of particular effects	2 Value of the environment affected (a) Acceptability of damage (to individuals and to community as a whole) (b) Impact of development on wider social attitudes and policies: 'knock on' political effects
3 Definition of the time frame of impacts	3 Responsibility for future generations
4 Definitions of reversibility of impact, and of effectiveness of alternative developments or rehabilitation methods	4 Preferences between alternatives
5 Definition of sensitivity of monitoring and capacity to intervene to counter unforseen change	5 Acceptability of uncertainty and risk: demand for precision in monitoring and control
6 Definition of probability of accident	6 Acceptability of risk of accident

*The views expressed in this paper are those of the author, and not necessarily those of the Department of the Environment or the Department of Transport.

from amongst choices for development and then to consider appropriate management strategies – quite a different goal.'

The process involves a comparison of unlikes: a blend of objective and subjective judgements (Table 8.9). The basic questions relate to the intellectual validity of the process and of the procedures adopted to carry it out, and the adequacy of the way in which it has been institutionalized. The scientific component of many environmental impact assessments has not been rigorous, and a major aim of research must be to improve its validity. But because EIA is a part of a process of developing public policy, the administrative component – the scale, context, and relationship to other legal and managerial processes – is likely to be a crucial determinant of what is actually done. One of the defects of many analyses of the subject is the tacit assumption that there can be some universal procedure suited to global, regional and local issues and to all national administrative systems.

The aim of an EIA process is to secure sound environmental management. Within this, EIA may seek to ensure: (a) protection of the health and safety of the public and the maintenance of environmental quality; (b) optimal balance of the economic benefits of a development against the environmental and other costs (including opportunity costs); (c) proper balancing of short-term and long-term environmental and social implications.

8.5.2 Limitations of EIA procedures

In judging the areas where research can help improve EIA, it is useful to analyse past experience (e.g. Holling, 1978; IIASA, 1979*a*, *b*). Holling and others have pointed out the problems that have arisen from misconceptions about the nature of the environmental impact assessment process and its relationship to ecology on the one hand and the institutions of planning and government on the other, and set these out in the shape of 12 alleged myths (summarized in Table 8.10).

These problems stem from three kinds of error: errors of perception, errors in response and errors in the institutional framework.

The problems stemming from errors of perception include:

(a) Failure to recognize that the EIA process must be sequential, beginning with the description of the structure of the systems concerned (including environmental and socioeconomic components) and then assessing alternative impacts from conceivable developments, presenting the results in a form suitable for political and managerial decision.
(b) Inadequate recognition of the spatial diversity of environmental systems.
(c) Inadequate recognition of the highly dynamic nature of many such systems.

Table 8.10 'Myths' of environmental impact analysis (modified from Holling, 1978)

Myth	*Truth*
A. *Wrong assumptions about the interaction between environment and society*	
1 The aim of EIA is to guide management to a stable, predictable end point	1 Neither stability nor precise prediction is a characteristic of dynamic social and environmental systems: the aim is to understand the likely changes, as they continue
2 Development programmes can be treated as set, extrinsic, factors superimposed on the environmental systems under study	2 Development programmes are just as subject to change as are environmental systems, and are integral parts of the system to be studied
3 Policies and social goals can be framed first and EIA then adopted to describe responses	3 Environmental understanding often helps to change social goals
4 Environmental concerns can be responded to only by institutional changes	4 There is often much latitude to adjust institutional policies to take account of environmental factors
B. *Wrong assumptions about the environment*	
5 All possible impacts of development must be considered in an EIA	5 Many impacts cannot be foreseen and forward planning must therefore provide for monitoring and adaptation
6 Each new assessment is unique	6 Ecological systems have many features and properties in common, and these can be reflected in parallels in the structure of assessments
7 Comprehensive surveys of environmental systems are an essential element in EIA	7 Comprehensive surveys generate excessive, often irrelevant data. What is needed is selective description of key parameters, with adequate attention to their dynamics
8 Any good scientific study must help an assessment	8 Much research not undertaken deliberately to help the assessment process will be useless. We need management models, not ecological ones
9 The right units for an assessment are either physical boundaries like a watershed or political ones like a district	9 The problem must define the boundaries: interactions may transgress preconceived frontiers

Table 8.10—(*cont.*)

Myth	*Truth*
10 Descriptive studies of parts of the present system can be integrated by systems analysis to provide understanding and prediction	10 Descriptions of what a complex system is doing rarely indicate what it would do under other conditions: the real question is what allowance must be made for our lack of predictive ability
11 Systems analysis allows the best option to be chosen	11 Systems analysis is not a reliable predictive tool, and the choice of alternatives depends on value judgements
12 Ecological analysis and impact assessment are designed to reduce uncertainty	12 Analyses must cater for inevitable uncertainty, but may indicate areas where it is greatest

(d) Failure to appreciate the nature of the relationship between ecological, developmental, and managerial components of the system.

The errors in response generally stem from this inadequate definition of the problem. The common approach (at least among scientists) when studying the environmental effects of development proposals is to attempt to proceed in four steps:

(a) To survey and analyse the environmental systems involved.
(b) To describe and analyse the developments proposed.
(c) To superimpose (b) on (a) and evaluate the likely consequences and the extent to which more environmental study will be likely to reduce uncertainty, and changes in development plans will be likely to reduce environmental damage.
(d) To monitor the real impact of the development, once sanctioned, and to provide for continuing adjustment through management, as it proceeds.

There is nothing inherently wrong with this approach. What has gone wrong is the way the various steps have sometimes been elaborated and components have been left out. In particular the following problems are common:

(a) Data overkill. Comprehensiveness is generally sought among those conducting EIAs (Munn, 1979), and often leads to the collection of substantial volumes of information of uneven validity and relevance, clogging the analytical process. This was particularly common in the early days of NEPA (Legore, Section 4.2).

(b) Merging of time differences. Results from surveys of various environmental components, made at different times, are often combined in a fashion that may actually obscure genuine fluctuations.
(c) Insufficient gearing to the wider socioeconomic processes involved. It is fallacious to assume that the environmental systems and the developmental proposals can reasonably be separated for analysis when the whole point of EIA is in fact to bring these dissimilar elements together.
(d) Insufficient interaction between the actors in the management process must also be avoided. Environmental scientists, developers and representatives of the potential regulatory authorities must come together from the outset. Workshops beginning at the stage of problem definition, are a key feature (Holling, 1978; IIASA, 1979*a*, *b*).

Errors in the institutional framework have four main forms:

(a) Imposition of excessively broad – or unduly rigid – guidelines and directives about the content of environmental impact statements, thereby hampering streamlining or adaptation to specific environments or administrative systems. For example the original guidelines laid down by the CEQ (1971) specified that environmental impact statements should include ten widely defined components. It has now been found necessary to issue revised guidelines so as to accelerate the process, make the results more useful and intelligible, and avoid the substantial burden on the staff reviewing the assessments (CEQ, 1979).
(b) Required methodology, imposed regardless of its appropriateness. Four particular, not necessarily mutually exclusive, methods have commonly been adopted in the EIA process. These are matrices, overlays, check-lists and simulation models. They vary in their comprehensiveness, selectivity, capacity to show interactions, handling of uncertainty and cost. No one method is ideal for all kinds of issue, and all influence the nature of the output to a considerable degree. Consequently the uniform adoption (e.g. by the World Bank) of a particular checklist could lead to a distortion in the subsequent judgements (Munn, 1979).
(c) Assumptions that satisfying an EIA procedure will automatically lead to a sound decision. EIA procedures need to be applied within the context of wider land-use strategies, to be related to procedures for integrated resource planning and management and to particular development control practices, and to be recognized as a tool for the achievement of these policies rather than an end in themselves.
(d) Saturation of the institutional system through a failure to screen out the many cases where elaborate EIA procedures are not required. For example, under the United States 1969 *National Environmental*

Policy Act, 6946 environmental impact statements had been filed by the end of 1975 (Lee and Wood, 1978), and one staff member of the office of Environmental and Health Affairs in the World Bank is reported to have reviewed 100 projects in 45 days. Ways of relating the complexity of the analysis more closely to the size of the problem are clearly necessary, and this was one reason why UK evaluation of the potential for EIA led to the recommendation that it be applied only to major developments (Catlow and Thirlwall, 1977).

The consensus among those who have recently reviewed the EIA system is that there is a role for a process of probability evaluation, so as to allow judgements about likely changes in the environment that a development will cause, the associated costs and benefits, and the best ways of guarding against waste. It must be emphasized, however, that this process is an input to the decision, not a substitute for it. The process needs to be as simple as possible or it will lose credibility. At the same time it needs to be as accurate as possible without causing excessive costs.

8.5.3 The role of research

The purpose of research in the area of environmental impact assessment is to upgrade the analytical process. The technique has proved valuable in many environmental and political contexts, but the bounds of its application also need definition.

It is equally evident that it is not the purpose of research in this field to increase our understanding of basic ecology. One major fallacy of the past has been the contention that *any* improvement in ecological knowledge must of necessity improve environmental impact evaluation procedures. There has also been confusion over the kind of model most useful in an EIA context. Ecosystem models are not required; the need is for management models incorporating development: environment interactions and the modifying influences of social choice.

The research needs fall into two groups: the first concerned with the conduct of EIA and the second with methods for the evaluation of the resulting environmental impact statement (Munn, 1979). The first group includes work on:

(a) Methodology to decide the type of EIA appropriate to a given situation. This will demand a capacity to analyse the uses to be made of the assessment and relate these to available techniques, and the whole is a process of defining and satisfying 'customer need'.
(b) Methodology for defining the environmental systems involved, and especially for recognizing the key variables within them, and the

degree of resolution needed both in space and time, as an aid to selective information gathering.

(c) Methodology for defining the environmental effects of social and institutional programmes and other non-construction activities.

(d) Methods for quantifying the effects of both developments and environmental changes on the wider socioeconomic system. This, which involves the comparison of unlikes to a considerable degree, is one of the least well catered for areas in this process.

(e) Methodology for building management models (not ecosystem models) at the level of sophistication required. Holling (1978) stresses the need for simple but effective simulation models and Munn (1979) emphasizes that they must incorporate physical, biological and socioeconomic parameters.

(f) Methodology (including improved workshops) for information exchange between the human components of the EIA process (including developers, environmental scientists and managers, and between different kinds of specialist and non-specialists).

Research may also be required on the following aspects of evaluation of EIA:

(1) Criteria for environmental quality. Munn (1979) lists this, but the requirement may go wider, incorporating the need for critical study of standards, goals, and indicators of environmental quality which may be used as a framework against which to evaluate the likely consequences of a planned development.

(2) Methodology for quantifying value judgements about the relative worth of various components of environmental quality. This implies a capacity to balance such unlikes as agricultural productivity, wildlife conservation, landscape beauty and amenity, full employment, and personal mobility: many indices developed so far have been of doubtful value.

(3) Methodology for monitoring of the actual impact of a development, and evaluating the adequacy of the initial environmental impact assessment.

8.5.4 An illustration of the problems

The ways in which scientific information may limit an EIA process can be illustrated by reference to one of the major potential impacts on one of the worlds least-disturbed series of ecosystems: the possible consequences of mineral exploration and exploitation in the Antarctic (Holdgate and Tinker, 1979).

The projected development (which at present has not gone beyond the earliest stages of scientific exploration) might involve the drilling for oil at sea on the continental shelves around the Antarctic continent, or the exploitation of hard rock minerals on land. The basic problem is what such exploitation might do to components of the Antarctic environment, to international scientific co-operation in the region, and to world economics.

The first question is whether an impact analysis process would be useful in approaching this situation and if so, in what way.

The objective of such an analysis would be to define what national or international regulatory procedures were needed in order to control development, avoid unnecessary environmental damage, and monitor the effectiveness of such action. To be justified, a formal EIA procedure would need to provide more reliable and economical insights than could come from less structured scientific analysis and intergovernmental discussions.

It is clearly unrealistic to attempt a single analysis for so vast an area. The Antarctic continent has a total area of some 14 million km^2 (one tenth of the world's land surface). The continental shelves, depressed by ice loading to depths often above 500 m also cover a very large total area: at maximal extent some 20 million km^2 of sea are covered by ice shelves or pack ice.

The first requirement for research is therefore to narrow the field of concern to areas where there is a reasonable likelihood of mineral development (a first step in defining the problem). So far, there have been no geophysical studies directly searching for oil fields at sea, but some preliminary drilling suggests there may well be hydrocarbons in the Ross Sea basin, and they could also occur in the Bellingshausen and Weddell Seas. On land geological exploration is more advanced, and 14 minerals which are of commercial value elsewhere in the world have been located in the Antarctic. However, no deposit that is economically worth extracting has been found there, and while calculations suggested that there might, by analogy with other continents of similar geological structure and history, be more than 900 major mineral deposits in the Antarctic, only 20 or so of these could be expected to be located in the less than 2% of area which is free of ice cover.

The other crucial element in defining the problem is how far engineering technology exists to permit exploration or exploitation of deposits under the extreme environmental and climatic conditions of this region. It is now possible to drill at sea in depths of over 1500 m: by the mid 1980s drilling rigs are likely to be capable of working down to 5000 m. Sea depth alone is, therefore, unlikely to prove a bar to drilling for oil on the Antarctic continental shelf or even off it. Nor is floating ice an insuperable barrier to exploration, since vessels now exist with a capacity of disconnecting from a well and moving away when threatened by icebergs. Exploitation of any finds, however, would be technically difficult and might require permanent installation of well-head equipment on or below the sea bed in positions where iceberg scour (which is known to occur down to 500 m in some Arctic areas) could not threaten their integrity.

Existing geological research techniques will clearly allow more precise definition of possible zones of exploitation. Whether or not these prove feasible, however, will depend on an evaluation of the technology and its cost, and this is an area in which engineering development continues to proceed rapidly, and where predictions of the future are very doubtful. The incentive to development will be provided by world economic trends, so that the speculation about future energy costs, and the relative value set upon oil as opposed to that set upon renewable or nuclear energy sources or coal will be crucial.

The most plausible development is for oil on the continental margins. The maximum environmental impacts would come at sea if, through accident, oil was released in quantity into the surface waters, or on-shore if oil pollution reached the coasts, or if large areas of the limited ice free ground were taken for land based installations.

It will be noted that only at this stage of problem definition do ecological questions (as commonly understood) arise. What are the most probable impacts on living organisms? Oil released at sea would be likely to affect the pelagic krill (*Euphausia superba*) which is important both as a resource (it is the subject of one of the worlds most rapidly expanding and potentially largest fisheries) and because it occupies a central place in the whole marine ecosystem, accounting for about 50% of the zooplankton biomass and providing a major food resource for fish, seals, birds and whales. Oil would be likely also directly to affect some of these consumers of krill, especially sea birds. As in other oceans, direct fouling of sea birds and marine mammals and of coastal communities is likely to be the major impact of oil spills. However, it is difficult to assess the size of such impacts, especially when water circulation, the effect of ice in confining the spread of oil, and the rate of natural breakdown of oil are not properly known. Even if, therefore, it is assumed that the essential system to quantify is that relating oil to krill and to higher organisms, research would be needed on the following parameters:

(a) The rate of dispersion and likely distribution of spilled oil in particular parts of the Antarctic.
(b) The interaction of the oil with the pack ice and with the plants attached to the ice.
(c) The rate of bacterial breakdown of oil in the Antarctic seas.
(d) The rates of uptake of oil-derived hydrocarbons by krill and other organisms and the impact of this on their populations and on the krill fisheries.
(e) The likely scale of impact of spilled oil on sea birds and seals, and on their consequent populations.

It is essential that these analyses allow the probabilities of different changes in response to different impacts to be assessed and an evaluation of the consequences of various margins of uncertainty. Only (b), (c), (d) and (e) of these would commonly be regarded as the province of the ecologist.

Table 8.11 Some components of the system affecting oil exploration and exploitation in Antarctic seas (from Holdgate and Tinker, 1979)

Factors influencing site selection
Geological indications of oil potential
Bathymetry and structure of sea bed
Iceberg and pack-ice condition
Currents (surface and deep)
Climate
Special biological features of site (and alternatives)
Cost of operation
Technological factors influencing impacts
Use of seismic explosives
Drilling processes
Nature of rigs or platforms
Nature of wellhead completion structure (e.g. on or below sea bed)
Nature of oil storage system
Nature of transport system
Nature of pollution control system (including management)
Nature of emergency procedures (including training)
Environmental factors affecting impacts
Types and concentrations of pollutants, especially oil, released
Rates of physical and biological degradation of oil and other pollutants
Rates and directions of dispersion of oil and pollutants
Relative location of emissions and living targets (benthos, phytoplankton, krill and other zooplankton, fish, seabirds, seals, whales, etc.)
Sensitivity of living targets to oil and other pollutants and to substances used in clean up

United States continental shelf experience suggests that about one well in 4000 can be expected to have a serious blow-out, and prior to the 1979 Gulf of Mexico incident the maximum volume of oil thus released was about 12 000 tonnes. The risk of oil pollution by a blow-out may therefore be much less than that from a shipping accident (the *Amoco Cadiz* in 1978 released 150 000 tonnes of oil, and because of their advantages in ice breaking, very large tankers could be expected to be used in the Antarctic so that a shipping disaster might involve as much as 250 000 to 500 000 tonnes of oil).

Single spills even of massive dimensions are unlikely to have a permanent impact on the very large sea bird and seal populations in the Antarctic, and still less impact on the plankton ecosystems. However, repeated spills could well build up a substantial chronic effect and it will be necessary to analyse the likely interaction in different areas between these and the populations at greatest risk and to develop ways of monitoring population changes.

Table 8.12 Some components of the system involved in hard rock mining on land (from Holdgate and Tinker, 1979)

Factors affecting scale of development

- Ore quality
- Size of ore body
- Availability of water and energy
- Location of suitable sites and construction materials, including harbours and overland transportation

Development features affecting scale of environmental impacts

- Location of development
- Size of mining installation (land required)
- Volume of waste and spoil generated, and location of dumping area
- Volume of ice/freshwater required
- Nature of energy source used
- Nature of transportation system used for mineral concentrate
- Nature and volume of chemical pollution released to air and water

Environmental features affecting scale of impacts

- Landscape features at sites of development
- Biological characteristics of sites of development (pattern and rarity of soil, vegetation, fauna)
- Biological characteristics (sensitivity) of plant and animal communities exposed to development

Environmental management affecting development impacts

- Constraints on location to avoid impact on ecologically or geomorphologically valued sites
- Grading, stabilization of dumped spoil to facilitate recolonization by vegetation
- Direct re-establishment of vegetation on bared ground or spoil (very difficult in the Antarctic)
- Removal of chemical pollutants in emissions to air and water
- Removal of installations once no longer required
- Costs of all these measures

The impacts of onshore developments associated with oil are easier to quantify. There are few points on the Antarctic coastline which are ice-free in summer and offer large enough areas of bare ground for the building of substantial installations. These are also the places where the impoverished Antarctic land vegetation and invertebrate fauna are most abundant, and are generally occupied by breeding colonies of sea birds, while the ice that is fast about them may support seal populations. Many of these areas have been scheduled as specially protected areas under the internationally agreed measures for the conservation of flora and fauna in the Antarctic (Holdgate,

1970). Research would need to quantify the likely scale of disruption, and evaluate its significance for the species concerned.

Tables 8.11 and 8.12 summarize the key components which would need to be quantified if a model were to be built as a foundation for environmental impact analysis in this region. But this information would be of little value unless the organizational framework existed for drawing upon the findings in a fashion that led towards a rational and responsible management policy. The key question, after all, is whether a predicted change, consequent upon a proposed development, really matters – and if so, what can be done to prevent or mitigate it, and these issues go beyond the competence of the ecologist alone. Because of the special international arrangements operating in the Antarctic under the Antarctic Treaty, considerable negotiation would be required in order to establish a regulatory authority which was compatible with the differing national attitudes to sovereignty and hence to the ownership of the Antarctic resources. If some form of 'authority' were established under a treaty or convention, this 'authority' would need to draw upon the expertise of geologists, technologists, biologists, climatologists, oceanographers and specialists in sea ice in securing information essential for the establishment of a management regime. Table 8.13 sets out the central questions which these various groups may be expected to ask of one another within this context, and it is from these that on-going research requirements would stem.

This example illustrates some of the crucial factors in determining the research requirements in support of EIA:

(a) The research can only be specified once the nature of the likely development and the ways in which it may be managed have been defined.

(b) The research must be closely geared to the increasingly precise definition of the nature and location of the impacts and of the targets affected.

(c) The aim of the research must be to quantify interactions to the degree required, rather than go into the fundamental ecology of all component species.

(d) Only some of the necessary research is ecological in a strict sense, and even that must be related to a wider socioeconomic and technological context.

(e) The key questions – which are concerned with the significance of predicted change and the possibility of mitigation – can only be worked out through the interaction of a range of scientists and managers. No one component – ecologist, planner, developer or administrator – can operate alone.

(f) This process of interaction must be a continuing one. Just as Antarctic mineral development is likely to go through successive phases of scientific exploration, commercial exploration, exploratory drilling and full exploitation, so the research required in order to determine the

Table 8.13 Questions likely to be asked in assessing the environmental impact of oil development in the Antarctic (note that the geologist is seen as a respondent rather than a questioner) (from Holdgate and Tinker, 1979)

Questioner	*Respondent*				
	'The authority'	Climatologist, oceanographer, and sea-ice specialist	Biologist	Technologist	Geologist
'The authority'		What is the probable dispersion of oil?	What is the nature and significance of the environmental impacts foreseen?	What are the chances of finding oil? What operational methods and plans for safe containment are there?	What are the characteristics of development sites?
Climatologist, oceanographer, and sea-ice specialist				What are the properties of the oil? Where are you located?	
Biologist		What is the likely dispersion pattern of the oil?		Where are you located? What is the risk of an oil spillage? What are your land support needs? What are the properties of your oil?	
Technologist	What constraints are you imposing on my operations? Where can I drill? What information must I supply? What permits must I seek?	What are the sea, seabed, tide, weather and ice conditions I may meet?	What do you consider the impacts of my operations will be? (Can I argue that 'the authority' is being unreasonable?)		Where should I explore for oil? Where can I find good conditions on shore?

responsible limits to be put on each step in the process will change. Monitoring of the impact of each stage is to be an essential preliminary to the regulation of the successive one.

The most profitable approach appears to involve the use of workshops, bringing together representatives of all the likely interests at both the stage of problem definition, and then at the stage of construction and testing of a model incorporating the best available information about the key interactions between the principal variables (Holling, 1978; IIASA 1979*a*, *b*; Holdgate and Tinker, 1979). By convening a series of such workshops, the assessment can be developed and adjusted as knowledge increases, and the development can be adjusted in the light of environmental understanding and evaluations of the socioeconomic effects. Workshops are also a convenient means of determining monitoring requirements, and ways in which management regimes may need adjustment. Where the development is a large one, there may be need for continuity through a 'core group' of specialists responsible for model building and testing. Such elaborate arrangements are, however, likely to be appropriate only where we are concerned with very large development, or with a series of smaller-scale impacts on a system like a major river.

8.5.5 Conclusions

The general term environmental impact assessment encompasses a wide range of activities, addressing problems of enormously variable scale. The first crucial question is how most economically to conduct that level of analysis that is really required. Two general guidelines can be suggested: (a) the process must be integrated within wider national (or where appropriate international) planning and regulatory practices; (b) it must take all relevant interests and factors into account, but be no more complex that the nature of the problem renders essential.

Accordingly the process depends on effective definition of the problem, in a fashion that allows its true significance to be assessed, and a decision taken on the degree of elaboration of analysis that is appropriate. The IIASA-based workshop methodology appears adequate for larger projects, but more research may be needed on the most effective way of analysing smaller-scale developments which are not suited to either a major workshop or an ex- extremely elaborate series of guidelines, and which must be catered for within national and local administration.

The crucial question is how far universally applicable logic and methodology can fit such disparate developments and situations. Check lists, overlays, matrices and simulation models all have their uses, especially to ensure that significant parameters are not overlooked, but none is perfect and it may be that the right thing is to provide a series of alternative methods, with clear guidelines on what each is most useful for. In this case research will need to concentrate on the framework for analysis and the diagnostic pointers towards appropriate methods as much as on the methods themselves.

There has been a common confusion over the degree of ecological analysis required for EIA. Elaborate ecosystem models are commonly not needed, but a high order of ecological insight is needed in order to select the variables of first importance in a particular situation and to describe the essential components and dynamics of the system. Research on how best to do this remains a high priority.

Finally, the process of information exchange is crucial and needs further study. Environmental impact statements are generally evaluated by non-scientists as well as scientists, and need to be cast in readily comprehended language which does not, however, distort the underlying rationale. They lead to decisions on questions of the acceptability of impact which neither ecologists nor planners nor administrators are capable of answering alone: there has to be interaction. Workshops are effective because they force interactions that indicate, and resolve, misunderstandings. But much EIA (in simplified form) goes on in the context of local or national planning decisions and in formulating industrial plans, and insufficient has been done on how to make a scientific input comprehensible and effective under such circumstances.

The central dilemma remains. Is EIA to be thought of as a formal process, needing administrative systems to accommodate it (as in the USA, and to a lesser degree, in Australia) or a concept to be fitted into a diversity of management systems, with a degree of elaboration determined by the circumstances? The answer to this question will have a significant impact on research priorities.

REFERENCES

Aponso, M.C.D.V. (1976) *Report on the Reappraisal of Cow Green and Derwent Reservoirs*, MSc thesis, University of Newcastle upon Tyne.

Armitage, P.D. (1976) *Freshwat. Biol.*, **6**, 229–40.

Armitage, P.D. & Capper, M.H. (1976) *Freshwat. Biol.*, **6**, 425–32.

Baker, J.M. (1971) in *The Ecological Effects of Oil Pollution on Littoral Communities* (ed. E.B. Cowell), Applied Science Publishers, Barking.

Baker, J.M. (1976) in *Marine Ecology and Oil Pollution* (ed. J.M. Baker), Applied Science Publishers, Barking.

Baker, J.M., Hiscock, S., Hiscock, K., Levell, D., Bishop, G., Precious, M., Collinson, R., Kingsbury, R. and O'Sullivan, A.J. (1981) *Irish Fisheries Investigations* Series B (Marine) **23**, 1–27.

Bayfield, N.G. & Burrow, G.C. (1976) *Biol. Conserv.*, **9**, 267–92.

Bayne, B.L., Widdows, J. & Thompson R.J. (1976) in *Marine Mussels and their Ecology and Physiology* (ed. B.L. Bayne), Cambridge University Press, Cambridge.

Bines, T.J. (1977) *Proc. N. England Soil Discussion Group*, **12**, 29–35.

Boydell, P. (1966) Evidence given to the Select Committee of the House of Lords on the Tees Valley and Cleveland Water Bill, 23 November 1966.

Bradshaw, M.E. & Doody, J.P. (1978) *Biol. Conserv.*, **4**, 223–42.

Bradshaw, M.E. & Jones, A.V. (1976) *Phytosociology in Upper Teesdale*, Trustees of the Teesdale Trust, Farnworth.

Buchanan, J.B., Sheader, M. & Kingston, P.F. (1978) *J. Mar. Biol. Ass. UK*, **58**, 191–209.

Catlow, J. & Thirlwall, G.C. (1977) *Environmental Impact Analysis in the United Kingdom*, Department of the Environment, London.

CEQ (1971) In *Environmental Quality: the Second Annual Report of the Council on Environmental Quality*, US Government Printing Office, Washington DC.

CEQ (1979) in *Environmental Quality: the Tenth Annual Report of the Council on Environmental Quality*, US Government Printing Office, Washington DC.

Chappell, G., Ainsworth, J.F., Cameron, R.A.D. & Redfern, M. (1971) *J. Appl. Ecol.*, **8**, 869–82.

Clapham, A.R. (1978) *Upper Teesdale, The Area and its Natural History*, Collins, London.

Countryside Commission (1978) *Digest of Countryside Recreation Statistics* (*CCP* 86) Countryside Commission, Cheltenham.

Crapp, G.B. (1970) *The Biological Effects of Marine Oil Pollution and Shore Cleaning* PhD thesis, University of Wales.

Crapp, G.B. (1973) *Irish Fisheries Investigations* Series B, No. 9.

Crawford, A.R. & Liddle, M.J. (1977) *Biol. Conserv.*, **12**, 135–42.

Crisp, D.T. & Mann, R.H.K. (1977*a*) *Fish. Mgmt.*, **8**, No. 2.

Crisp, D.T. & Mann, R.H.K. (1977*b*) *Fish Mgmt.*, **8**, No. 4.

Crisp, D.T., Mann, R.H.K. & McCormack, J.C. (1974) *J. Appl. Ecol.*, **11**, 969–96.

Cross, T.F., Southgate, T. & Meyers, A.A. (1979) *Mar. Poll. Bull.*, **10**, 104–7.

Cullinane, J.B., McCarthy, P. & Fletcher, A. (1975) *Mar. Poll. Bull.*, **6**, 173–6.

Dicks, B. (1976) in *Marine Ecology and Oil Pollution* (ed. J.M. Baker), Applied Science Publishers, Barking.

Dicks, B. & Iball, K. (1981) in *Proc. 1981 World Oil Spill Conference*, Atlanta, API/EPA/USCG, pp. 361–74.

Doody, J.P. (1975) *Studies in the Population Dynamics of Some Teesdale Plants*, PhD Thesis, University of Durham.

Dudley, G. (1976) in *Marine Ecology and Oil Pollution* (ed. J.M. Baker), Applied Science Publishers, Barking.

Godwin, H. & Walters, S.M. (1967) *Proc. Bot. Soc. Brit. Is.*, **6**, 348–9.

Gregory, R. (1971) *The Price of Amenity: Five Studies in Conservation and Government*, Macmillan, London.

Gregory, S. & Smith, K. (1967) *Weather*, **22**, 497–505.

Harding, R.J. (1978) *J. Inst. Water Engineers & Sci.*, **33**, 252–4.

Holdgate, M.W. (1970) in *Antarctic Ecology*, Vol. 2 (ed. M.W. Holdgate), Academic Press, London and New York.

Holdgate, M.W. & Tinker, J. (1979) *Oil and Other Minerals in the Antarctic*, Scott Polar Research Institute, Cambridge.

Holling, C.S. (1978) *Adaptive Environmental Assessment and Management. IIASA International Series on Applied Systems Analysis*, Vol. 3, J. Wiley & Sons, Chichester.

IIASA (1979*a*) *Adaptive Environmental Assessment and Management. Summary report of the First Policy Seminar* 18–21 June 1979, International Institute for Applied Systems Analysis, Laxenburg.

IIASA (1979*b*) *Expect the Unexpected. An adaptive Approach to Environmental Management*, Executive Report 1, International Institute for Applied Systems Analysis, Laxenburg.

Jeffrey, D.W. & Pigott, C.D. (1973) *J. Ecol.* **61**, 85–92.

Johnson, G.A.L. & Dunham, K.C. (1963) *The Geology of Moor House*, Nature Conservancy Monograph No. 2, HMSO, London.

Johnson, G.A.L., Robinson, D. & Hornung, M. (1971) *Nature, Lond.*, **232**, 453–6.

Jones, W.E., Fletcher, A., Bennell, B.J., McConnell, B.J., Richards, A.V.L. & Mack-Smith, J. (1979) in *Monitoring of the Marine Environment* (ed. D. Nichols), Institute of Biology, London.

Kennard, M.F. & Knill, J.L. (1969) *Inst. Water Engineers*, **23**, 87–136.

Lee, N. & Wood, C.M. (1978) *J. Environ. Mgmt.*, **6**, 57–71.

Levell, D. (1977) Southampton Water-Benthic Surveys 1975–1976, *Oil Poll. Res. Unit*, unpublished report.

Manley, G. (1952) *Climate and the British Scene*, New Naturalist series, Collins, London.

Marshall, B.S. (1977) *Stability of Reservoir Banks in Saccharoidal Limestone at Cow Green Reservoir, Co. Durham* MSc thesis, University of Newcastle upon Tyne.

Ministry of Town and Country Planning (1947) *Conservation of Nature in England and Wales*: Report of the Wild Life Conservation Special Committee (England and Wales), (Command 7122), HMSO, London.

Munn, R.E. (ed.) (1979) *Environmental Impact Assessment*, SCOPE 5, 2nd ed., J. Wiley & Sons, Chichester.

Myers, A.A., Cross, T.F. & Southgate, T. (1978) *Bantry Bay Survey*, Zoology Dept., First Annual Report University College, Cork.

Myers, A.A., Cross, T.F. & Southgate, T. (1979) *Bantry Bay Survey*, Zoology Dept., Second Annual Report University College, Cork.

Nelson-Smith, A. (1964) *Some Aspects of the Marine Ecology in Milford Haven*, PhD thesis, University of Wales.

Nelson-Smith, A. (1967) *Field Studies* **2**, 435–77.

NERC (1966) Evidence given to the Select Committee of the House of Lords on the Tees Valley and Cleveland Water Bill, 23 November 1966. (Statement in the report of the Secretary of State for Education and Science.)

O'Sullivan, A.J. (1975) *Mar. Poll. Bull.*, **6**, 3–4.

Parry, J.M. & Al-Mossawi, M.A.J. (1979) in *Monitoring of the Marine Environment* (ed. D. Nichols) Institute of Biology, London.

Pearson, T.H. (1971) *J. Exp. Mar. Biol. Ecol.*, **6**, 211–33.

Pigott, C.D. (1956) *J. Ecol.*, **44**, 545–86.

Praeger, R.L. (1934) *The Botanist in Ireland*, Hodges, Figgis & Co., Dublin.

Ratcliffe, D.A. (1977) *A Nature Conservation Review*, Cambridge University Press, London.

Rees, E.I.S., Nicholaidou, A. & Laskaridou, P. (1977) in *Biology of Benthic Organisms* (eds B.F. Keegan, P. O'Ceidigh & P.J.S. Boaden), Pergamon Press, London.

Robinson, D. (1971) *Contr. Mineral & Petrol.*, **32**, 245–50.

Royal Society of Edinburgh (1981) *Proc. R. Soc. Edin. Section B* (*Biol. Sci.*) (eds T.H. Pearson & S.O. Stanley), **80**, Parts 1–4, p. 367.

SVEAG (1976) *Oil Terminal at Sullom Voe, Environmental Impact Assessment*, Thuleprint Ltd, Sandwick, Shetland.

Whitby, M.C. & Willis, K.G. (1978) *Rural Resource Development*, Methuen, London.

CHAPTER NINE

Conclusions

Both ecologists and planners are involved in management of the environment although they operate at rather different levels. Environmental management is also carried out by a range of other disciplines including agriculture, forestry, water resource management, pollution control etc. However, since ecology is the science concerned with the functioning of organisms and their relationship with the environment, and planning provides the central, co-ordinating decision-making framework, they are both of central importance in environmental management.

The science of ecology relates to planning in two ways. The first is the application of ecological principles to planning situations. These have been incorporated into agriculture and forestry for a very long period, in relation to, for example, the maximization of carrying capacity and productivity. More recently, they are also being applied to pollution control in the development of standards for controlling pollutant doses to man or animals. Ecological principles relating to ecosystem development, restoration and management of land, tree planting and landscape design, conservation of species and habitats, etc., all have application to planning process.

The second way in which ecology relates to planning, and it is this area which has been the major concern in this book, is through the application of ecological techniques. The methodologies most pertinent to planning are those associated with: (a) the collection of base-line data about a site or sites; (b) prediction of the likely consequences of alternative courses of action and the identification of mitigation measures and (c) evaluation and communication of information gained from (a) and (b).

The papers in this book appear to indicate that techniques for base-line survey are available and, in general, are adequate for most planning requirements. There may, however, be a need for a continued process of 'fine-tuning'. The challenge for ecologists is not to produce a scientific treatise, but rather to ensure that the final decision is taken in the light of best available knowledge, within the time available. Ecologists should not be afraid of best practicable means as applied to their own work provided that the information is received and acted upon as such, and that opportunities and financial support are available for post-auditing exercises in order that future attempts may be improved. Similarly, planners, perhaps, need to appreciate that natural environments are complex and that in these cases adequate answers can only be achieved by painstaking research over a long period of time. Clearly forward planning and anticipation is required in these situations.

The papers in this book also seem to indicate that techniques for evaluation and communication of ecological information have developed rapidly in recent years. The problem here is that planning involves comparisons of unlikes. Consequently there is a tendency to convert data into a common form, usually for ecological parameters through the use of numerical indices, charts or matrices. Within the planning system, these are often interpreted as having some degree of precision so that numerical values may, for example, be summed or subtracted to obtain a total value index. Biological systems rarely lend themselves to such precisions.

Furthermore, the derivation of numerical indices involves value judgements about the bases for weightings, etc. The indices by themselves will be of limited use, or susceptible to mis-interpretation, if the basis for the value judgements are subsequently changed. The resolution of this problem will be difficult and represents a major challenge for planners and ecologists alike.

There does seem also to be some controversy over who should undertake these evaluations. Planners appear to be afraid of undertaking evaluations of ecological data. The resolution to this problem must involve a slow educational process and ecologists may have to continue bearing most of the evaluation burden as an interim measure. It is, however, clearly important that the basis of the value judgement is spelt out and that the information is provided in such a way that re-evaluation can take place if the bases of the judgement change.

Perhaps the most difficult tasks for ecologists lies within the areas of predicting the likely consequences of alternative courses of action and developing prescriptions to achieve specific objectives. Charts and matrices are not methods for predicting likely outcomes despite their widespread use, but, rather, are approaches for summarizing and communicating information on impact predictions. The most profitable methodologies for quantitative predictions appear to be through modelling exercises. Most current models are, however, of only limited application or value and further developments of models developed from information on biological systems must be welcomed. In addition, the need for post-auditing exercises cannot be overemphasized provided that they are carried out with the specific aim of improving our ability to make future predictions.

Within certain limits, ecological methodologies are available to satisfy many of the planners requirements and there are numerous examples in this book of the positive benefits resulting from the application of these methods to specific planning requirements. However, to achieve these ends, it is imperative that the planner should have a clear idea of their requirements.

Ecologists involved in environmental management systems at the research level cannot be expected to know what the decision-makers requirements are and more importantly, what they may be in the future. Their role is to judge whether the requirements are achievable and how to satisfy them. The resolution of the problem, can only be achieved by continued dialogue between ecologists and planners. It is unfortunate that such dialogue has proved difficult in the past largely due to differences in their training. Planners deal in aggregation and evaluation of environmental matters whereas ecologists generally have a quantitative background which leaves little room for subjective judgements. Clearly, considerable effort, tolerance and education will be required to overcome these perceived differences.

In addition to the conflicts arising from poor dialogue between ecologists and planners, there are a number of other causes of what is felt by some groups to be poor environmental management. The planners influence is not all-embracing and much of the responsibility for management of the rural environment lies with Central Government agencies (such as MAFF, Forestry Commission, Countryside Commission, Nature Conservancy Council, etc. or nationalized industries). Considerable space has been devoted in this book to the conflicts which arise, and the changes in legislation which could achieve a more reasonable reconciliation of objectives. The present trend is towards ensuring better consultation between these agencies but if this fails then the next stage could be an extension of planning control into rural areas or the adoption of a rural land-use strategy as strongly advocated by many conservation organizations.

The consideration of environmental matters in the UK is covered by a variety of legislation. It has been argued that the division of environmental controls into distinct elements (development planning on the one hand and effluent control on the other) can lead to omissions or overlap. However, the flexibility in the UK system does allow for reconciliation of amenity and economic interests in the local context. Flexibility, can, however, lead to inconsistencies as shown by the stark contrast of the depth of environmental considerations given in planning the oil-related developments on the Mersey Marshes and Canvey Island. There are also inconsistencies with the many agencies which are presently outside planning control or policing developments made by the planning authorities themselves where they act as advocate and judge at the same time.

It is argued that these inconsistencies may be overcome by the adoption of more formalized procedures such as environmental impact assessment as carried out under the *American National Environmental Protection Act*. This debate has been brought to the forefront by the issuance of a draft Directive by the EEC for the 'Environmental Assessment of Projects'. This empowers the developer to produce an EIS which gives a description of the project and the environment, an assessment of the effects, procedures for amelioration of effects, a description of compliance with local plans and pollution control legislation and where the effects are appreciable, to discuss alternatives to the development. The EIS would be issued alongside the planning application and the local authority would review the document with all bodies involved in environmental control and the public. The views expressed herein suggest that these formal procedures could lead to long and irrelevant assessments, delays and increased costs unless rigorous controls are exercised. There is also concern that (a) the definition of developments to be subjected to an EIA would not allow for differences in the sensitivity of an area, (b) publication of an EIS would lead to problems of confidentiality and (c) the discussion of alternative sites could lead to unnecessary anxiety or the blighting of areas under consideration. On the other hand, views are expressed here that EIA would force the consideration of environmental matters and lead to a more explicit decision-making process. It could also strengthen the existing planning process and save time and costs. Indeed, some feel that EIA should also be applied to major changes in rural land use and also plan-making and policy assessment.

Whatever the differences regarding procedural matters all parties seem to agree upon the need for and early consideration of environmental matters in developmental planning. Concern for the delays caused by extended public inquiries has also led to preparation of and EIS prior to the inquiry by many developers and the introduction of pre-inquiry review meetings. The way ahead for major controversial or novel developments may be shown from the preparation of an environmental assessment of the Severn Barrage Scheme prior to consideration of detailed engineering factors. It is also clear that an EIS is only a 'snapshot' and environmental considerations should also be incorporated throughout the design and construction phases.

Additional retrospective analysis of the effectiveness of planning procedures or the accuracy of predictions is essential for continued improvement in the validity of environmental impact assessments. However, the EIA process is more often limited by poor communication than by inadequate ecological information. It is to be hoped that the dialogue between planners and ecologists entered into in this book will be continued and that communications and perceived differences will be resolved by mutual education.

Index